Molecular Approaches to Ecology and Evolution

Edited by
Rob DeSalle
Bernd Schierwater

Birkhäuser · Boston · Berlin

Editors

Dr. Rob DeSalle
American Museum of
Natural History
Central Park West at 79th Street
New York, NY 10024-5192
USA

Dr. Bernd Schierwater
Zoologisches Institut
J.W. Goethe-Universität
Siesmayerstr. 70
D-60054 Frankfurt
Germany

Die Deutsche Bibliothek – CIP-Einheitsaufnahme

Molecular approaches to ecology and evolution / ed. by Rob
DeSalle ; Bernd Schierwater. - Basel ; Boston ; Berlin : Birkhäuser
1998
    ISBN 3-7643-5725-8 (Basel ...)
    ISBN 0-8176-5725-8 (Boston)

Library of Congress, Cataloging-in-Publication Data

Molecular approaches to ecology and evolution / edited by Rob DeSalle,
    Bernd Schierwater.
        p.   cm.
    Includes bibliographical references and index.
    ISBN 0-8176-5725-8 (pbk. : alk. paper)
    1. Molecular ecology. 2. Molecular evolution. I. DeSalle, Rob.
  II. Schierwater, B. (Bernd), 1958 –   .
  QH541.15.M63M625 1998
  572.8'38 —dc21   98-15347
                                             CIP

© 1998 Birkhäuser Verlag, PO Box 133, CH-4010 Basel, Switzerland
Printed on acid-free paper produced from chlorine-free pulp
Cover design: Markus Etterich, Basel
Printed in Germany
ISBN 3-7643-5725-8
ISBN 0-8176-5725-8
9 8 7 6 5 4 3 2 1

Contents

List of contributors

George Amato, Science Resource Center, Wildlife Conservation Society, 2300 Southern Blvd., Bronx, New York, NY 10460, USA
E-mail: Gamato1@aol.com

Peter Brazaitis, Science Resource Center, Wildlife Conservation Society, 2300 Southern Blvd., Bronx, New York, NY 10460, USA

Gustavo Caetano-Anollés, Department of Biology, University of Oslo, P.O. Box 1045, Blindern, N-0316 Oslo, Norway

James M. Cheverud, Department of Anatomy & Neurobiology, Washington University School of Medicine, St. Louis, MO 63110, USA
E-mail: cheverud@thalamus.wustl.edu

Timothy M. Collins, Department of Biological Sciences, Florida International University, University Park, Miami, FL 33199, USA
E-mail: CollinsT@FIU.EDU

Clifford W. Cunningham, Zoology Department, Duke University, Durham, NC 27708, USA
E-mail: cliff@acpbu.duke.edu

M. N. Dawson, Department of Biology, University of California, 405 Hilgard Ave., Los Angeles, CA 90095-1606, USA

Amos S. Deinard, School of Veterinary Medicine, University of California, Davis, CA 95616, USA

John Gatesy, Department of Ecology and Evolution, University of Arizona, Tucson, AZ 85721, USA

Paul Z. Goldstein, American Museum of Natural History, Department of Entomology, Central Park West at 79th Street, New York, NY 10024, USA
E-mail: zoltan@amnh.org

Eduardo A. Groisman, Howard Hughes Medical Institute, Department of Molecular Microbiology, Washington University School of Medicine, St. Louis, MO 63110, USA
E-mail: groisman@borcim.wustl.edu

Heike Hadrys, Abt. Ökologie und Evolution, J. W. Goethe-Universität, Siesmayerstr. 70, D-60054 Frankfurt, Germany, and Department of Ecology and Evolution, Yale University, 165 Prospect St., New Haven, CT 06511, USA
E-mail: Hadrys@zoology.uni-frankfurt.de

Jody Hey, Department of Ecology, Evolution, and Natural Resources, Nelson Biological Labs, Rutgers University, P. O. Box 1059, Piscataway, NJ 08855, USA
E-mail: hey@mbcl.rutgers.edu

David K. Jacobs, Department of Biology, University of California, 405 Hilgard Ave., Los Angeles , CA 90095-1606, USA
E-mail: djacobs@ucla.edu

Allan Larson, Department of Biology, Washington University, St. Louis, MO 63130-4899, USA
E-mail: larson@wustlb.wustl.edu

Shannon E. Lee, Department of Biology, University of California, 405 Hilgard Ave., Los Angeles, CA 90095-1606, USA

Howard Ochman, Department of Biology, University of Rochester, Rochester, NY 14627, USA
E-mail: hoch@uhura.cc.rochester.edu

Josephine Pemberton, Institute of Cell, Animal and Population Biology, University of Edinburgh, West Mains Road, Edinburgh, EH9 3JT, UK
E-mail: J.Pemberton@ed.ac.uk

Kevin A. Raskoff, Department of Biology, University of California, 405 Hilgard Ave., Los Angeles , CA 90095-1606, USA

Howard C. Rosenbaum, Molecular Systematics Laboratory, American Museum of Natural History, 79th St. & CPW, New York, NY 10024-5192, USA, and Dept. of Biology, Yale University, PO Box 6666, New Haven, CT 06510, USA

Eric Routman, Department of Biology, San Francisco State University, 1600 Holloway Ave., San Francisco, CA 94132, USA
E-mail: routman@sfsu.edu

Christian Schlötterer, Institut für Tierzucht und Genetik, Veterinärmedizinische Universität Wien, Josef-Baumann Gasse 1, A-1210 Wien, Austria
E-mail: christian.schloetterer@vu-wien.ac.at

Michelle Pellissier Scott, Department of Zoology, University of New Hampshire, Durham, NH 03824, USA
E-mail: mps@christa.unh.edu

Michael T. Siva-Jothy, Department of Animal and Plant Sciences, University of Sheffield, Sheffield S10 2TN, UK
E-mail: m.siva-jothy@sheffield.ac.uk

Chelsea D. Specht, Department of Biology, New York University, New York, NY 10003, USA

Joseph L. Staton, Department of Biology, University of California, 405 Hilgard Ave., Los Angeles , CA 90095-1606, USA

Alan R. Templeton, Department of Biology, Washington University, St. Louis, MO 63130-4899, USA
E-mail: temple_a@biodec.wustl.edu

Alfried P. Vogler, Department of Entomology, The Natural History Museum, Cromwell Road, London, SW7 5BD, UK, and Department of Biology, Imperial College at Silkwood Park, Ascot, Berkshire, SL5 7PY, UK
E-mail: a.vogler@nhm.ac.uk

John Wakeley, Department of Organismic and Evolutionary Biology, Harvard University, Cambridge, MA 02138, USA
E-mail: jwakeley@oeb.harvard.edu

Michael S. Webster, Department of Biological Sciences, State University of New York at Buffalo, Buffalo, NY 14260, USA
E-mail: mwebster@acsu.buffalo.edu

David F. Westneat, Center for Evolutionary Ecology, T. H. Morgan School of Biological Sciences, 101 Morgan Building, University of Kentucky, Lexington, KY 40506-0225, USA
E-mail: biodfw@ukcc.uky.edu

Ward Wheeler, Department of Invertebrates, American Museum of Natural History, Central Park West @ 79th Street, New York, NY 10024, USA
E-mail wheeler@amnh.org

Scott M. Williams, Department of Microbiology, Immunology and Genetics, Meharry Medical College, Nashville, TN 37208, USA
E-mail: swilliams@ccvax.mmc.edu

Preface

Four years ago we edited a volume of 36 papers entitled *Molecular Approaches to Ecology and Evolution* (Schierwater et al., 1994), in which we attempted to put together a diverse array of papers that demonstrated the impact that the technological revolution of molecular biology has had on the field of evolutionary biology and ecology. The present volume borrows from that theme but attempts to focus more sharply on the impact that molecular biology has had on our understanding of different hierarchical levels important in evolutionary and ecological studies. Because DNA sequence variation is at the heart of every paper in the present volume, we feel it necessary to examine how DNA has affected study at various levels of biological organization. The majority of the chapters in the present volume follow themes established in the earlier volume; all chapters by authors in the previous volume are either fully updated or entirely new and expand into areas that we felt were important for a more complete understanding of the impact of DNA technology on ecology and evolution.

The collection of papers in this volume cover a diverse array of ecological and evolutionary questions and demonstrates the breadth of coverage molecular technology has imparted on modern evolutionary biology. There are also a broad range of hierarchical questions approached by the 17 papers in this volume. These topics range from identification of individuals in paternity studies to problems that systematists encounter in analyzing DNA sequence data such as alignment and the so-called big data set problem. The breadth of chapters in this volume is also mirrored by the proliferation of journal articles on the topic in the last 15 years. In addition to long-standing journals such as *Evolution, Journal of Molecular Evolution* and *Genetics* that have had a tradition of publication on strong molecular evolutionary topics, several new journals have been innitiated to deal with the steady production of studies focusing on the application of DNA technology to evolutionary topics. Journals such as *Molecular Ecology, Molecular Phylogenetics and Evolution, Molecular Marine Biology, Ancient Biomolecules* and *Molecular Biology and Evolution* have specifically been created to deal with this recent interest in molecular evolutionary topics. Specialty journals in the systematics field have also begun to focus more and more on molecular topics in evolutionary biology, with journals such as *Cladistics*, *Systematic Biology* and the *Annals of the Entomological Society of America* routinely dedicating substantial proportions of issues to molecular biology issues. The field of developmental biology has also gotten into the act with molecular evolution topics appearing in journals such as *Developmental Biology*, *Genes and Development* and *Development*. Even the revered and seemingly timeless *Wilhelm Roux's Archive for Developmental Biology* has gone through a name change to *Development and Evolution* to reflect the importance of the evolutionary perspective.

This book is divided into three major parts that focus on three distinct levels in evolutionary and ecological biology. The lowest level is of individuals and populations and requires the fine-scale analysis of DNA sequences from large numbers of individuals. Analysis of animal mating systems tends to be the major objective of these kinds of studies, and DNA sequence technologies such as restriction fragment length polymorphism (RFLP) methods and microsatellite techniques have been used to delve into this area. The upper level concerns systematics and how DNA sequence data are used in the interpretation of the "tree of life". Many problems and controversies have arisen since the incorporation of DNA techniques into systematics. We attempt to present some of these controversies in chapters that take a head-on approach to some of the problems. Sandwiched between these upper and lower sections of the book is a section on species and how molecular biology has altered our approach to understanding species and how we approach defining species. We suggest that any bridge (if bridges do indeed exist) across the upper and lower levels discussed above can only be found through the understanding of change at speciation. Finally, let us not forget that any attempts to project structural boundaries into the continuum of the evolution of life must necessarily be artificial.

We have assembled a broad array of techniques, approaches, applications and controversies. Indeed, we think that this volume will complement a growing collection of summary volumes on the use of molecular techniques such as Hillis et al. (1997), Zimmer et al. (1993), Avise (1994), Avise and Hamrick (1995), Ferraris and Palumbi (1996), Karp et al. (1998), Miyamoto and Cracraft (1991), Burke et al. (1992) etc. All of the chapters in this volume have up-to-date reference lists, and in addition we have attempted to cover material only slightly touched upon in the chapters in our introductory chapters to the major sections, which are also fully referenced. We hope that these in-depth bibliographies will allow the book to be used as a starting reference source for current literature in the field. Our introductory chapters also address future directions at the three hierarchical levels we have targeted. We intend for the volume to be used by researchers as well as graduate students and advanced undergraduates in seminar classes.

Finally, we thank the authors for their cooperation in the production of the chapters. Their response to our request for new or revised material was prompt and extremely enthusiastic, and we greatly appreciate their understanding in the review process. We are grateful to the same individuals listed in the previous volume for reviews and criticism on early drafts of chapters. These reviewers waded through literally hundreds of manuscript pages and greatly enhanced the final product. We also thank Janine Kern, our editor in this project at Birkhäuser, for her encouragement and understanding during the process of putting this volume together. Ultimately, though, we as editors assume responsibility for the final product, including its mistakes, presentation and organization. We hope the reader will find this volume accessible, enjoyable and helpful.

January 1998   Rob DeSalle, American Museum of Natural History
                                Bernd Schierwater, University of Frankfurt

References

Avise, J. C. (1994) *Molecular Markers, Natural History and Evolution*, Chapman and Hall, New York.

Avise, J. C. and Hamrick, J. L. (1995) *Conservation Genetics: Case Histories from Nature*, Chapman and Hall, New York.

Burke, T., Dolf, G., Jeffreys, A. J. and Wolff, R. (eds) (1992) *DNA Fingerprinting Approaches and Applications*, Birkhäuser Verlag, Basel.

Ferraris, J. D. and Palumbi, S. R. (eds) (1996) *Molecular Zoology: Advances, Strategies and Protocols*, Wiley-Liss, New York, pp. 65–194.

Hillis, D. M., Moritz, C. and Mable, B. K. (eds) (1996) *Molecular Systematics*, Sinauer Associates, Sunderland, MA.

Karp, A., Isaac, P. G. and Ingram, D. S. (eds) (1998) *Molecular Tools for Screening Biodiversity*, Chapman and Hall, London.

Miyamato, M. M. and Cracraft, J. (1991) *Phylogenetic Analysis of DNA Sequences*, Oxford University Press, Oxford.

Schierwater, B., Streit, B., Wagner, G. P. and DeSalle, R. (eds) (1994) *Molecular Ecology and Evolution: Approaches and Applications*, Birkhäuser, Basel.

Smith, T. B. and Wayne, R. (eds) (1997) *Molecular Genetic Approaches in Conservation*, Oxford University Press, Oxford.

Zimmer, E. A., White, T. J., Cann, R. L. and Wilson, A. C. (eds) (1993) *Methods in Enzymology 224. Molecular Evolution: Producing the Biochemical Data*, Academic Press, San Diego.

Part 1. Population biology, kinship and fingerprinting

Bernd Schierwater and Rob DeSalle

Molecular methods have become indispensable tools for studying the dynamics and evolutionary factors acting on natural populations. In former days biologists were dependent on morphological and behavioral mutants to study genetic variation in populations. Today, the limited number of phenotypically variable systems and their unknown relationship to genotypes have been replaced by almost unlimited numbers of molecular variants of unknown relationship to a phenotype. And for the future we might see the characterization of variation directly for the genes contributing to the phenotypic diversity of features and adaptation.

The first section of this volume deals with the *status quo* of population-level studies and illustrates the power of recent molecular technologies for screening individuals and populations in order to determine genetic variation, genetic distance, gene flow, population size, kinship, fitness and other related subjects. Many of these questions were not approachable a few years ago but may now be asked because of the advent of new technologies (for a historical review, see Powell, 1994). The selection of the following chapters aims (i) to illustrate the nature of recent questions being asked – and likely be pursued in the near future – and (ii) to review the potential of state-of-the-art technologies to give answers to these questions. The focus is on the level of individuals and populations; the potential for other studies will only briefly be mentioned here.

Some 30 years ago, a novel and relatively simplistic methodology, starch-gel electrophoresis of allozymic proteins, became available (Lewontin and Hubby, 1966; Harris, 1966), which revolutionized research in population biology. Because of the ease of obtaining Mendelian data, screening allozymes was the dominant method for the analysis of genetic variation in natural populations for more than two decades. Owing to low cost, safety and ease of use, allozyme electrophoresis may still be the method of choice for some applications (for refs and review, see Hartl et al., 1994; Powell, 1994). Present and future research, however, is and will continue to be dominated by technologies that investigate variation in the DNA molecule directly. At the level of DNA, variation between individuals, or any two operational taxonomic units, is practically unlimited; and due to constant progress in the development of genetic tools, this variation has become more and more easily detectable.

R. DeSalle, B. Schierwater (eds) Molecular Approaches to Ecology and Evolution
©1998, Birkhäuser Verlag Basel

Ever since the detection of restriction fragment length polymorphisms (RFLPs) and the invention of "DNA fingerprinting" in the early to mid-1980s (Wyman and White, 1980; Jeffreys et al., 1985), technological progress has consistently increased the ease, accuracy and resolving power of available DNA techniques for screening genomes. At the same time, the number of DNA-screening techniques has increased, while the average life span of these techniques has decreased. Traditional DNA fingerprint techniques, which are based on Southern blot hybridizations, seem to have gone out of vogue already and have been widely replaced by polymerase chain reaction (PCR)-based techniques, that can amplify DNA loci at high to very high degrees of polymorphism. In contrast, RFLP analyses have remained indispensable for certain applications, especially for the screening of uniparentally inherited genetic markers in population and conservation genetics (e.g. Schroth et al., 1996). PCR-typing techniques require only nanogram amounts of DNA, are not labor-intensive, and are particularly suited for high analytical throughputs. Since the publication of our preceding volume (Schierwater et al., 1994), the number of microsatellite applications and newly developed PCR-based arbitrarily amplified DNA (AAD) techniques, including their use in empirical studies, has increased dramatically at the expense of studies using, for example, allozyme and traditional (Southern blot-based) multilocus DNA fingerprinting (cf. volumes by Hoelzel, 1992; Ferraris and Palumbi, 1996). This trend is reflected in the selection of the chapters in this book.

PCR-based techniques are available for both, the amplification of multilocus and single-locus DNA sites. Presently, AAD (Schierwater, 1995) techniques dominate the field of multilocus and microsatellite typing and the field of single-locus DNA typing. AAD techniques are based on the amplification of several to numerous DNA fragments by means of arbitrary primers. These techniques do not require any sequence information of the genome under investigation and are immediately applicable to any genome of interest. Furthermore, they are rapid and cost-effective and require only very basic molecular skills. None of these techniques, however, can directly generate allelic information, and data interpretation is not necessarily straightforward. In contrast, single-locus microsatellite typing reveals easy-to-interpret allelic markers, but it cannot be readily applied to anonymous genomes. In addition, this technique requires the isolation and characterization of polymorphic loci beforehand, which can be time- and labor-intensive, and requires more advanced molecular skills. A fruitful link between AAD and microsatellite markers was recently established that not only facilitates the isolation of polymorphic microsatellites but also allows their isolation from nanogram amounts of DNA of any unknown genome under investigation. The so-called RAMS (randomly amplified microsatellites) technique uses AAD fingerprint profiles, instead of genomic libraries, as the DNA source for isolation of microsatellites (see Ender et al., 1996, and the chapter by Siva-Jothy and Hadrys).

In choosing a research tool to answer a specific question, the researcher has to find a feasible method (or combination of methods) that can answer the question.

If a question can be answered by one method, then most likely there will also be other methods available to answer the question. Thus cost-efficiency, prior experience, laboratory setup, and also "personal taste" may become additional criteria for deciding on the "optimal" method. For example, if the determination of paternity is the issue, a researcher might decide to use microsatellite typing because he has the primers for the study organism already at hand, and he might argue that multilocus DNA fingerprinting by means of an AAD technique is less suited because of a high number of potential fathers in the population. Another researcher, however, might decide to use a multilocus AAD technique like RAPD, AP-PCR, DAF or AFLP fingerprinting (see Hadrys et al., 1992, 1993 and chapter by Caetano-Anollés for overviews), because of their ease of use, the steadily increasing number of successful applications and perhaps his own former positive experience. He might argue that the number of potential fathers in a population is not a major limitation, since he can easily generate any desired number of polymorphic AAD markers necessary to resolve paternity. Yet another researcher might decide to use traditional, Southern blot-based, multilocus DNA fingerprinting because the DNA is not limited and because of the existing laboratory setup. Finally, a researcher may combine and exploit the advantages of two or more methods, for example using RAPD fingerprinting for statistical paternity assignments in large offspring clutches and to simultaneously isolate polymorphic microsatellite loci, which then are applied to certain subproblems (cf. chapter by Siva-Jothy and Hadrys). Thus it is part of the healthy soul of science that researchers differ not only in the questions they ask but also in the tools they prefer.

Studies of birds have played an important role in demonstrating the importance of genetic relatedness as a key factor in behavioral ecology, and evolutionary and conservation biology. The determination of kinship is central, for example, to detect and quantify extrapair fertilization and intraspecific brood parasitism, which likely have been important selective forces in the evolution of avian mating systems, morphology and behavior. In their chapter, Webster and Westneat outline the importance of birds as model systems in behavioral ecology and review several interesting questions in avian population genetics. They include a review of the potential of and limitations to a variety of molecular methods that have been applied to answer these questions.

Some of the most promising model systems for studying the mechanisms of sexual selection, the evolution of mating systems and the evolution of sociality are found in insects. Here the complexity of behavioral and morphological adaptations reaches a peak and challenges the power of genetic methods to unravel the underlying selective forces. Fitness estimations may become very difficult because of high degrees of polygamy, sperm competition mechanisms, complex reproductive morphologies, large offspring clutches, sociality and an enormous spectrum of inter- and intrasexual competition in many systems (Thornhill and Alcock, 1983). Furthermore, behavioral ecologists discuss the existence of "cryptic" female choice, that is whether females might be capable of selectively using the sperm of a certain male after collecting and storing sperm from different mates. In the chapter by Siva-Jothy and Hadrys, this most modern and challenging question is discussed in detail along with molecular approaches that seem suited to demonstrate cryptic female choice. A more general discussion of molecular approaches applied to measure the key parameter "fitness" in insects is given in the chapter by Scott and Williams.

While the focus of the first three papers in this section is on questions investigated in certain model systems by means of appropriate molecular tools, the focus of the final two chapters is on the molecular tools themselves, and discusses the potential and application of the most prominent techniques currently in use. Recently, several new AAD techniques, or modifications and improvements of older ones (note, "old" in this context means 8 years or less!), have been developed that have just begun to be used in the fields of population biology, conservation genetics and behavioral and applied ecology. The chapter by Schloetterer and Pemberton illustrates the usefulness of PCR-based microsatellite analyses for genetic analysis of natural populations, and reviews application studies in different areas of behavioral ecology and population genetics. Rosenbaum and Dienard examine several theoretical questions relevant to the use of microsatellites in evolutionary studies. By detailing what is known about mutational processes and the patterns of microsatellite evolution, these authors discuss the benefits and pitfalls of microsatellite use. The chapter by Caetano-Anollés reviews the potentials of the variety of the PCR-based AAD techniques. Since many of these techniques are less than 4 years old, the author reviews their potential not only for the analysis of populations and individuals but for screening genomes in general. This way the reader might detect new possibilities for genetically analyzing populations and individuals.

It is interesting and intriguing that the first five chapters discuss at least one common point, the dependency of the deciphering of selective forces that affect population structures on the determination of kinship. Although the methods detailed in these chapters overlap, all the authors hold different attitudes toward the usefulness of different techniques for generating and analyzing the genetic data. In the future we may expect deeper insights into the structure of conserved and variable regions of DNA in a variety of organisms, which will be helpful in further improving the generation and analysis of evolutionary data (including the theoretical understanding of PCR; cf. Schierwater et al., 1996). Clearly, the applications of microsatellites and AAD techniques to the analysis of individuals and populations have not peaked yet. We may expect a rapid increase in the isolation of microsatellite loci from a variety of organisms, and a further increase in the number of new technologies, particularly in the area of AAD-related techniques. Sooner or later we will begin to use RNA screening to study variation in gene expression in natural populations. AAD techniques may provide a means for RNA typing (see chapter by Caetano-Anollés) as a first step in directly characterizing variation of genes contributing to phenotypic variation and adaptation in natural populations.

References

Ender, A., Streit, B., Städler, T., Schwenk, K. and Schierwater, B. (1996) RAPD identification of microsatellites in *Daphnia*. *Molec. Ecol.* 5: 437–441.

Ferraris, J. D. and Palumbi, S. R. (eds) (1996) *Molecular Zoology: Advances, Strategies and Protocols*, Wiley-Liss, New York, pp. 65–194.

Hadrys, H., Balick, M. and Schierwater, B. (1992) Applications of random amplified polymorphic DNA (RAPD) in molecular ecology. *Molec. Ecol.* 1: 55–63.

Hadrys, H., Schierwater, B., DeSalle, R., Dellaporta, S. and Buss, L. W. (1993) Determination of paternities in dragonflies by Random Amplified Polymorphic DNA fingerprinting. *Molec. Ecol.* 2: 79–87.

Harris, H. (1966) Enzyme polymorphism in man. *Proc. R. Soc. Lond. B* 164: 298–310.

Hartl, G. B., Willing, R. and Nadlinger, K. (1994) Allozymes in mammalian population genetics and systematics: indicative function of a marker system reconsidered. *In*: Schierwater, B., Streit, B., Wagner, G, P. and DeSalle, R. (eds) *Molecular Ecology and Evolution: Approaches and Applications*, Birkhäuser, Basel, pp. 299–231.

Hoelzel, A. R. (ed.) (1992) *Molecular Genetic Analysis of Populations. A Practical Approach*, Oxford University Press, Oxford.

Jeffreys, A. J, Wilson, V. and Thein, S. L. (1985) Hypervariable "minisatellite" regions in human DNA. *Nature* 314: 67–72.

Lewontin, R. C. and Hubby, J. L. (1966) A molecular approach to the study of genic heterozygosity in natural populations. II. Amount of variation and degree of heterozygosity in natural populations of *Drosophila pseudoobscura*. *Genetics* 54: 595–609.

Powell J. R. (1994) Molecular techniques in population genetics: A brief history. *In*: Schierwater, B., Streit, B., Wagner, G, P. and DeSalle, R. (eds) *Molecular Ecology and Evolution: Approaches and Applications*, Birkhäuser, Basel, pp. 131–156.

Schierwater, B. (1995) Arbitrarily amplified DNA in systematics and phylogenetics. *Electrophoresis* 16: 1643–1647.

Schierwater, B., Streit, B., Wagner, G. P. and DeSalle, R. (eds) (1994) *Molecular Ecology and Evolution: Approaches and Applications*, Birkhäuser, Basel.

Schierwater, B., Metzler, D., Krüger, K. and Streit, B. (1996) The effects of nested primer binding sites on the reproducibility of PCR: mathematical modeling and computer simulation studies. *J. Comp. Biol.* 2: 235–251.

Schroth, W., Streit, B. and Schierwater, B. (1996) Evolutionary handicap for sea turtles. *Nature* 384: 521–522.

Thornhill, R. and Alcock, J. (1983) *The Evolution of Insect Mating Systems*, Harvard University Press, Cambridge.

Wyman, A. R. and White, R. (1980) A highly polymorphic locus in human DNA. *Proc. Natl. Acad. Sci. USA* 77: 6754–6758.

The use of molecular markers to study kinship in birds: techniques and questions

Michael S. Webster[1] *and David F. Westneat*[2]

[1] Department of Biological Sciences, State University of New York at Buffalo, Buffalo, NY 14260, USA
[2] Center for Evolutionary Ecology, T. H. Morgan School of Biological Sciences, 101 Morgan Building, University of Kentucky, Lexington, KY 40506-0225, USA

Summary

Questions of kinship have long been central to behavioral, ecological and evolutionary studies of birds, and recently developed molecular genetic methods have opened a new frontier to answering these questions. We present an overview of these questions, and we explore the advantages and disadvantages of various molecular techniques for answering them. Parent-offspring relationships are presently a major focus of attention. Extra-pair fertilizations (EPF) and intraspecific brood parasitism (IBP) are known to be important aspects of avian mating systems, yet relatively little beyond estimates of frequencies of these events is known. Multilocus techniques are currently in wide use, and have important advantages for identifying those offspring that result from EPF and IBP. Multilocus techniques, though, have serious limitations when it comes to searching for and identifying the actual parents of such offspring. Single locus techniques avoid most of these limitations and are therefore most appropriate for studies in which true parents must be identified. For the field biologist, speed of development and analysis are critical factors determining the scope of a study. Unfortunately, no technique at present combines speedy development with rapid analysis times; some evidence suggests that polymerase chain reaction analysis of microsatellite loci may be a valuable compromise. Many studies may combine single locus and multilocus techniques to efficiently identify actual parents. Relatively few studies of birds have analyzed relatedness beyond parent-offspring relationships, yet multilocus and single locus techniques can provide estimates of average relatedness within subgroups of a population. Although it is unlikely that any current technique will allow determination of relatedness between dyads of individuals with much precision, some techniques do allow general estimates of relatedness in such cases. Finally, recently developed markers hold great promise for the study of avian sex ratios. Our review uncovered many unanswered questions about kinship in birds; the most exciting and successful research integrates speedy and precise molecular tools with extensive behavioral and experimental studies of free-living individuals.

R. DeSalle, B. Schierwater (eds) Molecular Approaches to Ecology and Evolution
©1998, Birkhäuser Verlag Basel

Introduction

Kinship refers to the descent of two or more individuals from a common ancestor, and this concept plays a central role in modern evolutionary and ecological theory. Indeed, a full understanding of cooperation, mating behavior, parental care, mate choice, dispersal and sexual selection all depend at some level on information about kinship. Moreover, kinship forms the basis for describing patterns of population structure, and is a necessary consideration in designing breeding and reintroduction plans for endangered species (e.g. Schroth et al., 1996). It is no surprise, then, that determining genetic relatedness is a critical task facing researchers in the fields of behavioral, evolutionary and conservation biology.

Studies of birds have played a key role in the development of these fields (Konishi et al., 1989), and kinship in birds has been an important area of such research for some time. Until recently, kinship in birds was determined primarily through observations of reproductive behavior. When an adult male and female were seen associating with each other, they were assumed to be "mated" to each other, and any young produced were assumed to be their offspring. Genetic studies have now clearly shown that, in many species, offspring often are not descendent from the male or female at the nest (Westneat et al., 1990; Birkhead and Møller, 1992; Westneat and Webster, 1994). This makes it clear that genetic relatedness (i.e. kinship) cannot confidently be inferred from social relationships.

Recent developments in molecular technology have provided an array of tools that can be used to assess kinship. These tools have cleared the way to answering a number of previously inaccessible questions, and have generated a flood of new

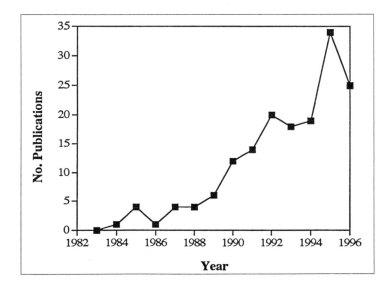

Figure 1. The number of studies that have used molecular genetic techniques to assess kinship in birds vs. year of publication. Data were gathered by searching Bioabstracts.

questions. As a consequence, in recent years we have seen an explosion of research projects employing molecular techniques to investigate kinship in birds (Fig. 1). These tools, though, have also created new burdens, generated new problems and perhaps even constrained the types of questions being asked.

In this chapter, we review studies of kinship in birds that have used molecular tools. Our intent is not to exhaustively review the available techniques, as this is done elsewhere, nor is it to review all of the studies that have used molecular genetic techniques. Rather, in this chapter we focus on the match between questions of kinship and the techniques available to answer those questions. We restrict our review and discussion to questions of genetic relatedness among individuals in the same population and do not consider molecular techniques to determine relationships among different populations or species (i.e. phylogenetics). Similarly, we do not provide detailed descriptions of the precise molecular or statistical techniques involved, as many of these issues are discussed in other chapters of this volume and elsewhere.

General description of techniques

Molecular genetic techniques can usefully be divided into two major groups: those that assay many loci simultaneously and those that assay only one locus at a time (Tab. 1). The distinction is fundamental to choosing the proper technique for the question at hand. Multilocus techniques are able to survey more genetic variation at once, but the inability to separate among loci can limit the scope of the information they yield. In contrast, single locus markers are usually easy to interpret (locus-specific genotypes and allele frequencies are easy to determine), but the variation screened is usually less than needed, making repeated screenings necessary. The decision to use multilocus techniques, single locus techniques or a combination of both will depend on several factors, many of which we outline in Table 1 and discuss below.

Multilocus techniques

Minisatellites

Minisatellite DNA is composed of tandem repeats of a core sequence, generally 15–30 bases in length. Because many minisatellite regions scattered throughout the genome are constructed from the same core sequence, probes homologous to that sequence usually survey multiple loci and generate complicated banding patterns that, in many populations, are individually unique (Jeffreys et al., 1985a, b; Burke and Bruford, 1987; Westneat, 1990). A comparison of band sharing between two

Table 1. Update of Burke et al. (1991); comparison of molecular genetic techniques available to study kinship in birds[*]

	Multi-locus minisatellite	RAPD	Protein electrophoresis	Single-locus minisatellite	RFLP	PCR-amplified microsatellites
Development time[†]	1 month	1–2 weeks	2–4 weeks	1–3 months	1–4 months	2–6 months
Genomic specificity	from plants to humans	from plants to humans	from plants to humans	species-genus, possibly family	unknown, likely species only	typically species only, some primers work between sub-orders
Processing time[‡]	2–4 weeks[§]	1 week[§]	2 days	2–4 weeks[§]	2–4 weeks[§]	2–4 days[§]
Genetic variation	high	medium	low	3–20+ alleles	2–4+ alleles	3–15+ alleles
Risk of anomalous results	low-moderate (due to insufficient number of scorable bands)	high (due to dominant alleles, sensitive PCR conditions)	low (Some rare null alleles)	low	low	low-moderate (due to modest frequency of null alleles)
Ease of scoring	moderate-difficult	difficult	easy-moderate	easy	easy	easy-moderate

[*]Based on experience of authors with proteins, multi-locus minisatellites, RAPDs and PCR-amplified microsatellites, and from references. [†]Assumes lab is using the technique successfully on another species; if not, add 6–12 months minimum for all the DNA techniques, 2–6 months for proteins in order to learn and develop correct procedures. [‡]For a set of 15–20 samples from collected tissue to bands ready to be scored (includes 1–4 days to isolate and resuspend DNA for DNA techniques). [§]Time is for one probe, primer, or locus. Additional probe/loci likely to be needed and so additional time necessary (might not require doubling).

individuals gives an estimate of genetic similarity and, once population-level variation is taken into account, kinship (Reeve et al., 1990, 1992).

At present, multilocus minisatellite analyses are the most popular technique being used to study kinship in birds (Westneat and Webster, 1994). This popularity derives from the facts that (i) the amount of genetic variation that can be detected is very high, and (ii) it has low genomic specificity, meaning that probes developed for one species often work on many other, distantly related species. Indeed, nearly all the bird studies using multilocus minisatellite techniques have used probes developed from mammals or M13 phage, and even arbitrary oligonucleotide probes often yield useful DNA profiles (Epplen et al., 1991). There are, however, three drawbacks to multilocus minisatellite analyses. First, the analyses can take a long time to complete (Burke, 1989). Second, the presence of many bands, often separated by very small distances on the gel, can make scoring difficult. As a consequence, comparisons between individuals run far apart on the same gel are not easy (Piper and Rabenold, 1992), and comparisons between individuals run on different gels are extremely difficult. Finally, minisatellite analyses require a relatively large amount of high-quality DNA, making analysis difficult for small or difficult-to-capture organisms.

RAPD

Randomly amplified polymorphic DNA (RAPD) profiles are produced by using a single short, arbitrary primer to amplify small sections of DNA using polymerase chain reaction (PCR) (see Caetano-Anollés, this volume). The overwhelming advantage of RAPD analysis is its speed and technical simplicity (Tab. 1); because the primers used are arbitrary, relatively little need be known about the genome of the study organism. A second advantage is that, like other PCR-based techniques, very little DNA is needed from each individual surveyed, so very small samples can be used. Three difficulties with the RAPD procedure are that (i) spurious bands may be produced as lab artifacts or as the result of PCR particularities (e.g. Levitan and Grosberg, 1993; Schierwater and Ender, 1993; Grosberg et al., 1996; Schierwater et al., 1996), (ii) interpretation of the banding patterns can be difficult and (iii) variability is only moderate. However, one approach for analyzing parentage that appears to be successful is to focus on diagnostic bands known to vary in the population (Lewis and Snow, 1992; Levitan and Grosberg, 1993). Although they have been used to study kinship in invertebrates (e.g. Hadrys et al., 1993; Levitan and Grosberg, 1993; Jones et al., 1994; O'Donnell, 1996; Siva-Jothy and Hadrys, this volume), very few studies of birds have used RAPD markers (for a notable exception, see Haig et al., 1994b; Haig, 1996).

Single-locus techniques

Protein electrophoresis

This procedure was the first genetic technique to gain widespread use in analyses of population structure, parentage and kinship. As a consequence, the statistical techniques necessary for analysis are well developed (e.g., Pamilo and Crozier, 1982; Pamilo, 1984; Westneat et al., 1987; Wrege and Emlen, 1987). The major advantages of protein electrophoresis are that it is fast, the techniques have become routine and scoring large numbers of individuals is relatively easy. The primary disadvantage, particularly in birds, is that the genetic variability assayed is usually quite low (Barrowclough and Corbin, 1978; Avise and Aquadro, 1982; Zink, 1982; Westneat et al., 1987), dramatically reducing the ability to determine kinship.

Single-locus minisatellites (SLM)

This technique is identical to the multilocus minisatellite analysis, except that the probe used recognizes a single minisatellite locus rather than many (e.g. Nakamura et al., 1987; Hanotte et al., 1991a). Thus, each individual will show only one band for each allele at the locus. This simplified banding pattern greatly facilitates scoring and allows banding patterns to be compared among gels. Individual minisatellite loci show a high degree of variability (e.g. Gyllensten et al., 1990; Hanotte et al., 1991b; Kempenaers et al., 1992), so a combination of only a few SLM probes should allow for powerful analyses of kinship. Unfortunately, SLM analyses have two major disadvantages: (i) the time necessary to process a set of samples and produce results for one probe is similar to that necessary for multilocus minisatellites; and (ii) SLM probes are likely to be genome-specific, although some probes might be useful for phylogenetically closely related species (Gyllensten et al., 1990; Burke et al., 1991; Hanotte et al., 1992). Even with recent improvements in the technique (e.g. Hanotte et al., 1991b), the process of isolating probes can be quite lengthy (potentially months).

Restriction fragment length polymorphism (RFLP)

Although RFLP sometimes refers to a general set of techniques which includes both single and multilocus minisatellites, here we use the term to refer specifically to techniques that assay variation at restriction sites. The basic technique is identical to minisatellite analyses (Southern analysis), except here a segment of DNA from the study organism, which has a polymorphic restriction site located within it, is isolated and used as a probe. The resulting banding patterns show which individuals are homozygous or heterozygous for the restriction site. The advantage of RFLP analysis is the ease of scoring and interpreting the patterns for a large number of individuals. The disadvantages are (i) the processing time for RFLP analysis is similar to that for minisatellite analyses, (ii) considerable time is necessary to develop the probes and (iii) each probe detects much lower levels of genetic variability than most of the other procedures outlined in this chapter. Due to these disadvantages, this technique will be of limited applicability and will be most useful in cases where a polymorphic restriction site is already known to exist (e.g. Quinn et al., 1987, 1989).

Microsatellites

Microsatellites, or simple sequence repeats (SSR), are similar to minisatellites in that alleles differ in the number of repeats of a core sequence, but the core sequence is only a few base pairs in length, making them short enough to be PCR-amplified. Microsatellite analyses hold much promise for field researchers (see Schlötterer and Pemberton, this volume), as they combine a high level of detectable genetic variation with the fast processing time of PCR (Tab. 1). There are two general shortcomings of using microsatellites to determine kinship. First, the time needed to develop a microsatellite marker system is the longest for all the techniques surveyed here. Second, null alleles (alleles that do not amplify, leading to the false classification of heterozygotes as homozygotes) are sometimes present (Paetkau and Strobeck, 1995; Pemberton et al., 1995; Primmer et al., 1995), so results must be interpreted carefully.

Several studies of non avian taxa have demonstrated the utility of microsatellite markers for kinship studies (e.g. Amos et al., 1993; Evans, 1993; Morin et al., 1994; Taylor et al., 1994; Kellogg et al., 1995; Altmann et al., 1996; Dow and Ashley, 1996; Paxton et al., 1996; Richard et al., 1996). However, despite the potential power of these markers, to date very few kinship studies of birds have used microsatellites. The reason for this appears to be related to the small size of the avian genome and its relative lack of "junk" DNA (Tiersch and Wachtel, 1991; Hughes and Hughes, 1995); clones bearing SSR loci appear to be more than an order of magnitude less frequent in birds than in genomic libraries constructed for other taxa (M. S. Webster and D. F. Westneat, unpublished data). This makes the search for useful SSR loci a slow and frustrating process in most cases. Nevertheless, some researchers have used standard cloning techniques to isolate polymorphic microsatellites from the avian genome (Ellegren, 1992; Hanotte et al., 1994; Primmer et al., 1995; Neumann and Wetton, 1996; Bensch et al., 1997; Piertney and Dallas, 1997), and some fascinating studies of kinship in birds have utilized such markers (McDonald and Potts, 1994; Ellegren et al., 1995; Gibbs et al., 1996, 1997; Sheldon and Ellegren, 1996).

Although few studies of birds have used microsatellite markers, two recent developments hold great promise for future studies. First, several techniques have been developed to enrich a genomic library for microsatellites before the library is screened. Although details of the techniques vary, most utilize hybridization to membranes (Armour et al., 1994; Kandpal et al., 1994) or to biotinylated target molecules (Karagyozov et al., 1993; Kijas et al., 1994; Edwards et al., 1996; Hammond et al., 1997) to isolate genomic fragments containing microsatellites. Our own experiments have shown that these techniques can increase the frequency of microsatellites in a genomic library by two or more orders of magnitude, and that most of the SSR so isolated are relatively long and polymorphic (M. S. Webster, unpublished data). Other methods to facilitate the isolation of microsatellites have also been suggested, including a rapid means of isolating polymorphic microsatellite loci from nanogram amounts of DNA (Ender et al., 1996). Second, although microsatellites are generally thought to have high species specificity, at least some SSR loci have conserved flanking regions (e.g. Schlötterer et al., 1991; Bowcock et al., 1994; Engel et al., 1996), suggesting that some primers will have wide taxonomic applicability. An extensive survey of bird microsatellites (Primmer et al., 1996) found that approximately half of the microsatellites isolated from a given species were present and polymorphic in other members of the same family, and one primer pair was found that amplified an SSR in virtually all species surveyed. This suggests that a "universal set" of primers that amplify polymorphic loci in a wide variety of birds might be developed eventually.

Specific questions and appropriate techniques

Which technique to use depends on the question being asked. Here we review some of the questions that researchers have addressed using molecular techniques and briefly summarize the most successful applications to answering each. An earlier chapter (Westneat and Webster, 1994) gives a detailed review of molecular studies of kinship in birds. In this chapter we focus almost exclusively on studies that have been published since that earlier review, highlight those questions that have yet to be adequately answered and offer suggestions on the most useful approaches to answering them. We examine two general categories of questions: those that concern parentage and those that concern relatedness outside of parent-offspring relationships.

Questions of parentage

In birds, alternative mating strategies are frequent, making actual patterns of descent much different from that implied by social behavior. Questions about parentage can be divided into three main types, which roughly correspond to the amount of information sought about a system.

1. Do social associations reflect mating relationships, and approximately how often are they wrong (crude exclusion studies)?

    This question has already been answered in its broadest sense. Thanks to a large number of detailed molecular studies (Birkhead and Møller, 1992; Westneat and Webster, 1994), we now know that extra-pair fertilizations (EPF) are a common feature of avian mating systems, and that intraspecific brood parasitism (IBP) is frequent in at least some species. These results have transformed the study of mating systems and sexual selection; it is now foolish to study the social system of any bird without at least considering the possible effects of these reproductive strategies.

    Although this basic question has now been answered, questions regarding the taxonomic distribution and consequences of such strategies have not. The frequency of EPF varies dramatically across species, with some studies showing little or no evidence for EPF (Hill et al., 1994; Warkentin et al., 1994; Hasselquist et al., 1995; Haydock et al., 1996) and others showing exceptionally high EPF rates (Brooker et al., 1990; Dixon et al., 1994; Mulder et al., 1994). What ecological and/or historical factors account for this variation? Similarly, the frequency of IBP varies among species, but on the whole appears to be much less common than EPF (Westneat and Webster, 1994). On the surface this is puzzling, as the benefits of brood parasitism to females are potentially quite high. What ecological and/or historical factors promote IBP, and why is this female reproductive tactic so rare relative to extra-pair copulations?

    One approach to answering questions such as these is to use general estimates of the frequency of EPF and IBP in comparative studies. For example, an interspecific comparison of parentage in passerines found the frequency of EPF to be associated with the degree of female breeding synchrony (Stutchbury and Morton, 1995). Similarly, behavioral studies suggest that the frequency of EPF may be associated with degree of coloniality (Møller and Birkhead, 1993b), but a rigorous interspecific examination of the association between EPF and nesting dispersion remains to be performed (Birkhead and Møller, 1992). Some genetic studies of colonial nesting birds also have suggested that dense nesting aggregations may promote IBP (Wrege and Emlen, 1987; Brown and Brown, 1988; Birkhead et al., 1990), but IBP also occurs in noncolonial species (Quinn et al., 1987; Price et al., 1989; McKitrick, 1990; McRae and Burke, 1996) and this hypothesis requires a rigorous comparative test with proper phylogenetic controls. Interspecific comparisons also can be used to examine the consequences of EPF and IBP. For example, across species the frequency of EPF is correlated with reduced male parental care (Møller and Birkhead, 1993a), suggesting that a reduction in paternal investment is one evolutionary consequence of EPF. Similarly,

EPF frequency is related to male plumage coloration (Møller and Birkhead, 1994) and relative gonad size (Møller and Briskie, 1995), supporting the hypothesis that sexual selection operating through EPF is an important evolutionary force shaping male morphology.

Clearly, many more species need to be studied if we are to answer comparative questions about the evolution and ecology of these reproductive strategies. Furthermore, researchers beginning a new study will often wish to obtain a rough estimate of the frequencies of EPF and IBP before investing more time and effort into a study of their details. All the methods described above can be used to estimate the frequency of EPF and IBP, but because these estimates do not require that the parentage of every offspring be identified, the best technique could sacrifice precision to gain increased speed and ease of analysis. For example, protein electrophoresis has been successfully employed in a number of studies (e.g. Gowaty and Karlin, 1984; Westneat, 1987; Wrege and Emlen, 1987). However, although it is fast and development time is minimal, the level of variation is often very low, leading to a need for large sample sizes, difficulties in distinguishing IBP from EPF (e.g. Westneat et al., 1987; Romagnano et al., 1989) and large error limits on estimates of IBP and EPF rates (Westneat et al., 1987). A preliminary survey for variable loci should be undertaken first, in order to calculate the probability of detecting EPF or IBP and thus estimate sample sizes that might be necessary to achieve an acceptable estimate.

Although all of the above techniques can produce an estimate of the frequency of EPF and IBP in a population, the technique of choice to answer this question is multilocus minisatellite analysis. This technique offers a high level of resolution, with probabilities of detection being greater than 0.99 (e.g., Westneat, 1990, 1993; Decker et al., 1993). This high level of resolution allows EPF to be easily separated from IBP. Furthermore, development time is not substantial for a lab set up to do DNA work. Given the low development time and high resolution, it is no surprise that studies using multilocus probes are so popular. In the future it is possible that PCR-amplified microsatellites will be used to estimate the frequency of EPF and IBP, as a relatively small number of highly variable loci can yield a great deal of information. However, until a set of universal primers is developed, the excessive development time of this technique for individual species makes it an unwise choice for those interested in a general estimate of the frequency of alternative mating strategies.

2. Which social associations do not reflect actual mating relationships, why and what are the consequences (precise exclusion studies)?

This question moves one step beyond the previous one; it requires that most or all individual young who have resulted from EPF or IBP be identified. This is the prerequisite to a host of specific questions about the factors influencing rates of these behaviors and their effects on other aspects of reproductive behavior.

For example, studies of the factors influencing the effectiveness of male mate-guarding behavior (Birkhead and Møller, 1992) and the influence of parentage on parental care (Westneat and Sherman, 1993; Whittingham et al., 1993) often require that the parentage of individual offspring be known.

Several early studies used protein electrophoresis to answer these questions (e.g. Westneat, 1988; Bollinger and Gavin, 1991; Gowaty and Bridges, 1991). However, because protein analyses often have difficulty separating EPF from IBP and can detect some but not all cases of these phenomena, exceptionally large sample sizes are generally needed and conclusions drawn are somewhat limited. With a few exceptions (e.g. Gyllensten et al., 1990; Dixon et al., 1994), the single locus techniques summarized above have not been used extensively to answer questions at this level of resolution, most likely because the lengthy development time has been prohibitive. Those studies which have used microsatellites or SLM to identify EPF and IBP were designed to answer questions at a greater level of resolution and are discussed below.

At present, multilocus minisatellites are the technique of choice for this level of question (but see Wetton and Parkin, 1997). When all individuals from the same family can be run together on a single gel and the actual parents of excluded young do not have to be identified, multilocus probes combine high statistical power with relatively short development time. Using this technique, researchers have been able to examine male and female behaviors leading to variation in frequency of extra-pair young (Dunn et al., 1994; Sheldon and Burke, 1994; Stutchbury et al., 1994; Owens et al., 1995; Webster, 1995; Burley et al., 1996; Gray, 1996; Krokene et al., 1996; Wagner et al., 1996a), social factors affecting mating patterns (Westneat, 1993; Mulder et al., 1994; Alatalo et al., 1996; McRae, 1996), and male traits affecting paternity (Hill et al., 1994; Rätti et al., 1995).

Many interesting correlative results have come from these studies. For example, several recent studies have found that males of many species do not seem to adjust their paternal output in relation to the number of EPF in their brood (e.g. Wagner et al., 1996b; Yezerinac et al., 1996), whereas other studies have found that males do adjust in accordance with their paternity (e.g. Dixon et al., 1994; Weatherhead et al., 1994; Westneat, 1995) but do not preferentially feed their own young within the brood (Westneat et al., 1995). Interestingly, Freeman-Gallant (1996) found that female savannah sparrows engage in EPC less frequently if their mates provide high levels of paternal care to the young, and suggested that females use paternal feeding rates as a measure of male genetic quality. Finally, for the cooperatively breeding superb fairy wren, EPF are most common in groups with helpers (Mulder et al., 1994), but breeding males appear to tolerate helpers because the males are then freed to pursue EPC with other females (Green et al., 1995).

Although correlative studies such as these have been extremely useful for evaluating hypotheses, the power of multilocus minisatellites has best been

shown in experimental studies. For example, by experimentally inducing bright-
ly colored male pied flycatchers to nest on territories adjacent to dull-colored
males, Rätti et al. (1994) were able to show that plumage color plays little role
in male EPF success in this species. It also might be possible to directly manipulate plumage traits to test this hypothesis (e.g. Smith et al., 1991; Ellegren et al.,
1995), though this has rarely been done. Burley et al. (1996) experimentally manipulated male attractiveness to show that unattractive males are cuckolded
more often than attractive males, and that unattractive males invest more in
parental care, whereas attractive males invest more in the pursuit of EPF.
Experimental removals have also been used to test the effectiveness of male mate
guarding (Davies et al., 1992; Lifjeld and Robertson, 1992; Westneat, 1994;
Macdougall-Shackleton et al., 1996).

More experiments such as these are clearly needed to test the wide variety of
hypotheses to explain why females engage in EPC, the ecological factors influencing EPF rates and the effects of EPF on other aspects of social behavior. In addition, there have been very few quantitative studies of IBP using multilocus minisatellites (e.g. Birkhead et al., 1990; McRae and Burke, 1996). Clearly, additional correlative and experimental studies using molecular techniques on species
known to engage in IBP will be useful. Because they require little development
time and are relatively precise, multilocus minisatellites will continue to be a
dominant tool for answering these types of questions in the near future. The
lengthy processing time and quantity of DNA required, though, limit the scope
of some studies. If a set of PCR-amplified microsatellites with wide taxonomic
applicability are developed, such that the probability of detection of EPF or IBP
approaches 0.95–0.99, then they will likely replace multilocus probes at this level.

3. Who is mating with whom, why and what are the consequences (assignment
studies)?

Many issues in behavioral ecology, particularly those dealing with reproductive strategies and sexual selection, require a precise knowledge of who is mating with whom. For example, understanding the action of sexual selection on
male traits now requires measuring components of fitness that include success at
extra-pair matings (Webster et al., 1995), and testing ideas about the potential
benefits of EPF to females requires knowing which males fertilized their eggs. In
addition, an understanding of the factors influencing IBP will require uncovering
which females lay parasitic eggs and the conditions under which they do so. To
address these gaps in our knowledge, the genetic parents of a nestling must be
identified from the pool of potential parents.

Multilocus DNA fingerprinting can be used to answer questions at this level
of resolution, but the applicability of this method may be somewhat limited. In
some cases, particularly polyandrous species (Burke et al., 1989; Oring et al.,

1992; Owens et al., 1995) and some cooperative breeders (Rabenold et al., 1990), the set of potential sires is more or less limited to a small number of males from within the social group. In these cases it is relatively easy to run all potential sires on the same gel with the offspring and their mother(s). In most avian systems, however, the genetic parents of a nestling could potentially be any of a large number of adults in the population. This raises problems for standard multilocus minisatellite analyses because (i) searching through all adults increases the probabilities of finding a match purely by chance, and (ii) it is not possible to run a nestling and all of its potential parents on the same gel. This latter problem is a serious drawback, as complex multilocus patterns are difficult to compare among gels. Thus, new gels with the nestling and a suite of potential parents must be run, increasing the already immense amount of effort as well as the amount of DNA needed from each individual.

Despite these limitations, a few studies have assigned paternity based solely on data from multilocus ministatellite analyses. For the most part, these have been studies in which extra-pair sires came from adjacent territories (e.g. Westneat, 1993; Stutchbury et al., 1994; Hasselquist et al., 1996). The use of inlane standards and computer scoring of band sizes has also been used to identify the sires of EPF young from among a large pool of potential sires (Mulder et al., 1994; Weatherhead and Boag, 1995). Although this approach can work with some effort, substantial resolution may be lost, reducing the available information and correspondingly increasing the probability that a match can arise by chance alone.

It is at this level of genetic resolution that the power and utility of single-locus DNA markers become evident, because single locus markers yield genotypes that can be easily compared among individuals run on different gels. Some studies have used such markers to determine the subset of males in the population that are potential sires of an EPF offspring, and then used multilocus DNA fingerprinting to identify the actual sire from this subset. For example, by screening the population with just one single locus probe, Gibbs et al. (1990) were able to reduce the pool of potential sires from 21 to 3 or fewer for each EPF nestling. RAPD markers might also be useful for such screening, as the technique of matching diagnostic alleles might reduce the number of likely parents to a manageable few (Lewis and Snow, 1992; Levitan and Grosberg, 1993).

An alternative, and promising, approach is to use single locus markers alone to determine the parentage of young. In particular, PCR-amplified microsatellite analyses are fast (after useful loci have been isolated), and a small number of loci should provide the genetic resolution necessary to assign parentage precisely. Although microsatellite studies of birds have lagged behind those of other taxa (see above), some studies have demonstrated their utility (Ellegren, 1992; Primmer et al., 1995, 1996). For example, given a single microsatellite locus with 10 alleles at equal frequency ($q = 0.10$), the probability that a given individual might by chance match an excluded offspring carrying a diagnostic allele is 2pq + q2 or 0.19. The probability that at least one individual out of a set of 20 possible parents carries allele q by chance is 0.99 (1 – [Pr. of not matching]20, or 1-[1-0.19]20). However, the probability that a given individual would match at four such loci (if they assort independently) is 0.001, and the probability that at least 1 out of 20 potential parents matched by chance alone is 0.021. Analysis of IBP might require more loci, because two parents must be identified. The key to this procedure is the initial development of the microsatellite primers in the species of interest. Because of the eventual speed at which analyses can be done, this development time is probably worthwhile for researchers interested in questions at this level.

Of the few studies that have successfully identified the sires of extra-pair young, most have examined male traits that lead to success in obtaining EPF. Some studies have shown that females prefer males bearing particular morphological or behavioral traits, and that these traits appear to indicate male viability (Kempenaers et al., 1992; Weatherhead and Boag, 1995; Hasselquist et al., 1996), supporting the hypothesis that females copulate with males who contribute "good genes" (i.e. increased viability and/or attractiveness) to their offspring. Remarkably, in two of these studies extra-pair offspring had higher survivorship than within-pair offspring (Kempenaers et al., 1992; Hasselquist et al., 1996). These few studies that have identified sires of all offspring are clearly just the tip of the iceberg; we anticipate that future studies using microsatellites or a combination of molecular methods will generate similarly fascinating results.

Measuring relatedness

For many questions in behavioral ecology and evolutionary biology, such as studies of cooperative behavior, inbreeding and/or dispersal, knowing the relatedness between individuals will be important. In some cases, this can be accomplished by combining parentage analyses with behavioral observations to construct a pedigree (e.g. Haig et al., 1993). Such analyses, though, are not always possible, and so a more practical approach to estimating relatedness is preferable. Few studies of birds have used genetic techniques to measure relatedness, and this approach has found its widest application and development in studies of social insects and mammals. Here we review some of the molecular and statistical techniques available and indicate how they could be applied to studies of birds.

1. What is the average relatedness among a group of individuals?

As in analyses parentage, analyses of kinship fall into different categories of precision, depending on the level of relatedness information required. For some questions, specific estimates of the degree of relatedness (r) may not be needed, such that a qualitative assessment of average relatedness within a group may be adequate. For example, one might wish to know if males or females choose mates that are more related than expected by chance (Bateson, 1983), whether limited dispersal leads to population subdivision (Seitz et al., 1994), or whether subsets of individuals within a population showing a cooperative behavior are more closely related than randomly selected individuals (Packer et al., 1991; Pfennig and Reeve, 1993). For other questions a more precise assessment of the average relatedness within a group may be needed to test quantitative predictions. For example, studies of the mechanisms of kin recognition and analyses of kin selection often require that the average coefficient of relatedness (r) be estimated.

Fortunately, the same techniques that allow qualitative assessment of average relatedness also can be used to calculate the average coefficient of relatedness within the group. Indeed, any technique that estimates degree of genetic similarity, including protein electrophoresis, can be used to answer questions at this level of resolution. However, highly variable loci such as minisatellites will provide the most powerful method for comparing average relatedness within groups (Lynch, 1990, 1991; Jones et al., 1991; Haig et al., 1993; Blouin et al., 1996). Levels of band-sharing similarity can be used to qualitatively compare relatedness among individuals in different groups. Moreover, techniques originally derived for protein allozymes can be used to calculate the average degree of relatedness from data for any single locus marker (Pamilo and Crozier, 1982; Pamilo, 1984, 1989; Wilkinson and McCracken, 1985; Queller and Goodnight, 1989), and these analytical techniques have been extended to multilocus minisatellite (Reeve et al., 1992) and RAPD (Lynch and Milligan, 1994) markers. These approaches have been empirically demonstrated to give accurate estimates of average relatedness (Reeve et al., 1992; Webster, 1995; Blouin et al., 1996).

To date, relatively few studies of birds have used such molecular kinship information to test general hypotheses, and those that have fall into three general categories. First, a number of studies have used genetic information to examine population genetic structure, and a growing number of these are using DNA markers (e.g. Avise et al., 1992; Haig et al., 1993, 1994a; Haig, 1996). For example, Baker et al. (1995) found that limited dispersal has led to extreme population substructuring in flightless kiwis. Tegelstrøm and Sjoberg (1995) used DNA fingerprinting to show a dramatically low level of genetic variation in European populations of Canada geese, apparently as a consequence of repeated founder events. Some studies have used high-precision molecular markers to show a remarkable degree of small-scale population substructure based on kinship (Parker et al., 1995; Friessen et al., 1996). Similar approaches are being used to examine genetic variation and design effective conservation programs for several endangered species of bird (Brock and White, 1992; Geyer et al., 1993; Haig et al., 1994a, 1995; Signer et al., 1994).

21

Second, molecular markers have been used to assess the outcome of inbreeding (May et al., 1993a; Marin et al., 1994; Zhu et al., 1996) and potential mechanisms to avoid it. For example, some studies have shown that parents with high genetic similarity (as measured by multilocus minisatellite band sharing) have higher levels of reproductive failure than other mated pairs (Brock and White, 1992; Bensch et al., 1994; Kempenaers et al., 1996; see also McRae, 1996). These studies did not find that females used extra-pair copulations to avoid the deleterious effects of inbreeding (Brooker et al., 1990), although other studies have given some evidence in support of this hypothesis (Wetton and Parkin, 1991; Rätti et al., 1995).

Third, kinship is thought to play a central role in the evolution of cooperation, and several studies of social insects (Crozier et al., 1987; Strassmann, 1989; Ross, 1993; Evans, 1995; Heinze, 1995; Paxton et al., 1996) and mammals (Reeve et al., 1990; Packer et al., 1991; Lehman et al., 1992; Morin et al., 1994) have used molecular techniques to test this hypothesis. Relatively few studies of cooperative birds have tested this hypothesis using molecular markers. Molecular studies have demonstrated mixed parentage within the nests of several communally breeding birds (Jamieson et al., 1994; Millar et al., 1994; Quinn et al., 1994; Faaborg et al., 1995; Põldmaa et al., 1995; McRae, 1996). Similar studies of cooperatively breeding birds have shown that extra-pair copulations are usually rare or absent (Haig et al., 1994c; Dickinson et al., 1995; Haydock et al., 1996), such that helpers are usually raising close kin. Interestingly, EPF are extremely common in cooperatively breeding fairy wrens (Brooker et al., 1990; Mulder et al., 1994; Webster and Pruett-Jones, unpublished data), casting serious doubt on the importance of kinship in the evolution of helping behavior in this group (Dunn et al., 1995; see also Brooke and Hartley, 1995).

2. What is the exact relatedness between pairs of individuals (precise calculations of relatedness)?

In some cases it would be desirable to know the degree of relatedness between a specific pair of individuals. Because molecular genetic techniques provide a measure of genetic similarity, it is tempting to conclude that a measure such as the degree of band-sharing can be used to calculate relatedness. As described above, it is possible to calculate average relatedness among several individuals in a group, but as the number of individuals in the group decreases, confidence in the estimate of r also decreases. The reasons for this are presented in detail elsewhere (Lynch, 1988, 1990, 1991), and are empirically confirmed for birds (Westneat, 1990; Rabenold et al., 1991; Piper and Rabenold, 1992). In general, because molecular techniques examine only a portion of the genome, and because the degree of genetic similarity can vary among individuals in many kinship categories, measures of genetic similarity likely will overlap broadly for different kinship categories. Consequently, a measure of genetic similarity between two individuals does not necessarily correspond to a specific coefficient of relatedness between them.

Given these problems, it is not surprising that few studies have used molecular genetic techniques to calculate a coefficient of relatedness between specific individuals. In one interesting study, McDonald and Potts (1994) used microsatellites to calculate the degree of relatedness between cooperating pairs of male long-tailed manakins. Although the error limits were large, these estimates allowed the kin selection hypothesis for the evolution of cooperation to be rejected in the manakin system. Two other studies of birds were able to distinguish between first-order (parent-offspring, siblings) and second-order (cousins) relatives using band-sharing proportions alone (Jones et al., 1991; Piper and Rabenold, 1992), but most other avian studies have not achieved this level of resolution (Westneat, 1990; Haig et al., 1993).

In the future, studies requiring this level of genetic resolution may only be possible in special situations. If estimates of relatedness between individuals do not have to be extremely precise (as in the manakin study), then several of the techniques mentioned above may be applicable. Furthermore, a combination of genetic analyses and long-term behavioral observations (giving a putative pedigree) may allow relatedness between dyads to be assessed, but generally only in sedentary populations. Pedigrees are much more difficult to determine for populations where tracking offspring to adulthood is not possible, and calculating r between specific dyads of individuals in these circumstances will only be possible if several assumptions are made. At present the amount of work required for the necessary level of resolution is probably too great to make the endeavor worthwhile; scaled-back versions of hypotheses that require only estimates of average r among a group are much more feasible.

Sex ratio

Although not questions of kinship *per se*, sex ratios are potentially influenced by issues of kinship, as parental manipulation of sex ratios could be affected by relatedness as well as other ecological factors (Clutton-Brock, 1986; Gowaty, 1993). The determination of primary (hatching) sex ratio is extremely difficult in most birds, as it is often not possible to determine the gender of individuals until long after they have left the nest. This problem is particularly accute for migratory passerines that typically show strong sex-biased mortality and dispersal. Fortunately, methods for isolating molecular sex markers have been developed (Griffiths and Tiwari, 1993), useful markers have already been isolated for birds (Quinn et al., 1990; Griffiths et al., 1992; Millar et al., 1992; May et al., 1993b; Griffiths and Tiwari, 1995; Millar et al., 1996) and some of these markers appear to have nearly universal applicability (Ellegren, 1996; Griffiths et al., 1996). Thanks to these developments, a handful of recent studies have been able to address fascinating and fundamental questions regarding avian sex ratios (e.g. Longmire et al., 1991; Olsen and Cockburn, 1991; Griffiths, 1992; Ellegren et al., 1996; Nishiumi et al., 1996; Sheldon and Ellegren, 1996; Svensson and Nilsson, 1996; Komdeur et al., 1997). We anticipate an explosion of similar studies in the near future.

Conclusions and future directions

The above review clearly demonstrates the power that molecular techniques bring to the study of kinship in birds and other taxa. We fully expect the curve shown in Figure 1 to continue to rise steeply for some time, especially if PCR-based techniques become widely available. In our earlier review (Westneat and Webster, 1994), we listed four general directions for future molecular studies of kinship in birds. The recent studies outlined above have made some significant progress in these areas, but a number of interesting questions remain to be answered. First, molecular markers have found their widest application in the study of extra-pair paternity, and we predict that this will continue to be the case. The studies reviewed here have suggested that females are often in control of extra-pair paternity, that they may copulate with extra-pair males to obtain "good genes", and that extra-pair fertilizations are a strong selective force shaping male morphology and behavior. However, these studies are far from conclusive and much debate still exists regarding the function and consequences of EPF (e.g. Sheldon, 1994; Wagner, 1994; Weatherhead, 1994; Webster et al., 1995). Moreover, too few species have been studied to know how generalizable conclusions from one or two might be. We anticipate a number of future studies, employing a mixture of genetic markers and experimental manipulation, will investigate the mechanisms of EPF and sperm competition, ecological and social factors influencing the prevalence of EPF and the selective consequences of EPF. Second, we know less about IBP than we do about EPF. Molecular markers have the potential to give a detailed picture of the costs and benefits of this female reproductive strategy. This will be necessary before we can understand why IBP is so rare relative to EPF, and why it is seemingly common in some species. Third, the application of molecular markers to questions of kinship beyond that of parent-offspring is just now getting underway. The handful of avian studies that have been conducted show a surprising level of fine-grain kin structure in populations; these few studies have undoubtedly just scratched the surface. Moreover, although we have learned a great deal from studies that did not use molecular markers (Emlen, 1991), these markers will undoubtedly allow powerful, quantitative tests of competing hypotheses for the evolution of cooperation in birds, just as they have for studies of social insects. Finally, the recent addition of molecular tools for sex determination has opened the door for a suite of studies examining fundamental questions about sex ratio variation in birds. Like many of the other questions now being addressed with molecular markers, the ability to answer these questions was unthinkable just a few short years ago, and every new result seems to spark new questions.

As we stated in our earlier review, molecular methods are only one of several tools available to evolutionary biologists. In the end, careful integration of molecular tools into detailed observational and clever experimental studies will generate the most dividends. Hopefully our discussion here can help researchers accomplish this by balancing the benefit gained from a particular tool with the effort necessary to use it.

References

Alatalo, R. V., Burke, T., Dann, J., Hanotte, O., Höglund, J., Lundberg, A., Moss, R. and Rintamäki, P. T. (1996) Paternity, copulation disturbance and female choice in lekking black grouse. *Anim. Behav.* 52: 861–873.

Altmann, J., Alberts, S. C., Haines, S. A., Dubach, J., Muruthi, P., Coote, T., Geffen, E., Cheesman, D. J., Mututua, R. S., Saiyalel, S. N., Wayne, R. K., Lacy, R. C. and Bruford, M. W. (1996.) Behavior predicts genetic structure in a wild primate group. *Proc. Natl. Acad. Sci. USA* 93: 5797–5801.

Amos, B., Schlötterer, C. and Tautz, D. (1993) Social structure of pilot whales revealed by analytical DNA profiling. *Science* 260: 670–672.

Armour, J. A. L., Neumann, R., Gobert, S. and Jeffreys, A. J. (1994) Isolation of human simple repeat loci by hybridization selection. *Hum. Mol. Genet.* 3: 599–605.

Avise, J. C. and Aquadro, C. F. (1982) A comparative summary of genetic distances in the vertebrates. Patterns and correlations. *Evolut. Biol.* 15: 151–185.

Avise, J. C., Alisauskas, R. T., Nelson, W. S. and Ankney, C. D. (1992) Matriarchal population genetic structure in an avian species with female natal philopatry. *Evolution* 46: 1084–1096.

Baker, A. J., Daugherty, C. H., Colbourne, R. and McLennan, J. L. (1995) Flightless brown kiwis of New Zealand possess extremely subdivided population structure and cryptic species like small mammals. *Proc. Natl. Acad. Sci. USA* 92: 8254–8258.

Barrowclough, G. F. and Corbin, K. W. (1978) Genetic variation and differentiation in the Parulidae. *Auk* 95: 691–702.

Bateson, P. (1983) Optimal outbreeding. *In*: Bateson, P. (ed.) *Mate Choice*, Cambridge University Press, Cambridge, pp. 257–278.

Bensch, S., Haasselquist, D. and von Schantz, T. (1994) Genetic similarity between parents predicts hatching failure: nonincestuous inbreeding in the great reed warbler? *Evolution* 48: 317–326.

Bensch, S., Price, T. and Kohn, J. (1997) Isolation and characterization of microsatellite loci in a Phylloscopus warbler. *Molec. Ecol.* 6: 91–92.

Birkhead, T. R. and Møller, A. P. (1992) *Sperm Competition in Birds*, Academic Press, New York.

Birkhead, T. R., Burke, T., Zann, R., Hunter, F. M. and Krupa, A. P. (1990) Extra-pair paternity and intraspecific brood parasitism in wild zebra finches *Taeniopygia guttata*, revealed by DNA fingerprinting. *Behav. Ecol. Sociobiol.* 27: 315–324.

Blouin, M. S., Parsons, M., Lacailee, V. and Lotz, S. (1996) Use of microsatellite loci to classify individuals by relatedness. *Molec. Ecol.* 5: 393–401.

Bollinger, E. K. and Gavin, T. A. (1991) Patterns of extra-pair fertilizations in bobolinks. *Behav. Ecol. Sociobiol.* 29: 1–7.

Bowcock, A. M., Ruiz-Linares, A., Tomfohrde, J., Minch, E., Kidd, J. R. and Cavalli-Sforza, L. L. (1994) High resolution of human evolutionary trees with polymorphic microsatellites. *Nature* 368: 455–457.

Brock, M. K. and White, B. N. (1992) Application of DNA fingerprinting to the recovery program of the endangered Peurto Rican parrot. *Proc. Natl. Acad. Sci. USA* 89: 11121–11125.

Brooke, M. D. L. and Hartley, I. R. (1995) Nesting Henderson reed-warblers (*Acrocephalus vaughani taiti*) studied by DNA fingerprinting: unrelated coalitions in a stable habitat? *Auk* 112: 77–86.

Brooker, M. G., Rowley, I., Adams, M. and Baverstock, P. R. (1990) Promiscuity: an inbreeding avoidance mechanism in a socially monogamous species? *Behav. Ecol. Sociobiol.* 26: 191–199.

Brown, C. R. and Brown, M. B. (1988) Genetic evidence of multiple parentage in broods of cliff swallows. *Behav. Ecol. Sociobiol.* 23: 379–387.

Burke, T. (1989) DNA fingerprinting and other methods for the study of mating success. *Trends. Ecol. Evolut.* 4: 139–144.

Burke, T. and Bruford, M. W. (1987) DNA fingerprinting in birds. *Nature* 327: 149–152.

Burke, T., Davies, N. B., Bruford, M. W. and Hatchwell, B. J. (1989) Parental care and mating behaviour of polyandrous dunnocks *Prunella modularis* related to paternity by DNA fingerprinting. *Nature* 338: 249–251.

Burke, T., Hanotte, O., Bruford, M. W. and Cairns, E. (1991) Multilocus and single locus minisatellite analysis in population biological studies. *In*: Burke, T., Dolf, G., Jeffreys, A. J. and Wolff, R. (eds) *DNA Fingerprinting: Approaches and Applications*, Birkhäuser Verlag, Basel, pp. 154–168.

Burley, N. T., Parker, P. G. and Kundy, K. (1996) Sexual selection and extrapair fertilization in a socially monogamous passerine, the zebra finch (*Taeniopygia guttata*). *Behav. Ecol.* 7: 218–226.

Clutton-Brock, T. H. (1986) Sex ratio variation in birds. *Ibis* 128: 317–329.

Crozier, R. H., Smith, B. H. and Crozier, Y. C. (1987) Relatedness and population structure of the primitively eusocial bee *Lasioglossum zephyrum* (Hymenoptera: Halictidae) in Kansas. *Evolution* 41: 902–910.

Davies, N. B., Hatchwell, B. J., Robson, T. and Burke, T. (1992) Paternity and parental effort in dunnocks *Prunella modularis*: how good are male chick-feeding rules? *Anim. Behav.* 43: 729–745.

Decker, M. D., Parker, P. G., Minchella, D. J. and Rabenold, K. N. (1993) Monogamy in black vultures: genetic evidence from DNA fingerprinting. *Behav. Ecol.* 4: 29–35.

Dickinson, J., Haydock, H., Koenig, W., Stanback, M. and Pitelka, F. (1995) Genetic monogamy in single-male groups of acorn woodpeckers, *Melanerpes formicivorus*. *Molec. Ecol.* 4: 765–769.

Dixon, A., Ross, D., O'Malley, S. L. C. and Burke, T. (1994) Paternal investment inversely related to degree of extra-pair paternity in the reed bunting. *Nature* 371: 698–700.

Dow, B. D. and Ashley, M. V. (1996) Microsatellite analysis of seed dispersal and parentage of saplings in bur oak, *Quercus macrocarpa. Molec. Ecol.* 5: 615–627.

Dunn, P. O., Cockburn, A. and Mulder, R. A. (1995) Fairy-wren helpers often care for young to which they are unrelated. *Proc. R. Soc. Lond. B* 259: 339–343.

Dunn, P. O., Whittingham, L. A., Lifjeld, J. T. and Boag, P. T. (1994) Effects of breeding density, synchrony, and experience on extrapair paternity in tree swallows. *Behav. Ecol.* 5: 123–129.

Edwards, K. J., Barker, J. H. A., Daly, A., Jones, C. and Karp, A. (1996) Microsatellite libraries enriched for several microsatellite sequences in plants. *BioTechniques* 20: 758–760.

Ellegren, H. (1992) Polymerase-chain-reaction (PCR) analysis of microsatellites – a new approach to studies of genetic relationships in birds. *Auk* 109: 886–895.

Ellegren, H. (1996) First gene on the avian W chromosome (CHD) provides a tag for universal sexing of non-ratite birds. *Proc. R. Soc. Lond. B* 263: 1635–1641.

Ellegren, H., Lifjeld, J. T., Slagsvold, T. and Primmer, C. R. (1995) Handicapped males and extrapair paternity in pied flycatchers: a study using microsatellite markers. *Molec. Ecol.* 4: 739–744.

Ellegren, H., Gustafsson, L. and Sheldon, B. C. (1996) Sex ratio adjustment in relation to paternal attractiveness in a wild bird population. *Proc. Natl. Acad. Sci. USA* 93: 11723–11728.

Emlen, S.T. (1991) Evolution of cooperative breeding in birds and mammals. *In*: Krebs, J. R. and Davies, N. B. (eds) *Behavioral Ecology: An Evolutionary Approach*, Blackwell, Oxford, pp. 301–337.

Ender, A., Schwenk, K., Städler, T., Streit, B. and Schierwater, B. (1996) RAPD identification of microsatellites in *Daphnia. Molec. Ecol.* 5: 437–441.

Engel, S. R., Linn, R. A., Taylor, J. F. and Davis, S. K. (1996) Conservation of microsatellite loci across species of artiodactyls: implications for population studies. *J. Mammal.* 77: 504–518.

Epplen, J. T., Ammer, H., Epplen, C., Kammerbauer, C., Mitreiter, R., Roewer, L., Schwaiger, W., Steimle, V., Zischler, H., Albert, E., Andreas, A., Beyermann, B., Meyer, W., Buitkamp, J., Nanda, I., Schmid, M., Nürnberg, P., Pena, S. D. J., Pöche, H., Sprecher, W., Schartl, M., Weising, K. and Yassouridis, A. (1991) Oligonucleotide fingerprinting using simple repeat motifs: a convenient, ubiquitously applicable method to detect hypervariability for multiple purposes. *In*: Burke, T., Dolf, G., Jeffreys, A. J. and Wolff, R. (eds) *DNA Fingerprinting: Approaches and Applications*, Birkhäuser Verlag, Basel, pp. 50–69.

Evans, J. D. (1993) Parentage analyses in ant colonies using simple sequence repeat loci. *Molec. Ecol.* 2: 393–397.

Evans, J. D. (1995) Relatedness threshold for the production of female sexuals in colonies of a polygynous ant, *Myrmica tahoensis*, as revealed by microsatellite DNA analysis. *Proc. Natl. Acad. Sci. USA* 92: 6514–6517.

Faaborg, J., Parker, P. G., DeLay, L., de Vries, T., Bednarz, J. C., Maria Paz, S., Naranjo, J. and Waite, T. A. (1995) Confirmation of cooperative polyandry in the Galapagos hawk (*Buteo galapagoensis*). *Behav. Ecol. Sociobiol.* 36: 83–90.

Freeman-Gallant, C. R. (1996) DNA fingerprinting reveals female preference for male parental care in savannah sparrows. *Proc. R. Soc. Lond. B* 263: 157–160.

Friessen, V. L., Montevecchi, W. A., Gaston, A. J., Barrett, R. T. and Davidson, W. S. (1996) Molecular evidence for kin groups in the absence of large-scale genetic differentiation in a migratory bird. *Evolution* 50: 924–930.

Geyer, C. J., Ryder, O. A., Chemnick, L. G. and Thompson, E. A. (1993) Analysis of relatedness in the California condors, from DNA fingerprints. *Molec. Biol. Evol.* 10: 571–589.

Gibbs, H. L., Weatherhead, P. J., Boag, P. T., White, B. N., Tabak, L. M. and Hoysak, D. J. (1990) Realized reproductive success of polygynous red-winged blackbirds revealed by DNA markers. *Science* 250: 1394–1397.

Gibbs, H. L., Brooke, M. D. L. and Davies, N. B. (1996) Analysis of genetic differentiation of host races of the common cuckoo *Cuculus canorus* using mitochondrial and microsatellite DNA variation. *Proc. R. Soc. Lond. B* 263: 89–96.

Gibbs, H. L., Miller, P., Alderson, G. and Sealy, S. G. (1997) Genetic analysis of brown-headed cowbirds *Molothrus ater* raised by different hosts: data from mtDNA and microsatellite DNA markers. *Molec. Ecol.* 6: 189–193.

Gowaty, P. A. (1993) Differential dispersal, local resource competition and sex ratio variation in birds. *Amer. Naturalist* 141: 263–280.

Gowaty, P. A. and W. C. Bridges (1991) Behavioral, demographic, and environmental correlates of extrapair fertilizations in eastern bluebirds, *Sialia sialis*. *Behav. Ecol.* 2: 339–350.

Gowaty, P. A. and Karlin, A. A. (1984) Multiple maternity and paternity in single broods of apparently monogamous eastern bluebirds (*Sialia sialis*). *Behav. Ecol. Sociobiol.* 15: 91–95.

Gray, E. M. (1996) Female control of offspring paternity in a western population of red-winged blackbirds (*Agelaius phoeniceus*). *Behav. Ecol. Sociobiol.* 38: 267–278.

Green, D. J., Cockburn, A., Hall, M. L., Osmond, H. and Dunn, P. O. (1995) Increased opportunities for cuckoldry may be why dominant male fairy-wrens tolerate helpers. *Proc. R. Soc. Lond. B* 262: 297–303.

Griffiths, R. (1992) Sex-biased mortality in the lesser black-backed gull *Larus fuscus* during the nestling stage. *Ibis* 134: 237–244.

Griffiths, R. and Tiwari, B. (1993) The isolation of molecular genetic markers for the identification of sex. *Proc. Natl. Acad. Sci. USA* 90: 8324–8326.

Griffiths, R. and Tiwari, B. (1995) Sex of the last wild Spix's macaw. *Nature* 375: 454.

Griffiths, R., Tiwari, B. and Becher, S. A. (1992) The identification of sex in the starling *Sturnus vulagaris* using a molecular DNA technique. *Molec. Ecol.* 1: 191–194.

Griffiths, R., Daan, S. and Dijkstra, C. (1996) Sex identification in birds using two CHD genes. *Proc. R. Soc. Lond. B* 263: 1251–1256.

Grosberg, R. K., Levitan, D. R. and Cameron, B. B. (1996) Characterization of genetic structure and genealogies using RAPD-PCR markers: A random primer for the novice and nervous. *In*: Ferraris, J. D. and Palumbi, S. R. (eds) *Molecular Zoology: Advances, Strategies and Protocols*, Wiley-Liss, New York, pp. 67–95.

Gyllensten, U. B., Jakobsson, S. and Temrin, H. (1990) No evidence for illegitimate young in monogamous and polygynous warblers. *Nature* 343: 168–170.

Hadrys, H., Schierwater, B., Dellaporta, S. L., Desalle, R. and Buss, L. W. (1993) Determination of paternity in dragonflies by Random Amplified Polymorphic DNA fingerprinting. *Molec. Ecol.* 2: 79–87.

Haig, S. M. (1996) Population structure of red-cockaded woodpeckers in south Florida: RAPDs revisited. *Molec. Ecol.* 5: 725–734.

Haig, S. M., Belthoff, J. R. and Allen, D. H. (1993) Examination of population structure in red-cockaded woodpeckers using DNA profiles. *Evolution* 47: 185–194.

Haig, S. M., Ballou, J. D. and Casna, N. J. (1994a) Identification of kin structure among Guam rail founders: a comparison of pedigrees and DNA profiles. *Molec. Ecol.* 3: 109–119.

Haig, S. M., Rhymer, J. M. and Heckel, D. G. (1994b) Population differentiation in randomly amplified polymorphic DNA of red-cockaded woodpeckers *Picoides borealis. Molec. Ecol.* 3: 581–595.

Haig, S. M., Walters, J. R. and Plissner, J. H. (1994c) Genetic evidence for monogamy in the cooperatively breeding red-cockaded woodpecker. *Behav. Ecol. Sociobiol.* 34: 295–303.

Haig, S. M., Ballou, J. D. and Casna, N. J. (1995) Genetic identification of kin in micronesian kingfishers. *J. Hered.* 86: 423–431.

Hammond, R. L., Saccheri, I. J., Ciofi, C., Coote, T., Funk, S. M., McMillan, W. O., Bayes, M. K., Taylor, E. and Bruford, M. W. (1998) Isolation of microsatellite markers in animals. *In*: Karp, A., Isaac, P. G. and Ingram, D. S. (eds) *Molecular Tools for Screening Biodiversity: Plants and Animals*, Chapman and Hall, London, pp. 279–285.

Hanotte, O., Burke, T., Armour, J. A. L. and Jeffreys, A. J. (1991a) Cloning, characterization and evolution of Indian peafowl *Pavo cristatus* minisatellite loci. *In*: Burke, T., Dolf, G., Jeffreys, A. J. and Wolff, R. (eds) *DNA Fingerprinting: Approaches and Applications*, Birkhäuser Verlag, Basel, pp. 193–216.

Hanotte, O., Burke, T., Armour, J. A. L. and Jeffreys, A. J. (1991b) Hypervariable minisatellite DNA sequences in the Indian peafowl *Pavo cristatus. Genomics* 9: 587–597.

Hanotte, O., Cairns, E., Robson, T., Double, M. C. and Burke, T. (1992) Cross-species hybridization of a single-locus minisatellite probe in passerine birds. *Molec. Ecol.* 1: 127–130.

Hanotte, O., Zanon, C., Pugh, A., Greig, C., Dixon, A. and Burke, T. (1994) Isolation and characterization of microsatellite loci in a passerine bird: the reed bunting *Emberiza schoeniclus. Molec. Ecol.* 3: 529–530.

Hasselquist, D., Bensch, S. and von Schantz, T. (1995) Low frequency of extrapair paternity in the polygynous great reed warbler, *Acrocephalus arundinaceus. Behav. Ecol.* 6: 27–38.

Hasselquist, D., Bensch, S. and von Schantz, T. (1996) Correlation between male song repertoire, extra-pair paternity and offspring survival in the great reed warbler. *Nature* 381: 229–232.

Haydock, J., Parker, P. G. and Rabenold, K. N. (1996) Extra-pair paternity uncommon in the cooperativley breeding bicolored wren. *Behav. Ecol. Sociobiol.* 38: 1–16.

Heinze, J. (1995) Reproductive skew and genetic relatedness in Leptothorax ants. *Proc. R. Soc. Lond.* B 261: 375–379.

Hill, G. E., Montgomerie, R., Roeder, C. and Boag, P. (1994) Sexual selection and cuckoldry in a monogamous songbird: implications for sexual selection theory. *Behav. Ecol. Sociobiol.* 35: 193–199.

Hughes, A. L. and Hughes, M. K. (1995) Small genomes for better flyers. *Nature* 377: 391.

Jamieson, I. G., Quinn, J. S., Rose, P. A. and White, B. N. (1994) Shared paternity among non-relatives is a result of an egalitarian mating system in a communally breeding bird, the pukeko. *Proc. R. Soc. Lond.* B 257: 271.

Jeffreys, A. J., Wilson, V. and Thein, S. L. (1985a) Hypervariable "minisatellite" regions in human DNA. *Nature* 314: 67–73.

Jeffreys, A. J., Wilson, V. and Thein, S. L. (1985b) Individual-specific "fingerprints" of human DNA. *Nature* 316: 76–79.

Jones, C. S., Lessells, C. M. and Krebs, J. R. (1991) Helpers-at-the-nest in European bee-eaters (*Merops apiaster*): a genetic analysis. *In*: Burke, T., Dolf, G., Jeffreys, A. J. and Wolff, R. (eds) *DNA Fingerprinting: Approaches and Applications*, Birkhäuser Verlag, Basel, pp. 169–192.

Jones, C. S., Okamura, B. and Noble, L. R. (1994) Parent and larval RAPD fingerprints reveal outcrossing in freshwater bryozoans. *Molec. Ecol.* 3: 193–199.

Kandpal, R. P., Kandpal, G. and Weissman, S. M. (1994) Construction of libraries enriched for sequence repeats and jumping clones, and hybridization selection for region-specific markers. *Proc. Natl. Acad. Sci. USA* 91: 88–92.

Karagyozov, L., Kalcheva, I. D. and Chapman, V. M. (1993) Construction of random small-insert genomic libraries highly enriched for simple sequence repeats. *Nucl. Acid. Res.* 21: 3911–3912.

Kellogg, K. A., Markert, J. A., Stauffer, J. R. and Kocher, T. D. (1995) Microsatellite variation demonstrates multiple paternity in lekking cichlid fishes from Lake Malawi, Africa. *Proc. R. Soc. Lond. B* 260: 79–84.

Kempenaers, B., Verheyen, G. R., Van den Broeck, M., Burke, T., Van Broeckhoven, C. and Dhondt, A. A. (1992) Extra-pair paternity results from female preference for high-quality males in the blue tit. *Nature* 357: 494–496.

Kempenaers, B., Adriaensen, F., van Noordwijk, A. J. and Dhondt, A. A. (1996) Genetic similarity, inbreeding and hatching failure in blue tits: are hatched eggs infertile? *Proc. R. Soc. Lond. B* 263: 179–185.

Kijas, J. M. H., Fowler, J. C. S., Garbett, C. A. and Thomas, M. R. (1994) Enrichment of microsatellites from the citrus genome using biotinylated oligonucleotide sequences bound to streptavidin-coated magnetic particles. *BioTechniques* 16: 656–662.

Komdeur, J., Daan, S., Tinbergen, J. and Mateman, C. (1997) Extreme adaptive modification in sex ratio of the Seychelles warbler's eggs. *Nature* 385: 522–525.

Konishi, M., Emlen, S. T., Ricklefs, R. E. and Wingfield, J. C. (1989) Contributions of bird studies to biology. *Science* 246: 465–472.

Krokene, C., Anthonisen, K., Lifjeld, J. T. and Amundsen, T. (1996) Paternity and paternity assurance behaviour in the bluethroat, *Luscinia s. svecica. Anim. Behav.* 52: 405–417.

Lehman, N., Clarkson, P., Mech, L. D., Meier, T. J. and Wayne, R. K. (1992) A study of the genetic relationships within and among wolf packs using DNA fingerprinting and mitochondrial DNA. *Behav. Ecol. Sociobiol.* 30: 83–94.

Levitan, D. R. and Grosberg, R. K. (1993) The analysis of paternity and maternity in the marine hydrozoan *Hydractinia symbiolognicarpus* using randomly amplified polymorphic DNA (RAPD) markers. *Molec. Ecol.* 2: 315–326.

Lewis, P. O. and Snow, A. A. (1992) Deterministic paternity exclusion using RAPD markers. *Molec. Ecol.* 1: 155–160.

Lifjeld, J. T. and Robertson, R. J. (1992) Female control of extra-pair fertilization in tree swallows. *Behav. Ecol. Sociobiol.* 31: 89–96.

Longmire, J. L., Ambrose, R. E., Brown, N. C., Cade, T. J., Maechtle, T. L., Seegar, W. S., Ward, F. P. and White, C. M. (1991) Use of sex-linked minisatellite fragments to investigate genetic differentiation and migration of North American populations of the peregrine falcon (*Falco peregrinus*). *In*: Burke, T., Dolf, G., Jeffreys, A. J. and Wolff, R. (eds) *DNA Fingerprinting: Approaches and Applications*, Birkhäuser Verlag, Basel, pp. 217–229.

Lynch, M. (1988) Estimation of relatedness by DNA fingerprinting. *Molec. Biol. Evol.* 5: 584–599.

Lynch, M. (1990) The similarity index and DNA fingerprinting. *Molec. Biol. Evol.* 7: 478–484.

Lynch, M. (1991) Analysis of population genetic structure by DNA fingerprinting. *In*: Burke, T., Dolf, G., Jeffreys, A. J. and Wolff, R. (eds) *DNA Fingerprinting: Approaches and Applications*, Birkhäuser Verlag, Basel, pp. 113–126.

Lynch, M. (1994) Analysis of population genetic structure with RAPD markers. *Molec. Ecol.*

3: 91–99.

Macdougall-Shackleton, E. A., Robertson, R. J. and Boag, P. T. (1996) Temporary male removal increases extra-pair paternity in eastern bluebirds. *Anim. Behav.* 52: 1177–1183.

Marin, G., Marchesini, M., Tiloca, G. and Pagano, A. (1994) DNA fingerprinting fails to reveal inbreeding in a small, closed population of bearded tits (*Panurus biarmicus* L.). *Ethol. Ecol. Evol.* 6: 243–248.

May, C. A., Wetton, J .H., Davis, P. E., Brookfield, J. F. Y. and Parkin, D T. (1993a) Single-locus profiling reveals loss of variation in inbred populations of the red kite (*Milvus milvus*). *Proc. R. Soc. Lond. B* 251: 165–170.

May, C. A., Wetton, J .H. and Parkin, D T. (1993b) Polymorphic sex-specific sequences in birds of prey. *Proc. R. Soc. Lond. B* 253: 271–276.

McDonald, D. B. and Potts, W. K. (1994) Cooperative display and relatedness among males in a lek-mating bird. *Science* 266: 1030–1032.

McKitrick, M. C. (1990) Genetic evidence for multiple parentage in eastern kingbirds (*Tyrannus tyrannus*). *Behav. Ecol. Sociobiol.* 26: 149–155.

McRae, S. B. (1996) Family values: costs and benefits of communal nesting in the moorhen. *Anim. Behav.* 52: 225–245.

McRae, S. B. and Burke, T. (1996) Intraspecific brood parasitism in the moorhen: parentage and parasite-host relationships determined by DNA fingerprinting. *Behav. Ecol. Sociobiol.* 38: 115–129.

Millar, C. D., Lambert, D. M., Bellamy, A. R., Stapleton, P. M. and Young, E. C. (1992) Sex-specific restriction fragments and sex ratios revealed by DNA fingerprinting in the brown skua. *J. Hered.* 83: 350–355.

Millar, C. D., Anthony, I., Lambert, D. M., Stapleton, P. M., Bergmann, C. C., Bellamy, A. R. and Young, E. C. (1994) Patterns of reproductive success determined by DNA fingerprinting in a communally breeding oceanic bird. *Biol. J. Linn. Soc.* 52: 31–48.

Millar, C. D., Lambert, D. M., Anderson, S. and Halverson, J. L. (1996) Molecular sexing of the communally breeding pukeko: an important ecological tool. *Molec. Ecol.* 5: 289–293.

Møller, A. P. and Birkhead, T. R. (1993a) Certainty of paternity covaries with paternal care in birds. *Behav. Ecol. Sociobiol.* 33: 261–268.

Møller, A. P. and Birkhead, T. R. (1993b) Cuckoldry and sociality: a comparative study of birds. *Amer. Naturalist* 142: 118–140.

Møller, A. P. and Birkhead, T. R. (1994) The evolution of plumage brightness in birds is related to extrapair paternity. *Evolution* 48: 1089–1100.

Møller, A. P. and Briskie, J. V. (1995) Extra-pair paternity, sperm competition and the evolution of testis size in birds. *Behav. Ecol. Sociobiol.* 36: 357–365.

Morin, P. A., Moore, J. J., Chakraborty, R., Jin, L., Goodall, J. and Woodruff, D. S. (1994) Kin selection, social structure, gene flow and the evolution of chimpanzees. *Science* 265: 1193–1201.

Mulder, R. A., Dunn, P. O., Cockburn, A., Lazenby-Cohen, K. A. and Howell, M. J. (1994) Helpers liberate female fairy-wrens from constraints on extra-pair mate choice. *Proc. R. Soc. Lond. B* 255: 223–229.

Nakamura, Y., Leppert, M., O'Connell, P., Wolff, R., Holm, T., Culver, M., Martin, C., Fujimoto, E., Hoff, M., Kumlin, E. and White, R. (1987) Variable number of tandem repeat (VNTR) markers for human gene mapping. *Science* 235: 1616–1622.

Neumann, K. and J. H. Wetton (1996) Highly polymorphic microsatellites in the house sparrow *Passer domesticus*. *Molec. Ecol.* 5: 307–309.

Nishiumi, I., Yamagishi, S., Maekawa, H. and Shimoda, C. (1996) Paternal expenditure is related to brood sex ratio in polygynous great reed warblers. *Behav. Ecol. Sociobiol.* 39:
211–217.

O'Donnell, S. (1996) RAPD markers suggest genotypic effects on forager specialization in a
eusocial wasp. *Behav. Ecol. Sociobiol.* 38: 83–88.

Olsen, P. D. and Cockburn, A. (1991) Female-biased sex allocation in peregrine falcons and
other raptors. *Behav. Ecol. Sociobiol.* 28: 417–423.

Oring, L. W., Fleixcher, R. C., Reed, J. M. and Marsden, K. E. (1992) Cuckoldry through
stored sperm in the sequentially polyandrous spotted sandpiper. *Nature* 359: 631–633.

Owens, I. P. F., Dixon, A., Burke, T. and Thompson, D. B. A. (1995) Strategic paternity assurance in the sex-role reversed Eurasion dotteral (*Charadrius morinellus*): behavioral
and genetic evidence. *Behav. Ecol.* 6: 14–21.

Packer, C., Gilbert, D. A., Pusey, A. E. and O'Brien, S. J. (1991) A molecular genetic analysis of kinship and cooperation in African lions. *Nature* 351: 562–565.

Paetkau, D. and Strobeck, C. (1995) The molecular basis and evolutionary history of a microsatellite null allele in bears. *Molec. Ecol.* 4: 519–520.

Pamilo, P. (1984) Genotypic correlation and regression in social groups: multiple alleles, multiple loci and subdivided populations. *Genetics* 107: 307–320.

Pamilo, P. (1989) Estimating relatedness in social groups. *Trends. Ecol. Evolut.* 4: 353–355.

Pamilo, P. and R. H. Crozier (1982) Measuring genetic relatedness in natural populations:
methodology. *Theor. Pop. Biol.* 21: 171–193.

Parker, P. G., Waite, T. A. and Decker, M. D. (1995) Kinship and association in communally
roosting black vultures. *Anim. Behav.* 49: 395–401.

Paxton, R. J., Thoren, P. A., Tengo, J., Estoup, A. and Pamilo, P. (1996) Mating structure and
nestmate relatedness in a communal bee, *Andrena jacobi* (Hymenoptera, Andrenidae), using microsatellites. *Molec. Ecol.* 5: 511–519.

Pemberton, J. M., Slate, J., Bancroft, D. R. and Barrett, J. A. (1995) Nonamplifying alleles at
microsatellite loci: a caution for parentage and population studies. *Molec. Ecol.* 4:
249–252.

Pfennig, D. W. and Reeve, H. K. (1993) Nepotism in a solitary wasp as revealed by DNA fingerprinting. *Evolution* 47: 700–704.

Piertney, S. B. and Dallas, J. F. (1997) Isolation and characterization of hypervariable microsatellites in the red grouse *Lagopus lagopus scoticus. Molec. Ecol.* 6: 93–95.

Piper, W. H. and Rabenold, P. P. (1992) Use of fragment-sharing estimates from DNA fingerprinting to determine relatedness in a tropical wren. *Molec. Ecol.* 1: 69–78.

Põldmaa, T., Montgomerie, R. and Boag, P. (1995) Mating system of the cooperatively breeding noisy miner *Manorina melanocephala*, as revealed by DNA profiling. *Behav. Ecol.
Sociobiol.* 37: 137–143.

Price, D. K., Collier, G. E. and Thompson, C. F. (1989) Multiple parentage in broods of house
wrens: genetic evidence. *J. Hered.* 80: 1–5.

Primmer, C. R., Møller, A. P. and Ellegren, H. (1995) Resolving genetic relationships with
microsatellite markers: a parentage testing system for the swallow *Hirundo rustica.
Molec. Ecol.* 4: 493–498.

Primmer, C. R., Møller, A. P. and Ellegren, H. (1996) A wide-range survey of cross-species
microsatellite amplification in birds. *Molec. Ecol.* 5: 365–378.

Queller, D. C. and Goodnight, K. F. (1989) Estimating relatedness using genetic markers.
Evolution 43: 258–275.

Quinn, T. W., Quinn, J. S., Cooke, F. and White, B. N. (1987) DNA marker analysis detects multiple maternity and paternity in single broods of the lesser snow goose. *Nature* 326: 392–394.

Quinn, T. W., Davies, J. C., Cooke, F. and White, B. N. (1989) Genetic analysis of offspring of a female-female pair in the lesser snow goose (*Chen c. caerulescens*). *Auk* 106: 177–184.

Quinn, T. W., Cooke, F. and White, B. N. (1990) Molecular sexing of geese using a cloned Z chromosomal sequence with homology to the W chromosome. *Auk* 107: 199–202.

Quinn, J. S., Macedo, R. and White, B. N. (1994) Genetic relatedness of communally-breeding guira cuckoos. *Anim. Behav.* 47: 515–529.

Rabenold, P. P., Rabenold, K. N., Piper, W. H., Decker, M. D. and Haydock, J. (1991) Using DNA fingerprinting to assess kinship and genetic structure in avian populations. *In*: Dudley, E. C. (ed.) *The Unity of Evolutionary Biology*, Dioscorides Press, Portland, Oregon, pp. 611–620.

Rabenold, P. P., Rabenold, K. N., Piper, W. H., Haydock, J. and Zack, S. W. (1990) Shared paternity revealed by genetic analysis in cooperatively breeding tropical wrens. *Nature* 348: 538–540.

Rätti, O. (1994) Female reactions to male absence after pairing in the pied flycatcher. *Behav. Ecol. Sociobiol.* 35: 201–203.

Rätti, O., Hovi, M., Lundberg, A., Tegelstrom, H. and Alatalo, R. V. (1995) Extra-pair paternity and male characteristics in the pied flycatcher. *Behav. Ecol. Sociobiol.* 37: 419–425.

Reeve, H. K., Westneat, D. F., Noon, W. A., Sherman, P. W. and Aquadro, C. F. (1990) DNA "fingerprinting" reveals high levels of inbreeding in colonies of the eusocial naked mole-rat. *Proc. Natl. Acad. Sci. USA* 87: 2496–2500.

Reeve, H. K., Westneat, D. F. and Queller, D. C. (1992) Estimating average within-group relatedness from DNA fingerprints. *Molec. Ecol.* 1: 223–232.

Richard, K. R., Dillon, M. C., Whitehead, H. and Wright, J. M. (1996) Patterns of kinship in groups of free-living sperm whales (*Physeter macrocephalus*) revealed by multiple molecular genetic analyses. *Proc. Natl. Acad. Sci. USA* 93: 8792–8795.

Romagnano, L., McGuire, T. R. and Power, H. W. (1989) Pitfalls and improved techniques in avian parentage studies. *Auk* 106: 129–136.

Ross, K. G. (1993) The breeding system of the fire ant *Solenopsis invicta*: effects on colony genetic structure. *Amer. Naturalist* 141: 554–576.

Schierwater, B. and Ender, A. (1993) Different thermostable DNA polymerases may amplify different amplification patterns. *Nucl. Acid. Res.* 21: 4647–4648.

Schierwater, B., Metzler, D., Kruger, K. and Streit, B. (1996) The effects of nested primer binding sites on the reproducibility of PCR: Mathematical modeling and computer simulation studies. *J. Comp. Biol.* 2: 235–251.

Schlötterer, C., Amos, B. and Tautz, D. (1991) Conservation of polymorphic simple sequence loci in cetatean species. *Nature* 354: 63–65.

Schroth, W., Streit, B. and Schierwater, B. (1996) Evolutionary handicap for sea turtles. *Nature* 384: 521–522.

Seitz, A., Reh, W., Veith, M. and Wolfes, R. (1994) Gene flow between and within natural vertebrate populations. *Anim. Biotechnol.* 5: 155–168.

Sheldon, B. C. (1994) Male phenotype, fertility and the pursuit of extra-pair copulations by female birds. *Proc. R. Soc. Lond. B* 257: 25–30.

Sheldon, B. C. and Burke, T. (1994) Copulation behavior and paternity in the chaffinch. *Behav. Ecol. Sociobiol.* 34: 149–156.

Sheldon, B. C. and H. Ellegren (1996) Offspring sex and paternity in the collared flycatcher. *Proc. R. Soc. Lond. B* 263: 1017–1021.

Signer, E. N., Schmidt, C. R. and Jeffreys, J. (1994) DNA variability and parentage testing in captive Waldrapp ibises. *Molec. Ecol.* 3: 291–300.

Smith, H. G., Montgomerie, R., Põldmaa, T., White, B. N. and Boag, P. T. (1991) DNA fingerprinting reveals relation between tail ornaments and cuckoldry in swallows. *Behav. Ecol.* 2: 90–98.

Strassmann, J. E. (1989) Altruism and relatedness at colony foundation in social insects. *Trends. Ecol. Evolut.* 4: 371.

Stutchbury, B. J. and Morton, E. S. (1995) The effect of breeding synchorny on extra-pair mating systems in songbirds. *Behaviour* 132: 675–690.

Stutchbury, B. J., Rhymer, J. M. and Morton, E. S. (1994) Extrapair paternity in hooded warblers. *Behav. Ecol.* 5: 384–392.

Svensson, E. and Nilsson, J.-A. (1996) Mate quality affects offspring sex ratio in blue tits. *Proc. R. Soc. Lond. B* 263: 357–361.

Taylor, A. C., Sherwin, W. B. and Wayne, R. K. (1994) Genetic variation of microsatellite loci in a bottleneck species: the northern hairy-nosed wombat *Lasiorhinu krefftii. Molec. Ecol.* 3: 277–290.

Tegelström, H. and Sjoberg, G. (1995) Introduced Swedish Canada geese (*Branta canadensis*) have low levels of genetic variation as revealed by DNA fingerprinting. *J. Evol. Biol.* 8: 195–207.

Tiersch, T. R. and Wachtel, S. S. (1991) On the evolution of genome size in birds. *J. Hered.* 82: 363–368.

Wagner, R. H. (1994) Mixed mating strategies by females weaken the sexy son hypothesis. *Anim. Behav.* 47: 1207–1209.

Wagner, R. H., Schug, M. G. and Morton, E. S. (1996a) Condition-dependent control of paternity by female purple martins: implications for coloniality. *Behav. Ecol. Sociobiol.* 38: 379–389.

Wagner, R. H., Schug M. D. and Morton, E. S. (1996b) Confidence of paternity, actual paternity and parental effort by purple martins. *Anim. Behav.* 52: 123–132.

Warkentin, I. G., Carter, A. D., Curzon, R. E., Wetton, J. H., James, P. C., Oliphant, L. W. and Parkin, D. T. (1994) No evidence for extrapair fertilizations in the merlin revealed by DNA fingerprinting. *Molec. Ecol.* 3: 229–234.

Weatherhead, P. J. (1994) Mixed mating strategies by females may strengthen the sexy son hypothesis. *Anim. Behav.* 47: 1210–1211.

Weatherhead, P. J. and Boag, P. T. (1995) Pair and extra-pair mating success relative to male quality in red-winged blackbirds. *Behav. Ecol. Sociobiol.* 37: 81–91.

Weatherhead, P. J., Montgomerie, R., Gibbs, H. L. and Boag, P. T. (1994) The cost of extra-pair fertilizations to female red-winged blackbirds. *Proc. R. Soc. Lond. B* 258: 315–320.

Webster, M. S. (1995) The effects of female choice and copulations away from the colony on the fertilization success of male Montezuma oropendolas. *Auk* 112: 659–671.

Webster, M. S., Pruett-Jones, S., Westneat, D. F. and Arnold, S. J. (1995) Measuring the effects of pairing success, extra-pair copulations and mate quality on the opportunity for sexual selection. *Evolution* 49: 1147–1157.

Westneat, D. F. (1987) Extra-pair fertilizations in a predominantly monogamous bird: genetic evidence. *Anim. Behav.* 35: 877–886.

Westneat, D. F. (1988) Male parental care and extrapair copulations in the indigo bunting.

Auk 105: 149–160.

Westneat, D. F. (1990) Genetic parentage in the indigo bunting: a study using DNA finger-printing. *Behav. Ecol. Sociobiol.* 27: 67–76.

Westneat, D. F. (1993) Polygyny and extrapair fertilizations in eastern red-winged blackbirds (*Agelaius phoeniceus*). *Behav. Ecol.* 4: 49–60.

Westneat, D. F. (1994) To guard mates or go forage: conflicting demands affect the paternity of male red-winged blackbirds. *Amer. Naturalist* 144: 343–354.

Westneat, D. F. (1995) Paternity and paternal behaviour in the red-winged blackbird, *Agelaius phoeniceus*. *Anim. Behav.* 49: 31–35.

Westneat, D. F. and Sherman, P. W. (1993) Parentage and the evolution of parental behavior. *Behav. Ecol.* 4: 66–77.

Westneat, D. F. and Webster, M. S. (1994) Molecular analyses of kinship in birds: interesting questions and useful techniques. *In*: Schierwater, B., Streit, B., Wagner, G, P. and DeSalle, R. (eds) *Molecular Ecology and Evolution: Approaches and Applications*, Birkhäuser, Basel, pp. 91–126.

Westneat, D. F., Frederick, P. C. and Wiley, R. H. (1987) The use of genetic markers to estimate the frequency of successful alternative reproductive tactics. *Behav. Ecol. Sociobiol.* 21: 35–45.

Westneat, D. F., Sherman, P. W. and Morton, M. L. (1990) The ecology and evolution of extra-pair copulations in birds. *Curr. Ornithol.* 7: 331–369.

Westneat, D. F., Clark, A. B. and Rambo, K. C. (1995) Within-brood patterns of paternity and paternal behavior in red-winged blackbirds. *Behav. Ecol. Sociobiol.* 37: 349–356.

Wetton, J. H. and Parkin, D. T. (1991) An association between fertility and cuckoldry in the house sparrow, *Passer domesticus*. *Proc. R. Soc. Lond.* B 245: 227–234.

Wetton, J. H. and Parkin, D. T. (1997) A suite of falcon single-locus minisatellite probes: a powerful alternative to DNA fingerprinting. *Molec. Ecol.* 6: 119–128.

Whittingham, L. A., Dunn, P. O. and Roberson, R. J. (1993) Confidence of paternity and male parental care: an experimental study in tree swallows. *Anim. Behav.* 46: 139–147.

Wilkinson, G. S. and McCracken, G. F. (1985) On estimating relatedness using genetic markers. *Evolution* 39: 1169–1174.

Wrege, P. H. and Emlen, S. T. (1987) Biochemical determination of parental uncertainty in white-fronted bee-eaters. *Behav. Ecol. Sociobiol.* 20: 153–160.

Yezerinac, S. M., Weatherhead, P. J. and Boag, P. T. (1996) Cuckoldry and lack of parentage-dependent paternal care in yellow warblers: a cost-benefit approach. *Anim. Behav.* 52: 821–832.

Zhu, J., Nestor, K. E. and Moritsu, Y. (1996) Relationship between band sharing levels of DNA fingerprints and inbreeding coefficeints and estimation of true inbreeding in turkey lines. *Poultry Science* 75: 25–28.

Zink, R. M. (1982) Patterns of genic and morphologic variation among sparrows in the genera *Zonotrichia, Melospiza, Junco* and *Paserella*. *Auk* 99: 632–638.

A role for molecular biology in testing ideas about cryptic female choice

Michael T. Siva-Jothy[1] and Heike Hadrys[2,3]

[1] Department of Animal and Plant Sciences, University of Sheffield, Sheffield S10 2TN, UK
[2] Abt. Ökologie und Evolution, J. W. Goethe-Universität, Siesmayerstr. 70, D-60054 Frankfurt, Germany
[3] Department of Ecology and Evolution, Yale University, 165 Prospect St., New Haven, CT 06511, USA

Summary

If cryptic female choice exists, observations on mating behaviour will not reflect the actual patterns of paternity. Genetic profiling techniques are needed to identify patterns of paternity (and sperm storage) that deviate from expectations based on behavioural observations. Our chapter addresses this problem in two parts. The first provides a critical review of the concept of cryptic female choice leading to a set of criteria required to demonstrate its occurrence. Second, we explore the repertoire of molecular methods available to answer certain questions about cryptic female choice in insects. The potential and limitations of polymerase chain reaction-based single locus and multilocus DNA fingerprint techniques are discussed along with the different options of analysing the data. Two case studies illustrate that sperm storage and usage in odonates might enable females to control paternity.

Theoretical background

It is widely accepted that female choice is a widespread and important process (see reviews in Bradbury and Andersson, 1987; Andersson, 1994) and that it can persist because it provides the female with direct and/or indirect (i.e. genetic) benefits. There is growing evidence that female choice not only occurs before copulation ("overt" female choice) but also during, and after copulation (the "cryptic" choice of Thornhill (1983) and "copulatory courtship" of Eberhard (1991)). It is important to define the difference between cryptic and, by inference, overt female choice, partly because of the historical division between them, and partly because the choice processes are different. "Overt" choice results in the acceptance of a mating partner on the basis of somatic characters (i.e. expressed in the diploid somatic tissue) that convey signals prior to copulation. "Cryptic" choice is so called because it cannot be observed using the classic Darwinian criteria for identifying (overt) female choice (Eberhard, 1996). "Cryptic" choice results in the preferential "acceptance" (i.e. use) of a particular partner's sperm (gamete). Cryptic choice therefore operates on male gametes largely via anatomical and physiological processes in the female's genitalia. Choice may operate via pre-, syn- or even post-copulatory somatic signals (cf. Eberhard, 1996) modulated by the female's central nervous system, as well as via the female's genitalic environment operating on gametic traits (sperm membrane markers, or possibly seminal fluid components).

The only benefit females can get from exercising cryptic choice are indirect, and therefore genetic. So what might the indirect benefits of choice be to females? The intuitive answer is that females are attempting to choose the best-"quality" males. If, however, there is variance in male quality (there would be no point in females using this criterion to choose males unless there was), then only a few males will mate in every generation and the heritable variation for quality will be used up. At this point all the males in the population will be the sons of sons of... high-quality males, and there will be no benefit to female choice. Female choice may be costly (Reynolds and Gross, 1990), so choosy females will be selected against because selection, in the absence of substantive variation in male quality, will now favour females who follow the low-cost option of random mating (since all males are now high quality random mating will result in contact with high-quality males). This phenomenon has been dubbed "the lek paradox", and applies to any animal where a male's only contribution to his mate (and offspring) is sperm (e.g. Reynolds and Gross, 1990). These theoretical arguments are clearly sensitive to the perceived costs and benefits and, in light of the circumstantial evidence supporting the existence of cryptic choice, we need a good methodological approach to test the magnitude of these effects.

Sperm competition

We currently have a substantial theoretical and empirical base for understanding sperm competition (Parker, 1970). When females mate multiply and there is a delay between insemination and fertilisation, the outcome is nearly always nonrandom fertilisation success between different males. Nonrandom fertilisation is the outcome of sperm competition. Insects are predisposed to high levels of sperm competition, because most female insects mate with more than one male during a reproductive cycle and store the sperm of different males in discrete organs, creating the potential for intermale ejaculate overlap. Males are favoured by sexual selection if they can ensure that their sperm, and not a rival's, is used during fertilisation. Parker (1970) proposed that sperm competition resulted in two selection pressures acting on males, one that resulted in traits that enabled the preemption of stored ejaculates, and the other that produced traits which guarded against future preemption. The documentation of these adaptations has proliferated in recent years (see reviews in Thornhill and Alcock, 1983; Smith, 1984; Birkhead and Møller, 1998); the overwhelming pattern in insects is that the last male to mate with a female fertilises most of the eggs she lays in the subsequent bout of oviposition – there is last-male sperm precedence. However, although many examples of last-male sperm precedence exist in the literature, almost as common are reports of widespread intraspecific variation in the proportion of eggs fertilised by the last male (e.g. Schlager, 1960; Siva-Jothy and Tsubaki, 1989; Lewis and Austad, 1990; Hadrys et al., 1993; Eady, 1994). The mechanism by which males achieve sperm precedence is largely unknown (the exceptions are mainly from the Odonata), begging the question, What underlies the variance in fertilisation success between males?

Cryptic female choice

The same bias towards male aspects of precopulatory sexual selection that occurred after Darwin proposed the theory of evolution (when male-male competition was accepted and female choice resisted) has been repeated in the explosion of research on postcopulatory sexual selection.

The two preconditions for sperm competition are (i) that females mate multiply, and (ii) they store sperm from different males. Although both these preconditions are female traits, the role of the female in the operation and outcome of sperm competition has been largely neglected. Researchers are now correcting this bias, and the current shift is towards female perspectives (e.g. Rosenqvist and Berglund, 1992; Birkhead and Møller, 1993; LaMunyon and Eisner, 1993; Ward, 1993; Fincke, 1997; Simmons and Siva-Jothy, 1997). Gowaty (1994) has even suggested that a paradigm "riffle" towards female perspectives is occurring within behavioural ecology. As Rowe et al. (1994) point out, this "riffle" seems to have already occurred amongst insect researchers.

Recent evidence suggests that females are likely to play an important role in determining variation around the mean level of sperm precedence (Wilson et al., 1997). The nature of this variation is central to understanding the role of females in determining paternity after copulation (i.e. cryptic female choice). Villavaso (1975) was the first to investigate a role for female influences on variation in sperm precedence. He cut the nerve inervating the spermathecal muscle in a beetle, *Anthonomus grandis*, and consequently caused sperm "displacement" to fall from 65 to 21%. Female influences on sperm competition have also been emphasised by Walker (1980) and Simmons (1987), amongst others. Møller (1992) has suggested that females constrained from mating with males of the preferred phenotype might adjust their options through multiple mating. The most comprehensive theoretical framework for cryptic female choice has been proposed by Eberhard (1996). His ideas centre on a simple biological fact: once the male of any species with internal fertilisation has deposited sperm in the female's genital tract, she is largely in control of it. There are, of course, many ways in which males might continue to compete with each other inside the female [e.g. sperm removal by male odonates (Waage, 1979a), the transfer of "toxins" by male *Drosophila melanogaster* (Harshman and Prout, 1994; Clark et al., 1995)], but in general the sperm must migrate to, be stored in and be released from anatomical structures within the female. All these processes provide the female with the opportunity for exercising choice over whose ejaculate she uses for fertilisation.

How can we test whether females are exercising cryptic female choice? This is essentially a problem which requires the separation of male (e.g. sperm) and female effects. This is just as problematic as disentangling the same effects when they operate prior to copulation. The difficulty for empiricists trying to disentangle these effects operating after copulation results from the fact that females should be selected to utilise all avenues of choice available to them if such avenues provide a net benefit (e.g. Waage, 1996). Consequently, we might expect the females of a particular species to use a combination of overt and cryptic choice as well as active (female-controlled) and passive (selecting the winner from intrasexual competition) choice. Because selection will favour females that maximise net benefit (the balance between the payoff received from choice and the costs associated with choice), we expect different ratios of cryptic and overt choice in different taxa. At one end of the spectrum of female choice strategies will be species where it pays the females to let males compete in both overt and cryptic arenas, and then utilise the sperm from the winner (i.e. totally passive); at the other will be species where the females retain total control over reproductive decisions and exercise active choice (i.e. totally active). Spanning these extremes will be a host of situations where females use a combination of overt and cryptic, active and passive ways of increasing their net benefit. If this simple scheme is generally true, then identifying cryptic female choice empirically will be extremely difficult, and researchers would be best advised to hunt at one end of the female choice option spectrum. The catch 22 for empiricists is that we need to know where to look before we hunt, but we can't hunt until we know where to look.

Demonstrating the potential for cryptic female choice

There is potential for cryptic female choice in any mating system where the cost to females of resisting male copulatory attempts is greater than the cost of allowing the male to copulate, but where males cannot control when, or where, the female produces zygotes (e.g. Fincke, 1997). The last prerequisite is important since it provides the female with a means of avoiding any male-determined sperm precedence constraints on her ability to "select" sperm for fertilisation. This leads us to another important prerequisite. Cryptic female choice will generally be rare in mating systems where P_2 (or P_n) values are invariate. P_2 is the unit with which biologists measure the fertilisation outcome of controlled matings between a single virgin female and two males (see Boorman and Parker, 1974): the value represents the proportion of offspring fathered by the second male to mate. However, even in these mating systems females could bias paternity in favour of particular males by remating more frequently or by producing fewer zygotes after matings with "undesirable" males. These processes are, of course, overt rather than cryptic.

A good starting point for identifying cryptic female choice is to demonstrate that females have the potential to control sperm storage and usage, and that this female-based control results in variance in P_2 values. It is important to remember that (i) this approach does not constitute proof for the occurrence of cryptic female choice and (ii) a lack of evidence for the potential for cryptic female choice doesn't mean females can't exercise it. Why then bother to take this approach? We suggest it is useful because it enables us to focus attention on species and situations where more rigorous tests of ideas about cryptic female choice are more appropriate. We should focus on species for which the evidence in favour of a potential for exercising cryptic choice is strong, and in which the mating system characteristics suggest females may obtain a selective advantage from it.

Females may exercise cryptic choice through many routes (e.g. Eberhard, 1996), and supportive evidence for any would be useful: we suggest that when females directly influence sperm storage and usage there is considerable potential for them to control paternity. Carefully collected data from manipulations with *a priori* predictions about outcomes are needed to demonstrate that females have the potential to exercise cryptic choice. We will discuss ways in which molecular biological tools might help us to pinpoint such potential, and also consider the most likely alternative explainations for what we might expect to see.

Demonstrating cryptic female choice

There are four criteria that, if all met by a single species, would demonstrate the occurrence of cryptic female choice. First, sperm precedence values must be variable, otherwise females could not cryptically break out of the constraints imposed by sperm competition (they can, of course, break out of these constraints overtly, by remating sooner or fertilising fewer eggs between copulation bouts). Second, some of the variance in sperm precedence values must be attributable to female effects (and have a heritable basis). Third, the variance attributable to female effects must be correlated with somatic and/or gametic male trait(s). Fourth, female attributable shifts away from the population mean value of sperm precedence must result in higher female fitness (in the broadest sense – i.e. it includes genetic correlations between male traits and the female preference that increase offspring fitness over several generations) than if the female were constrained to use sperm at the population mean P_2 value or there must be a genetic correlation with the female preference. To date no single study has provided all four lines of evidence. Some studies have focussed largely on demonstrating a relationship between the bias in paternity over the population mean and some aspect of the successful male's phenotype in controlled mating situations (e.g. LaMunyon and Eisner, 1993; Ward, 1993). Unfortunately, these studies did not differentiate between female and male effects. Simmons et al. (1996) point out, the data in both studies are open to a more parsimonious interpretation based on ejaculate characteristics of the males, not on cryptic choice of males by the females.

In the rest of this chapter we will illustrate how certain molecular techniques have been applied to answering questions aimed at identifying the potential for cryptic female choice in insects, and will conclude by discussing the role we envisage molecular biology will play in identifying the mechanisms that enable females to exert cryptic choice.

The molecular methods of choice

An implicit assumption when considering cryptic female choice is that observed patterns of mating behaviour may not directly relate to the actual patterns of paternity. Behavioural ecologists therefore need genetic profiling techniques to identify deviations between patterns of realized paternity and expectations based on behavioural observations. When examining paternity in insects, the investigator is faced with three common problems when choosing an appropriate molecular technique: (i) small quantities of DNA from (ii) large numbers of offspring and (iii) no relevant genetic sequence information of the study organism (Hadrys and Siva-Jothy, 1994). The introduction of the polymerase chain reaction (PCR) (Saiki et al., 1988) made insects accessible to molecular genetic studies. To date two types of PCR-based DNA fingerprint techniques have become increasingly popular: arbitrarily amplified DNA techniques (AAD) (reviewed in Schierwater et al., 1997; Caetano-Anollés, this volume) and microsatellite typing (Tautz, 1989; reviewed in Schlötterer and Pemberton, this volume). The multilocus AAD techniques can be immediately and relatively easily applied to any insect system of interest, while the single-locus microsatellite typing requires relevant genetic sequence information beforehand.

AAD fingerprinting

AAD techniques have been developed into powerful tools for evolutionary and ecological research, as is demonstrated by more than 1000 publications (over the last 5 years) spanning a wide range of organisms to study genealogical relationships at different levels of complexity (reviewed in Hadrys et al., 1992; Williams et al., 1993; Caetano-Anollés, 1996; Grosberg et al., 1996; Schierwater et al., 1997). In insects AAD profiles have proven to be successful in identifying relatedness from parentage to a higher taxonomic level (Apostol et al., 1993; Hadrys et al., 1993; Smith et al., 1994; Schierwater, 1995).

An attractive feature of AAD techniques is that they do not require any sequence information of the genome under investigation. Instead, they define PCR priming conditions in which a single short arbitrary primer finds a large number of priming sites along any long genomic DNA template. A small fraction of such sites will be amplified in an exponential fashion (Schierwater et al., 1996). The resulting PCR products can be separated according to length by gel electrophoresis and visualised as a multilocus DNA profile of the template DNA. By comparing profiles from different genomes, a variable number of monomorphic as well as polymorphic fragments will be identified, depending on the degree of dissimilarity between the genomes (Hadrys and Siva-Jothy, 1994; Grosberg et al., 1996; Schierwater et al., 1997).

Three alternative AAD strategies were originally described: AP-PCR (Welsh and McClelland, 1990), RAPD (Williams et al., 1990) and DAF (Caetano-Anollés et al., 1991), which differ mainly by name, size of the primer and the number of markers generated per PCR reaction. Based on the three original protocols, a growing number of related techniques have been developed (reviewed in Caetano-Anollés, 1996). Probably the most prominent technique used for kinship analyses is the random amplified polymorphic DNA (RAPD)-Method. The RAPD technique has been used on insects to (i) demonstrate sperm competition (Hadrys et al., 1993; Reichardt and Wheeler, 1996; Siva-Jothy and Hooper, 1996); (ii) estimate last male sperm precedence (P_2) in offspring of multiply-mated females (Hadrys et al., 1993; Hooper and Siva-Jothy, 1996); (iii) identify reproductive success of different males (Fondrk et al., 1993; Scott and Williams, 1993; Hadrys, 1993); (iv) detect the presence of sperm from more than one male in female sperm stores (Siva-Jothy and Hooper, 1996; Hadrys and Robertz, in prep.; cf. case studies below); and (v) identify the sperm used during fertilisation (Hadrys and Robertz, in prep.; cf. first case study below).

Randomly amplified microsatellites

In cases where large pools of potential fathers need to be examined, dominant AAD markers have a resolution limit and codominant microsatellite markers might be more useful. By combining several microsatellite loci, discrimination between a large pool of potential fathers can be more easily achieved (for review, see Schlötterer and Pemberton, this volume). The benefit of using microsatellites is, however, purchased at one major cost. Identification of these loci for any new system can be labour- and time-intensive. Recently a new strategy has been developed to isolate polymorphic microsatellite loci based on RAPD fragments, randomly amplified microsatellites (RAMS) (Ender et al., 1996; Schierwater et al., 1997). This approach is comparatively fast, requires no prior sequence information and circumvents the nontrivial procedure of constructing and screening genomic libraries. RAPD fragments replace genomic libraries, and microsatellite loci are detected from RAPD profiles by Southern hybridization with labelled repeats (see Fig. 1). The RAMS technique has been used to isolate microsatellites from two dragonfly species – in combination with RAPD analyses – to unravel paternities and maternities in natural populations (Hadrys and Fincke, in prep.).

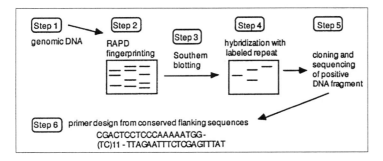

Figure 1. Strategy for the identification of RAMS (from Schierwater et al., 1997). Explanations are given in the text (for further details see Ender et al., 1996; Schierwater et al., 1997).

Analyses of molecular data

Another problem facing the empiricist trying to understand the nature of nonrandom paternity in mating systems concerns the interpretation of molecular genetic data sets. The analytical tools used to interpret molecular data are as diverse as the methods themselves. The profiles of dominant AAD markers, for example, are transformed into a 0/1 matrix (1 counts the presence of a particular marker and 0 the absence), and data interpretation is based on genetic "phenotypes". In contrast, data analysis of the codominant (locus-specific) microsatellites is based on the interpretation of allelic "genotypes".

To assess paternity from AAD profiles, one may use either (i) a small subset of informative diagnostic markers or (ii) all generated markers.

1. Assignment of paternity based on diagnostic markers by inclusion or exclusion is mainly nonstatistical. This approach is particularly sensitive to any distortion of Mendelian segregation patterns and any variation in the amplification protocol (e.g.

Lewis and Snow, 1992). Paternity exclusion is particularly critical, for it is often not possible to collect the entire pool of potential fathers (this is the case in many field studies on insects). However, Levitan and Grosberg (1993) demonstrated in a cnidarian system with known parent-offspring samples that based on as little as five diagnostic bands, nonparents could be correctly excluded for all but 1 of 30 offspring.

2. Rather than using a subset of diagnostic bands, as above, all generated markers may be used without preevaluation of whether a marker is informative or not. By comparing between two fingerprint profiles, coefficients are revealed which mirror the genetic distance between two individuals (Lynch, 1990). There are a variety of algorithms available to transform AAD markers into genetic distance measures. The software packages NTSYS-pc, version 1.60 (Rohlf, 1990), TREECON, version 1.1 (Van de Peer and Wachter, 1993) or PAUP, version 3.1 (Swofford, 1993), for example, provide aid in calculating appropriate distance measures for AAD profiles. For paternity analyses the distance measures can be used to group parents and offspring according to tree-building methods, for example: UPGMA-cluster-, neighbor-joining-, split-decomposition or maximum-parsimony analyses (e.g. Apostol et al., 1993; Fondrk et al., 1993; Hadrys and Siva-Jothy, 1994). The computer programs PHYLIP, version 3.75c (Felsenstein, 1995); TREECON, version 1.1 (Van de Peer and Wachter, 1993); SPLITSTREE, version 2.2 (Huson, 1997); or PAUP, version 3.1 (Swofford, 1993) provide appropriate tree-building algorithms.

As an alternative to tree-building methods, distance measures between groups or individuals may be compared statistically to separate relatives from nonrelatives in a quantitative manner. Reeve et al. (1992) obtained unambiguous estimates for full- and half-sibling relatedness by quantitative comparisons within and among broods, and Hadrys et al. (1992, 1993) have developed a scenario of testable expectations on paternity based on pairwise statistical comparisons of genetic distance measures.

Paternity assignment using microsatellite data can be performed by any of the methods applicable to AAD markers, but is more powerfully done by either paternity inclusion or exclusion in order to take full advantage of the allelic information (e.g. Schlötterer and Pemberton, this volume).

Case studies of sperm storage and utilisation in dragonflies

A good starting point for designing experiments to identify cryptic female choice is to demonstrate that the *potential* to exercise such choice exists in the mating system under investigation. In other words, females exercising cryptic choice should mate with more than one male, store sperm from different males simultaneously and produce changing patterns of sperm precedence that provide them with a net fitness benefit. The arena in which cryptic choice operates is, by definition, the only one where the female has the potential to influence paternity. In the case of insects, the female's genital tract and discrete sperm-storage organs are the vessels into which males deposit their ejaculate and/or manipulate the ejaculate of rivals, the site of sperm storage and maintenance by the female, and the point from which the female releases sperm to fertilise her eggs. The Odonata are perhaps the best-studied insect order with respect to their sexually selected anatomical, physiological and behavioural traits associated with copulation, sperm storage and usage. They are a good candidate group to search for cryptic female choice since females store the viable sperm of several mates for prolonged periods in multiple, often morphologically and histologically complex, sperm-storage organs (e.g. Miller, 1982; Waage, 1984; Siva-Jothy, 1987), show behavioural mechanisms that enable females to avoid the male sperm competition mechanism (e.g. Siva-Jothy and Hooper, 1995; Fincke, 1997; Hadrys and Robertz, in prep.), and show variation in sperm precedence patterns with time after copulation (e.g. Siva-Jothy and Tsubaki, 1989; Hadrys et al., 1993). The case studies we present here all show evidence that sperm storage and utilisation in odonates may provide females with the potential to control paternity independent of male sperm competition mechanisms.

Case study 1: Sperm storage and usage in female aeshnids

Female aeshnid dragonflies store sperm in multiple sperm-storage organs – a single bursa copulatrix and a pair of much smaller spermathecae. Males of most aeshnid species provide nothing more than sperm to their female mates. Males of the aeshnid *Anax parthenope* perform a costly guarding behaviour by remaining in contact with their female mate after copulation during oviposition (Rüppell and Hadrys, 1988). RAPD fingerprint analyses of the offspring have shown that the male which guards the female during oviposition secures 100% fertilisation success over prolonged periods of oviposition (Hadrys et al., 1993; Hadrys, 1993). To address the question whether the guarding male achieves sperm precedence by removing rival sperm during copulation, genetic analysis of the sperm composition in the female's sperm storage organs have been performed. The RAPD fingerprint analyses show that *A. parthenope* females store viable sperm from more than one mate and store the sperm from different ejaculates (mates) in separate storage compartments. Furthermore they selectively use sperm from the small, paired spermathecae for fertilisation (see Fig. 2A and B; Hadrys and Robertz, in prep.). This is in contrast to all known examples in other odonate species where females use the sperm from the bursa copulatrix to fertilise their eggs.

It has also been shown in other odonates that the spermathecal sperm cannot be removed by males (despite their ability to remove rival sperm from the bursa copulatrix) and contains a sperm composition of different mates (Siva-Jothy and Hooper, 1996). In contrast, *A. parthenope* females store the sperm from a single mate in the spermathecae, while the sperm stored in the bursa copulatrix represents a composi-

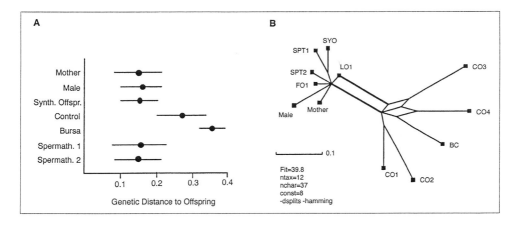

Figure 2. Inferring paternity from RAPD fingerprint profiles by means of (A) statistical comparisons of mean genetic distances and (B) split-decomposition analysis. Shown are the results for one A. parthenope *family. (A) Mean (±SD) pairwise genetic distance measures: Mother, offspring to the mother (known parental kinship); Male, offspring to mated male (putative first-degree kinship); Synth(etic) Offspr(ing), offspring to a mixture of equal amounts of DNA from the mother and putative father (see Hadrys et al., 1992, 1993); Control, offspring to unrelated control females and males (known non-first-degree kinship); Bursa, offspring to sperm stored in the bursa copulatrix; Spermath. 1,2, offspring to sperm stored in the two spermathecae. (B) Kinship tree derived from split-decomposition analyses (Huson, 1997) [same family as in (A)]. Shown here: the mother (Mother), the putative father (Male), unrelated individuals as controls (CO1–4), a sample of offspring from the first day (FO1) and the third day of oviposition (LO1) as well as the sperm stored in the bursa copulatrix (BC) and the paired spermathecae (SPT1,2).*

tion of at least two, and possibly all previous mates (see Fig. 2A). These results allow two alternative equally parsimonious explanations: (i) A. *parthenope* males are able to remove spermathecal sperm, (ii) males have a limited (or no) ability to remove rival sperm during copulation and the patterns of sperm partitioning between bursa copulatrix and the spermathecae is a result of female manipulation.

RAPD fingerprint analyses of the temporal distribution of the stored sperm [(i) during copulation, (ii) immediately after copulation] show that A. *parthenope* females did not store the sperm of the last male mate in the spermathecae immediatley after copulation. The sperm of the last male mate was identified as the sperm stored in the bursa copulatrix, while sperm of an anonymous male was found in both spermathecae and was used to fertilise the eggs (Hadrys and Robertz, in prep.). The results demonstrate that A. *parthenope* males are unable to remove rival sperm from the spermathecae during copulation, suggesting that the changing patterns of spermathecal sperm after copulation are likely the result of ejaculate manipulations by the females (Hadrys and Robertz, in prep.).

47

In sum, the multiple sperm-storage organs of *A. parthenope* females are demonstrated to function in sperm partitioning. The nature of sperm partitioning depends on the time after copulation and correlates with a specific male's fertilisation success. The data suggest that *A. parthenope* females might have the potential to manipulate sperm storage and usage.

Case study 2: Sperm storage and usage in female calopterygids

One consequence of complete sperm removal in the calopterygid damselfly *Calopteryx maculata* (Waage, 1979a) should be that females have no ability to manipulate ejaculates internally in order to produce nonrandom patterns of sperm precedence after copulation. However, studies of other calopterygids suggest that females are able to prevent complete sperm removal and are capable of using sperm from those "cached" ejaculates in later oviposition episodes (e.g. Siva-Jothy and Tsubaki, 1989; Siva-Jothy and Hooper, 1996). An ability to manipulate ejaculates may simply be the result of selection on females to cope with the logistical problems associated with maintaining viable sperm in storage. It may also enable females to choose (actively or passively) between mates.

Female calopterygid damselflies have multiple sperm-storage organs (Waage, 1979a; Waage, 1984; Siva-Jothy and Tsubaki, 1989; Siva-Jothy and Hooper, 1995) and can secure access to oviposition sites without mating with the resource-holding male (Waage, 1979b; Siva-Jothy and Tsubaki, 1989; Siva-Jothy and Hooper, 1995). In *C. splendens xanthostoma* the copulating male secures high levels of sperm precedence (Hooper and Siva-Jothy, 1996) in the eggs laid immediately by the female he guards after copulation. Oviposition secured in this manner is termed "mated and guarded" (MAG) oviposition; the copulating male achieves sperm precedence by removing sperm from the bursa copulatrix (Siva-Jothy and Hooper, 1995). The sperm in the smaller, paired spermathecae is not removed during copulation. In contrast to the aeshnid *A. parthenope*, RAPD analyses show that spermathecal sperm has a greater genetic diversity than bursal sperm (Siva-Jothy and Hooper, 1995) and probably represents a sperm cache from several, possibly all, previous mates. Female *C. s. xanthostoma* also secure oviposition without remating with a resource-holding male (termed "stealing a guard" (SAG) oviposition). Females conducting SAG oviposition have a significantly higher genetic diversity of sperm in the spermathecae than females arriving to conduct MAG oviposition (Siva-Jothy and Hooper, 1995) and, more interestingly, use sperm from this genetically diverse sperm cache to fertilise their eggs during such oviposition bouts (Siva-Jothy and Hooper, 1996). Few studies have considered the function of multiple sperm-storage organs in the context of a female's ability to partition ejaculates (e.g. Villavaso, 1975; Rodríguez, 1994; Walker, 1980), a function that is potentially very important when considering a female's ability to exercise cryptic choice.

In *C. s. xanthostoma* the nature of the partitioning appears to be related to a difference in the genetic diversity of stored sperm (Siva-Jothy and Hooper, 1995) as well as patterns of sperm usage associated with different female oviposition behaviours.

Conclusions

The evidence from these two families of odonates suggests that some subtle, sperm precedence-changing processes are occurring after copulation and within the female. We feel this is a good starting point from which to design experiments to test predictions from the theory of cryptic female choice. If we demonstrate the occurrence of this process, then the next logical step will be to identify the mechanism by which it occurs. Identifying the mechanisms underpinning cryptic female choice will, at its simplest level, involve identifying the spatial and temporal distribution of the viable spermatozoa of different males within the female's sperm-storage organs between insemination and fertilisation. Ideally, PCR-based DNA fingerprint techniques should identify individual sperm. After identifying cryptic female choice, unravelling what phenotypic criteria females use to base their choice on is central in understanding the operation of sexual selection and the way in which sexual conflicts of interest are resolved. Moreover, this research holds the exciting possibility of opening new vistas on the evolution of reproductive traits and the evolution of sex. In keeping with the reductionist nature of scientific investigation, we see the role of molecular biology as central in this endeavour.

Acknowledgements

Work was supported by grants from the German Science Foundation program Genetic Analyses of Social Mating Systems (DFG-Ha 1947/2-1; -Ha 1947/2-2) to HH. We thank Leo Buss and Ola M. Fincke for comments on the manuscript.

References

Andersson, M. (1994) *Sexual Selection*, Princeton University Press, Princeton.

Apostol, B. L., Black, W. C. IV, Miller, B. R., Reiter, P. and Beaty, B. J. (1993) Estimation of the number of full sibling families at an oviposition site using RAPD-PCR markers: applications to the mosquito *Aedes aegypti*. *Theor. Appl. Genet.* 86: 991–1000.

Birkhead, T. R. and Møller, A. P. (1993) Female control of paternity. *Trends Ecol. Evol.* 8: 100–104.

Birkhead, T. R. and Møller, A. P. (1998) *Sperm Competition and Sexual Selection*, Academic Press, London.

Boorman, E. and Parker, G. A. (1974) Sperm (ejaculate) competition in *Drosophila melanogaster*, and the reproductive value of females to males in relation to mating status. *Ecol. Entomol.* 1: 145–155.

Bradbury, J. W. and Andersson, M. B. (1987) *Sexual selection: testing the alternatives*. Wiley, Chicester.

Caetano-Anollés, G. (1896) Scanning of nucleic acids by *in vitro* amplification: new developments and applications. *Nat. Biotechnol.* 14: 1668–1674.

Caetano-Anollés, G., Bassam, B. J. and Gresshoff, P. M. (1991) DNA amplification fingerprinting using very short arbitrary oligonucleotide primers. *BioTechnology* 9: 553–557.

Clark, A. G. and Lanigan, C. M. S. (1993) Prospects of estimating nucleotide divergence with RAPDs. *Molec. Biol. Evol.* 10: 1096–1111.

Eady, P. E. (1994) Intraspecific variation in sperm precedence in *Callosobruchus maculatus*. *Ecol. Entomol.* 19: 11–16.

Eberhard, W. G. (1991) Copulatory courtship and cryptic female choice in insects. *Biol. Rev.* 66: 1–31.

Eberhard, W. G. (1996) *Female control: Sexual Selection by Cryptic Female Choice*. Princeton University Press, Princeton.

Ender, A., Schwenk, K., Städler, T., Streit, B. and Schierwater, B. (1996) RAPD identification of microsatellites in *Daphnia. Molec. Ecol.* 5: 437–441.

Felsenstein, J. (1995) PHYLIP: *Phylogenetic Inference Package, version 3.57c*. Department of Genetics, University of Washington, Seattle.

Fincke, O. M. (1997) Conflict resolution in the Odonata: implications for understanding female mating patterns and female choice. *Biol. J. Linn. Soc.* 60: 201–220.

Fondrk, M. K., Page, R. E. and Hunt, G. J. (1993) Paternity analysis of worker honeybees using random amplified polymorphic DNA. *Naturwissenschaften* 80: 226–231.

Gowaty, P. A. (1994) Architects of sperm competition. *Trends Ecol. Evol.* 9: 160–162.

Grosberg, R. K., Levitan, D. R. and Cameron, B. B. (1996) Characterization of genetic structure and genealogies using RAPD-PCR markers: a random primer for the novice and nervous. *In*: Ferraris, J. D. and Palumbi, S. R. (eds) *Molecular Zoology: Advances, Strategies and Protocols*, Wiley-Liss, New York, pp. 67–100.

Hadrys, H. (1993) Analysis of reproductive success in insect mating systems by RAPD fingerprinting. *Verhandlungen der Deutschen Zoologischen Gesellschaft* 86: 15–20.

Hadrys, H. and Siva-Jothy M. T. (1994) Unravelling the components that underlie insect reproductive traits using a simple molecular approach. *In*: Schierwater, B., Streit, B., Wagner, G, P. and DeSalle, R. (eds) *Molecular Ecology and Evolution: Approaches and Applications*, Birkhäuser, Basel, pp. 211–228.

Hadrys, H., Balick, M. and Schierwater, B. (1992) Applications of random amplified polymorphic DNA (RAPD) in molecular ecology. *Molec. Ecol.* 1: 55–63.

Hadrys, H., Schierwater, B., Dellaporta, S. L., DeSalle, R. and Buss, L. W. (1993) Determination of paternities in dragonflies by random amplified polymorphic DNA fingerprinting. *Molec. Ecol.* 2: 79–87.

Harshman, L. and Prout, T. (1994) Sperm displacement without sperm transfer in *Drosophila melanogaster*. *Evolution* 48: 758–766.

Hooper, R. and Siva-Jothy, M. T. (1996) Last male sperm precedence in a damselfly revealed by RAPD fingerprinting. *Molec. Ecol.* 5: 449–452.

Huson, D. H. (1997) *SPLITS TREE: A program for analyzing and visualizing evolutionary data*, version 2.3.2, huson@mathematik.uni-bielefeld.de, University of Bielefeld.

LaMunyon, C. W. and Eisner, T. (1993) Postcopulatory sexual selection in the arctiid moth (*Utetheisa ornatrix*). *Proc. Natl. Acad. Sci. USA* 90: 4689–4692.

Levitan, D. R. and Grosberg, R. K. (1993) The analysis of maternity and paternity in the marine hydozoan *Hydractinia symbiolongicarpus* using randomly amplified polymorphic DNA (RAPD) markers. *Molec. Ecol.* 2: 315–326.

Lewis, S. M. and Austad, S. N. (1990) Sexual selection in flour beetles: the relationship between sperm precedence and male olefactory attractiveness. *Behav. Ecol.* 5: 219–224.

Lewis, P. O. and Snow, A. A. (1992) Deterministic paternity exclusion using RAPD markers. *Molec. Ecol.* 1: 155–160.

Lynch, M. (1990) The similarity index and DNA fingerprinting. *Molec. Biol. Evol.* 5: 584–599.

Miller, P. L. (1982) Genital structure, sperm competition and reproductive behaviour in some African libellulid dragonflies. *Adv. Odonatol.* 1: 175–188.

Møller, A. P. (1992) Frequency of male copulations with multiple mates and sexual selection. *Amer. Naturalist.* 139: 1089–1101.

Parker, G. A. (1970) Sperm competition and its evolutionary consequences in the insects. *Biol. Rev.* 45: 525–567.

Reeve, H. K., Westneat, D. F. and Queller, D. C. (1992) Estimating within-group relatedness from DNA fingerprints. *Molec. Ecol.* 1: 223–232.

Reichardt, A. K. and Wheeler, D. E. (1996) Multiple mating in the ant *Acromyrmex versicolor*: a case of female control. *Behav. Ecol. Sociobiol.* 38: 219–225.

Reynolds, J. D. and Gross, M. R. (1990) Costs and benefits of mate choice: Is there a lek paradox? *Amer. Naturalist.* 136: 230–243.

Rodríguez, V. (1994) Function of the spermathecal muscle in *Chelymorpha alternans* Boheman (Coleoptera: Chrysomelidae: Cassidinae). *Physiol. Entomol.* 19: 198–202.

Rohlf, F. J. (1990) *NTSYS-pc: Numerical Taxonomy and Multivariate Analysis System*, version 1.60, Applied Biostatistics, Setauket, New York.

Rosenqvist, G. and Berglund, A. (1992) Is female sexual behaviour a neglected topic? *Trends Ecol. Evol.* 7: 177–176.

Rowe, L., Arnqvist, G., Sih, A. and Krupa, J. J. (1994) Sexual conflict and the evolutionary ecology of mating patterns: water striders as a model system. *Trends Ecol. Evol.* 9: 289–297.

Rüppell, G. and Hadrys, H. (1988) *Anax junius* (Aeshnidae) – Sexual male competion and oviposition behaviour. *Publ. Wiss. Film*, Göttingen, E 2998: 1–12.

Saiki, R. K., Gelfand, D. H., Stoffel, S., Scharf, S. J., Higuchi, R., Horn, G. T., Mullis, K. B. and Erlich, H. A. (1988) Primer-directed enzymatic amplification of DNA with a thermostable DNA polymerase. *Science* 239: 487–491.

Schierwater, B. (1995) Arbitrarily amplified DNA in systematics and phylogenetics. *Electrophoresis* 16: 1643–1647.

Schierwater, B., Metzler, D., Krüger, K. and Streit, B. (1996) The effects of nested primer binding sites on the reproducibility of PCR: mathematical modeling and computer simulation studies. *J. Comput. Biol.* 3: 235–251.

Schierwater, B., Ender, A., Schroth, W., Holzmann, H., Diez, A., Streit, B. and Hadrys, H. (1997) Arbitrarily amplified DNA in ecology and evolution. *In:* Caetano-Anollés, G. and Gresshoff, P. M. (eds) *DNA Markers: Protocols, Applications and Overviews*, Wiley, New York, pp. 313–330.

Scott, M. P. and Williams, S. M. (1993) Comparative reproductive success of communally-breeding burying beetles as assessed by PCR with randomly amplified polymorphic DNA. *Proc. Natl. Acad. Sci. USA* 90: 2242–2245.

Simmons, L. W. (1987) Sperm competition as a mechanism for female choice in the field cricket, *Gryllus bimaculatus. Behav. Ecol. Sociobiol.* 21: 197–202.

Simmons, L. W. and Siva-Jothy, M. T. (1998) Sperm competition in insects: mechanisms and the potential for selection. *In*: Birkhead, T. R. and Møller, A. P. (eds) *Sperm Competition,* Academic Press, London; *in press.*

Simmons, L. W., Stockley, P., Jackson, R. L. and Parker, G. A. (1996) Sperm competition or sperm selection: no evidence for female influence over paternity in yellow dung flies *Scatophaga stercoraria. Behav. Ecol. Sociobiol.* 38: 199–206.

Siva-Jothy, M. T. (1987) The structure and function of the female sperm-storage organs in libellulid dragonflies. *J. Insect Physiol.* 33: 559–567.

Siva-Jothy, M. T. and Hooper, R. (1995) The disposition and genetic diversity of stored sperm in females of the damselfly *Calopteryx splendens xanthostoma. Proc. R. Soc. Lond. B.* 259: 313–318.

Siva-Jothy, M. T. and Hooper, R. (1996) Differential use of sperm during oviposition in *Calopteryx splendens xanthostoma. Behav. Ecol. Sociobiol.* 39: 389–393.

Siva-Jothy, M. T. and Tsubaki, Y. (1989) Variation in copulation duration in *Mnais pruinosa pruinosa* (Calopterygidae; Odonata). I. Alternative mate-securing tactics and sperm precedence. *Behav. Ecol. Sociobiol.* 24: 39–45.

Smith, R. L. (1984) *Sperm Competition and the Evolution of Animal Mating Systems,* Academic Press, London.

Smith, J. L., Scott-Craig, J. S., Leadbetter, J. R., Bush, G. L., Roberts, D. L. and Fulbright, D. W. (1994) Characterization of random amplified polymorhic DNA (RAPD) products from *Xanthomonas campestris* and some comments on the use of RAPD products in phylogenetic analysis. *Mol. Phylogenet. Evol.* 3: 135–145.

Swofford, D. L. (1993) *PAUP: Phylogenetic Analysis Using Parsimony,* version 3.1, Illinois Natural History Survey, Champaign, IL.

Tautz, D. (1989) Hypervariability of simple sequences as a general source for polymorphic DNA markers. *Nucl. Acid. Res.* 17: 6463–6471.

Thornhill, R. (1983) Cryptic female choice and its implications in the scorpionfly, *Harpobittacus nigriceps. Amer. Naturalist.* 122: 765–788.

Thornhill, R. and Alcock, J. (1983) *The Evolution of Insect Mating Systems,* Harvard University Press, Cambridge.

Van de Peer, Y. and Wachter, R. D. E. (1993) TREECON: a software package for the construction and drawing of evolutionary trees. *Comput. Appl. Biosci.* 9: 177–182.

Villavaso, E. J. (1975) Functions of the spermathecal muscle of the boll weevil, *Anthonomus grandis. J. Insect Physiol.* 21: 1275–1278.

Waage, J. K. (1979a) Dual function of the damselfly penis: sperm removal and transfer. *Science* 203: 916–918.

Waage, J. K. (1979b) Adaptive significance of postcopulatory guarding of mates and nonmates by male *Calopteryx maculata* (Odonata). *Behav. Ecol. Sociobiol.* 6: 147–154.

Waage, J. K. (1984) Sperm competition and the evolution of odonate mating systems. *In*: Smith, R. L. (ed.) *Sperm Competition and the Evolution of Animal Mating Systems,* Academic Press, London, pp. 251–290.

Waage, J. K. (1996) Parental Investment – Minding the kids or keeping control? *In*: Gowaty, P. A. (ed.) *Feminism and Evolutionary Biology,* Chapman and Hall, New York, pp.

527–553.

Walker, W. F. (1980) Sperm utilisation strategies in non-social insects. *Amer. Naturalist.* 115: 780–799.

Ward, P. I. (1993) Females influence sperm storage and use in the yellow dung fly *Scathophaga stercoraria* (L.). *Behav. Ecol. Sociobiol.* 32: 313–319.

Welsh, J. and McClelland, M. (1990) Fingerprinting genomes using PCR with arbitrary primers. *Nucl. Acid. Res.* 19: 861–866.

Wilson, N., Tubman, S., Eady, P. E. and Robertson, G. W. (1997) Female genotype affects male success in sperm competition. *Proc. R. Soc. Lond. B.* 264: 1491–1495.

Williams, J. G. K., Hanafey, M. K., Rafalski, J. A. and Tingey, S. V. (1993) Genetic analysis using random amplified polymorphic DNA markers. *Methods Enzymol.* 218: 704–740.

Williams, J. G. K., Kubelik, A. R., Livak, K. J., Rafalski, J. A. and Tingey, S. V. (1990) DNA polymorphisms amplified by arbitrary primers are useful as genetic markers. *Nucl. Acid. Res.* 18: 6531–6535.

Molecular measures of insect fitness

Michelle Pellissier Scott[1] and Scott M. Williams[2, 3]

[1] Department of Zoology, University of New Hampshire, Durham, NH 03824, USA
[2] Division of Biomedical Sciences, Meharry Medical College, Nashville, TN 37208, USA
[3] Present address: Department of Microbiology, Immunology and Genetics, Meharry Medical College, Nashville, TN 37208, USA

Summary

Many questions concerning social behavior and reproductive ecology require an accurate measure of direct and/or indirect fitness. Recently developed molecular techniques that measure parentage and relatedness have made these basic questions more tractable. In this paper, we provide a summary of the methods and analyses of the most commonly used techniques for measuring fitness that have been applied to insect systems and discuss the strengths and limitations of each. The use of highly polymorphic microsatellite loci as genetic markers has been steadily increasing in the past few years, and we focus on their usefulness in the study of natural populations.

Introduction

Obtaining accurate measurements of fitness is critical to an evolutionary understanding of behavior. The traditional approach to this problem has been to estimate reproductive success, direct fitness from current reproduction, by observing copulation and egg laying. Using this method, reproductive tactics of insects can only be studied in species that are easily observed. However, observational data have drawbacks, because offspring are not always associated with their mother or the product of observed matings (Webster and Westneat, this volume). Female insects are also subject to sperm competition (Siva-Jothy and Hadrys, this volume). Because of these limitations, obtaining accurate information on reproductive success in natural populations was difficult before molecular techniques became available. Now, measuring reproductive success and thus assessing fitness is significantly easier.

The aims of measuring reproductive success vary. One may wish to identify parents and count the offspring of specific individuals, or to know the relative success of individuals with different behavioral strategies. In some cases, the central question of a study is answered by identifying offspring not produced by specific parents. For example, in monogamous, care-giving species it is important to know if some offspring are not the progeny of one or both of a pair, and to assess the consequences of brood parasitism or sperm competition. When studying social species, both direct and indirect fitness (i.e. inclusive fitness) must be evaluated. In these systems, individuals may forego personal reproduction, and their genes are passed on via relatives. Testing hypotheses for the evolution of social behavior requires not only the measurement of individual reproduction but the reproduction in relatives as well. For example, Hamilton (1964) proposed that inclusive fitness gains may predispose haplodiploid organisms to form cooperative societies because sisters are more closely related to each other than parents are to their offspring. Thus, testing this prediction by measuring the relatedness within social insect groups has become an important component of assessing fitness (Metcalf and Whitt, 1977a, b).

In this paper, we review and evaluate molecular approaches for measuring fitness in insects. We consider their appropriateness for use with small insects, their ease of use and expense as well as the quality and quantity of information obtained from them. Although molecular techniques can be used with insects to determine parentage, most attention has been on social insects and the measuring of inclusive fitness indirectly by determining relatedness or the number of males with whom a queen has mated. Most laboratory methods can be applied to both parentage and relatedness issues, requiring only different analyses that may be more or less satisfactory.

Techniques for measuring fitness

Allozyme analysis

The oldest molecular technique for assessing direct fitness is electrophoresis of allozymes. Early studies identified parentage using exclusion or likelihood analyses. For exclusion analysis genetic incompatibility eliminates some potential parents. Then a probability of exclusion is calculated on the basis of gene frequencies at several loci for the potential parents. Ideally, all but one parent of each sex can be eliminated, but this rarely happens in studies of natural populations where there are many potential parents. When the mother is known, an alternative approach is to calculate the likelihood of all males being the father based on the population-level frequencies of alleles and the probability of detecting paternity (LaMunyon, 1994). When there are many potential fathers (as with insect populations), the applicability of this approach is limited because the amount of variation is often too low to reduce ambiguity (Chakraborty et al., 1988). However, allozyme studies are still useful for some questions, for example, evaluating queen vs. worker contributions to reproduction (Visscher, 1996) or the last male advantage in sperm competition (LaMunyon, 1994).

Two other analyses with protein data have been useful in insect behavioral ecology. Pioneering studies used protein polymorphisms for maternity exclusion to estimate the minimum number of reproductive queens in *Polistes* wasps (Metcalf and Whitt, 1977a; Lester and Selander, 1981). These data helped to determine the relative reproductive success of females, queen replacement and multiple insemination in queens. Protein data can only identify a lower limit of the number of queens or inseminations. However, allozyme data have been useful in measuring the relative reproductive success in laboratory colonies of fire ants (Ross, 1988) and sweat bees (Kukuk and May, 1988) with known phenotypes. Studies have also used measures of relatedness obtained from allozyme data to estimate an upper limit on the number of male and female reproductive wasps in the field (Strassmann et al., 1991; Hughes et al., 1993).

Protein polymorphism was also used to address broader questions of inclusive fitness in social insects by estimating the genetic relatedness of colony members. Studies have been done in many social insects, including eusocial wasps (Metcalf and Whitt, 1977a, b; Lester and Selander, 1981; Queller et al., 1988, 1993a; Strassmann et al., 1989, 1991, 1994; Hughes et al., 1993), primitively eusocial bees (Crozier et al., 1987; Kukuk et al., 1987; Schwarz, 1987), ants (Crozier et al., 1984; Stille et al., 1991; Seppä, 1992; Seppä and Walin, 1996), and tent caterpillars (Costa and Ross, 1993). All of these studies used regression of colony allele frequency against that of the individual (0, 0.5, 1.0) (e.g. Pamilo and Crozier, 1982) or related techniques (Queller and Goodnight, 1989), but again resolution was limited by lack of variation.

Traditional DNA fingerprinting and other southern blot analyses

Minisatellite/multilocus probes

DNA markers can also be used to measure fitness, both directly through parent-offspring assessment and indirectly through relatedness. Several techniques have been used that address these issues, the first of which was DNA fingerprinting with minisatellite sequences (Jeffreys et al., 1985). Several probes are useful for many species and synthetic, species-specific probes can be developed as well. Due to the greater variability, minisatellite DNA fingerprinting offers an enormous advantage over allozyme studies to establish maternity and paternity (Burke, 1989; Burke et al., 1991; Weatherhead and Montgomerie, 1991). However, multilocus fingerprinting is rarely used to identify parentage in insects; exceptions are a study of sweat bees (Mueller et al., 1994) and another to determine sperm precedence in bushcrickets (Achmann et al., 1992). There have been a few studies that estimate relatedness in honeybees (Moritz et al., 1991; Blanchetot, 1991), solitary and primitively social bees (Blanchetot, 1992; Danforth et al., 1996) and a solitary wasp (Pfennig and Reeve, 1993).

Matching diagnostic paternal fragments to offspring was used to assign paternity in the laboratory study on sperm precedence in bushcrickets (Achmann et al., 1992). However, the predominant method of analysis in many studies determining parentage and relatedness with multilocus fingerprinting is band sharing, which compares the proportion of shared markers relative to known parent-offspring pairs or to the population at large (Pamilo and Crozier, 1982; Queller and Goodnight,1989). Band-sharing analyses have been criticized for estimating relatedness between specific pairs of individuals when "background" band sharing between unrelated individuals is high (Lynch, 1988, 1991), and Reeve et al. (1992) recommend excluding monomorphic bands from the analysis. The estimate of relatedness becomes less reliable with more distantly related individuals. Although estimates of relatedness between specific pairs of individuals (e.g. parent-offspring) cannot be accurate, the average within-group relatedness may be estimated satisfactorily using regression techniques (Reeve et al., 1992). However, nonindependence among pairwise comparisons (because each individual is used in multiple comparisons) further weakens band-sharing analyses. Subsampling to correct for nonindependence in several published data sets yielded larger standard errors, reducing or eliminating differences in band-sharing coefficients between relatives and nonrelatives (Danforth and Freeman-Gallant, 1996).

Multilocus fingerprinting has limitations and requires a reasonable amount of caution in its application. First, relatively large amounts of tissue may be required and are unavailable for small insect species. Second, unlike allozymes, different allelic states are impossible or at best difficult to detect (but see Reeve et al., 1990). This can confound some interpretations. Also, it is impossible to differentiate homozygotes from heterozygotes [although this may be handled in the analysis (Reeve et al., 1992)]. Third, it is important to know whether bands are inherited independently (i.e. are unlinked), especially if relatedness is being calculated using band-sharing measures. Nonindependence can bias interpretations. Fourth, as with other techniques, bands of identical length may not be the same fragments. This problem is probably minimized when extremely variable probes are used. Fifth, it is very difficult to compare banding patterns on different gels. This means that the size of a sample from which results can be measured is a function of gel size and not population size. But even when samples are on the same gel, scoring inaccuracy makes the assignment of paternity by direct inspection, for example, suspect. Finally, choice of enzyme may have a highly significant effect on fingerprint clarity and thus on band-sharing coefficients (Zeh et al., 1993).

Single-locus analysis

Single-locus fingerprinting analysis has not been applied to insects often, but may be useful. Two variants exist: random single-locus variation based on restriction fragment length polymorphism (RFLP) and single-locus minisatellites based on variable number of tandem repeats (VNTRs). In the first, RFLPs are detected by hybridizing Southern blots with single-copy probes constructed from a species-specific library (Quinn and White, 1987). A variant of this technique analyzes RFLPs of DNA amplified by the polymerase chain reaction (PCR) with specifc primers. This variation has been used to distinguish mitochondrial DNA (mtDNA) haplotypes to determine fitness of competing queens in fire ant colonies (Balas and Adams, 1996), but the analysis was limited by the lack of variation. Both the Southern blot- and PCR-based variants of this approach require the construction of probes or the *a priori* determination of specific DNA sequences. The PCR approach, however, is extremely easy to use once primer sequences have been determined. In both cases genetic variation is limited to the presence or absence of a restriction site, limiting the maximum heterozygosity to 0.50. Determining haplotypes for several sites simultaneously increases heterozygosity and hence the usefulness of this approach.

Single-locus minisatellite markers (VNTRs) can be identified by probes obtained from screening a library with standard minisatellite probes and isolating specific clones for Southern blot hybridizations (Nakamura et al., 1987). This technique avoids many of the problems of multilocus probes because bands are revealed by hybridization to nonrepetitive sequences flanking the VNTR. Thus, bands are allelic and homozygotes can be differentiated from heterozygotes. Single-locus probes are often quite variable with heterozygosities of 0.95 or more (Zeh and Zeh, 1994). This makes them useful for studies of parentage, sperm precedence and even to classify individuals by relatedness. If heterozygosity for each probe is 0.9, only six VNTR loci are needed to distinguish parent-offspring pairs from random pairs (Chakraborty and Jin, 1993).

PCR-based methods

RAPD

Randomly amplified polymorphic DNA (RAPD) (Williams et al., 1990; Welsh and McClelland, 1990) has become a popular alternative to traditional DNA analyses by allowing cheap, relatively easy analysis of nanogram quantities of DNA. This technique uses a single short primer (~10 bases) to amplify regions of the genome flanked by inverted repeats. PCR amplification of genomic DNA with such primers produces numerous fragments easily visualized on agarose or polyacrylamide gels. Band patterns can be exceedingly variable between individuals and have proven useful for analyses of parentage (Hadrys et al., 1993; Scott and Williams, 1993), sperm competition (Hooper and Siva-Jothy, 1996; Reichardt and Wheeler, 1996) and the evolution of sociality (Fondrk et al., 1993). RAPD markers have also been used to estimate the number of mates of social insects (Fondrk et al., 1993). The queen's genotype can be determined from banding patterns of the drones and the paternal contributions to the workers is determined by subtraction, and cluster analysis can divide workers into subfamilies (Apostol et al., 1993; Fondrk et al., 1993). The major limit to the amount of variation is the number of primers used, but an extremely large number of primers are available that can be quickly and inexpensively tested for their utility in any system.

There are several potential pitfalls in both the generation of data and analysis that must be recognized. First, RAPD can be sensitive to PCR reaction conditions. Standardization of temperature and concentrations of Mg^{2+}, template DNA and the specific polymerase used may be essential for reproducibility (Hadrys et al., 1992; Ellsworth et al., 1993; Schierwater and Ender, 1993; Smith et al., 1994). Second, nonparental bands may appear in offspring, and their frequency may need to be assessed, depending on the type of analysis used. They have been common in some taxa (Riedy et al., 1992) and rare in others (Scott et al., 1992). Third, contamination is always a problem with PCR, but especially so with whole-insect RAPD analysis, because DNA from others organisms such as parasites will be amplified as well. Fourth, RAPD data suffer from many of the problems of minisatellite analysis. Bands are probably independent, but band homology cannot be absolutely determined by inspection, partly because of band comigration (Smith et al., 1994; Lynch and Milligan, 1994). Also, RAPD markers usually segregate as dominant Mendelian alleles. Therefore, the presence of a given band in two individuals indicates that they share one identical allele, but it is impossible to tell if either or both is homozygous at this locus. This reduces the statistical power and accuracy of all analyses (Lynch and Milligan, 1994). Antolin et al. (1996) describe the use of single-strand conformation polymorphism analysis of RAPD markers to reveal codominant alleles that may help to overcome this problem. Also, codominant markers in haplodiploid species can be analyzed by segregation with less work than in diploid species (Hunt and Page, 1992; Shoemaker et al., 1994). Fifth, parentage analysis by exclusion has poor resolution because the fraction of randomly selected offspring that can be unambiguously assigned to a single male decreases with the number of potential fathers and increases with the number of loci and frequency of the recessive alleles. About 50 variable RAPD loci are needed for each offspring to identify paternity for 80% of the offspring (Lewis and Snow, 1992).

Many of the problems can be limited, if not eliminated, by careful laboratory procedures and/or analytical methods (Schierwater and Ender, 1993; Grosberg et al., 1996). These fixes, although not perfect, can produce strong and reliable conclusions, but they often require significant extra bench work such as Southern blot analysis of the bands or segregation analysis of the alleles. Since one of the main advantages of RAPD is the ease and speed with which data can be generated, this affects the tradeoff of ease vs. quality of data.

Despite limitations, the evaluation of parentage by inclusion has been successful (Scott and Williams, 1993; Reichardt and Wheeler, 1996). This method is best used only when DNA from all potential parents is available and the number of parental candidates is not too large. Only bands unique to one of the putative parents can be used, and the presence of these diagnostic bands in the offspring is then indicative of parentage. The absence of such bands in the offspring is uninformative. As long as the frequency of nonparental bands is moderately low, they should not confound the determination of parentage, especially if several diagnostic bands amplified from different primers are used. The probability of false inclusion of putative parents can be calculated from estimates of the populationwide frequency of the null allele (q) for each marker that identifies parentage in the offspring (see Lynch and Milligan, 1994).

The use of RAPDs for determination of parentage by band sharing, band matching (Apostol et al., 1993; Hooper and Siva-Jothy, 1996) or "synthetic offspring" (Hadrys et al., 1993) and for estimates of relatedness is probably rather limited for much the same reasons as the use of DNA fingerprints for these analyses (Lynch and Milligan, 1994). The degree of overlap in the distributions of relatedness, even for unrelated and parent-offspring combinations, may be broad, and more than 50 loci are probably needed to discriminate well (Lynch and Milligan, 1994). However, 50 or more RAPD loci can be realtively easy to amplify. Using RAPD bands with allelic frequencies between 0.1 and 0.6 improves the discriminatory power of band-matching analysis that scores both the presence and absence of bands (Apostol et al., 1993).

Microsatellite or simple sequence repeat markers

The most important new technique for measuring insect fitness is the use of microsatellite loci as genetic markers (Choudhary et al., 1993; Queller et al., 1993b; Strassmann et al., 1996a). These loci consist of runs of short repeating sequences, simple sequence repeats (SSRs), that can vary substantially in numbers between individuals. DNA amplified with unique sequences flanking these repeats is easily detectable as fragments of different lengths if individuals carry alleles with different numbers of repeats (Schlötterer and Pemberton, this volume). In addition, microsatellite fragments are codominant and inherited in Mendelian fashion, removing many of the problems with DNA fingerprinting and RAPD analyses. Microsatellites are generally highly variable, with average expected heterozygosities usually above 50% (Jarne and Lagoda, 1996; Strassmann et al., 1996b) and often above 90% (Evans, 1993). However, high mutation rates could, but usually do not, create problems resulting in false exclusion. This potential problem can be virtually eliminated when several primer pairs are used.

The PCR primers used in microsatellite analyses are species-specific. This specificity insures that alternative bands are allelic, but such sequence specificity required that primers be identified from genomic libraries and was time-consuming (Strassmann et al., 1996a). Newer methods, however, such as the randomly amplified microsatellites (RAMS) method, can greatly simplify this process (Ender et al., 1996; Schlötterer and Pemberton, this volume). Once loci are identified, data collection is fast and generally very informative for measurements of fitness.

Because of sequence variation among species, especially in the noncoding regions, where microsatellite loci are generally located, primers developed for one species may not work well on others. However, this is not always the case. Successful cross-species amplification depends on the evolutionary distance between the species, and about 40% of the primer sequences from one wasp species amplified variable fragments in other wasp species with a divergence time less than about 65 million years (Strassmann et al., 1996b). These results indicate that although many primers generated from a single species will only be useful for that species, others can be used in related groups. This cross-species possibility will make the use of microsatellites significantly easier once a large bank of these primers is developed.

Despite what was a relatively long lead time for generating markers, microsatellites are now being widely used to measure fitness in insects. In parentage analysis, even with a large pool of candidates, 10–20 loci should be able to eliminate all but one male and one female (Queller et al., 1993b), although 30 loci may be required to differentiate offspring of hymenopteran full sisters (Peters et al., 1995). For example, maternity and paternity analysis in a natural population of communally-breeding beetles with a moderate number of candidates (e.g. 10) succeeded in identifying female brood parasites and male extra-group fertilizations with only three loci (Scott et al., in press).

A large proportion of insect studies using microsatellites analyze the evolution of reproductive cooperation and conflict in social insects. In most cases, the queen's genotype is identified through sons (which are haploid) or workers, and paternal alleles are deduced by subtraction. Levels of polyandry and particularly the effective number of patrilines present in the colony at any one time have been widely investigated in honey bees (Estoup et al., 1994; Moritz et al., 1995, 1996; Oldroyd et al., 1995, 1996) and ants (Sundström et al., 1996). Maternity assignment and estimates of the minimum number of mates can even be made from analysis of stored sperm (Peters et al., 1995). Attention has also focused on mating structure in hymenopteran societies, the number of queens, and their relatedness to each other (Hamaguchi et al., 1993; Peters et al., 1995; Evans, 1995, 1996a, 1996b; Paxton et al., 1996).

Estimates of relatedness use the same analysis as for allozyme data (Pamilo and Crozier, 1982; Queller and Goodnight, 1989) and may be aided by the computer program RELATEDNESS 4.2 (Queller, Rice University). Cluster analysis can be used to identify family groups or full siblings from half siblings (Estoup et al., 1995). The coefficient of relatedness may also be calculated directly from the relative proportions of each patriline in a monogynous colony (Laidlaw and Page, 1984).

Conclusions

High-resolution molecular techniques have made it possible for behavioral ecologists to address problems that were formerly intractable. Accurate determination of parentage is vital, and we now have a choice of several methodologies (Tab. 1). In some cases, such as with laboratory studies where individuals of known allozyme phenotypes can be manipulated, the simplest and cheapest method to estimate reproductive success may still be protein analysis. In others, RAPD analyses may be useful because it is easy to master quickly. For those insects from which large amounts of DNA can be obtained, fingerprinting can also be useful. However, in

Table 1. Genetic techniques for measuring reproductive success

	Allozymes	Multilocus minisatellites	Single-locus probes	RAPD	micro satellites
Applications					
Direct fitness					
Parental inclusion	−	+	−	++	+++
Parental exclusion	+	+	+++	−	+++
Indirect fitness					
Relatedness of particular individuals	−	−	−	−	−
Average within-group relatedness	+	++	++	+	+++
Attributes					
Amount of variation	low	high	high	high[*]	high
Identification of allelic states	yes	no	yes	no	yes
Differentiation of homozygotes and heterozygotes	yes	no	yes	no	yes
Inference of parentage without all potential parents	−	−	++	−	++
Number of potential parents	few	moderate[†]	many	many	many
Ease of use	+++	+	++	+++	++
Quantity of DNA needed for analysis	NA	ng–μg	ng–μg	pg–ng	pg–ng

[*]With some important exceptions, e.g. dampwood termites (S. M. Williams, unpublished data). [†]Limited by resolution and the size of the gel.

most situations we think that microsatellites are the most powerful and useful tools for measuring fitness in insects. Markers are highly variable and codominant, and because this technique uses PCR, little tissue is needed. Methods to make development of microsatellite primers are being improved rapidly. As microsatellite use becomes more common, markers for more taxa will be identified, making them available for other studies of the same or related species.

Acknowledgements

This work was supported by grants from the National Science Foundation (DEB 9222148 and IBN 9628832) to MPS. SMW was supported by NIH-RCMI 5G12RR-AI03032-32-12 and K14-HL03321-01A1.

References

Achmann, R., Heller, K.-G. and Epplen, J. T. (1992) Last-male sperm precedence in the bushcricket *Poecilimon veluchianus* (Orthoptera, Tettigonioidea) demonstrated by DNA fingerprinting. *Molec. Ecol.* 1: 47–54.

Antolin, M. F., Bosio, C. F., Cotton, J., Sweeney, W., Strand, M. R. and Black W. C. IV (1996) Intensive linkage mapping in a wasp (*Bracon hebetor*) and a mosquito (*Aedes aegypti*) with single-strand conformation polymorphism analysis of random amplified polymorphic DNA markers. *Genetics* 143: 1727–1738.

Apostol, B. L., Black, W. C. IV, Miller, B. R., Reiter, P. and Beaty, B. J. (1993) Estimation of the number of full sibling families at an oviposition site using RAPD-PCR markers: applications to the mosquito *Aedes aegypti*. *Theor. Appl. Genet.* 86: 991–1000.

Balas, M. T. and Adams, E. S. (1996) The dissolution of cooperative groups: mechanisms of queen mortality in incipient fire ant colonies. *Behav. Ecol. Sociobiol.* 38: 391–399.

Blanchetot, A. (1991) Genetic relatedness in honeybees as established by DNA fingerprinting. *J. Hered.* 82: 391–396.

Blanchetot, A. (1992) DNA fingerprinting analysis in the solitary bee *Megachile rotundata*: variability and nest mate genetic relationships. *Genome* 35: 681–688.

Burke, T. (1989) DNA fingerprinting and other methods for the study of mating success. *Trends Ecol. Evol.* 4: 139–144.

Burke, T., Hanotte, O., Bruford, M. W. and Cairns, E. (1991) Multilocus and single locus minisatellite analysis in population biological studies. *In*: Burke, T., Dolf, G., Jeffreys, A. J. and Wolff, R. (eds) DNA *Fingerprinting: Approaches and Applications*, Birkhäuser Verlag, Basel, pp. 154–168.

Chakraborty, R and Jin, L. (1993) Determination of relatedness between individuals using DNA fingerprinting. *Hum. Biol.* 65: 875–895.

Chakraborty, R., Meagher, T. R. and Smouse, P. E. (1988) Parentage analysis with genetic markers in natural populations. I. The expected proportion of offspring with unambiguous paternity. *Genetics* 118: 527–536.

Choudhary, M., Strassmann, J. E., Solís, C. R. and Queller, D. C. (1993) Microsatellite variation in a social insect. *Biochem. Genetics* 31: 87–96.

Costa, J. T. III and Ross, K. G. (1993) Seasonal decline in intracolony genetic relatedness in eastern tent caterpillars: implications for social evolution. *Behav. Ecol. Sociobiol.* 32: 47–54.

Crozier, R. H., Pamilo, P. and Crozier, Y. C. (1984) Relatedness and microgeographic genetic variation in *Rhytidoponera mayri*, an Australian arid-zone ant. *Behav. Ecol. Sociobiol.* 15: 143–150.

Crozier, R. H., Smith, B. H. and Crozier, Y. C. (1987) Relatedness and population structure of the primitively eusocial bee *Lasioglossum zephyrum* (Hymenoptera: Halictidae) in Kansas. *Evolution* 4: 902–910.

Danforth, B. N. and Freeman-Gallant, C. R. (1996) DNA fingerprinting data and the problem of non-independence among pairwise comparisons. *Molec. Ecol.* 5: 221–227.

Danforth, B. N., Neff, J. L. and Barretto-Ko, P. (1996) Nest mate relatedness in a communal bee, *Perdita texana* (Hymenoptera: Andrenidae) based on DNA fingerprinting. *Evolution* 50: 276–284.

Ellsworth, D. L., Rittenhouse, K. D. and Honeycutt, R. L. (1993) Artifactual variation in randomly amplified polymorphic DNA banding patterns. *Biofeedback* 14: 214–217.

Ender, A., Schwenk, K., Städler, T., Streit, B. and Schierwater, B. (1996) RAPD identification of microsatellites in *Daphnia*. *Molec. Ecol.* 5: 437–441.

Estoup, A., Solignac, M. and Cornuet, J.-M. (1994) Precise assessment of the number of patrilines and of genetic relatedness in honeybee colonies. *Proc. R. Soc. Lond. B* 258: 1–7.

Estoup, A, Garnery, L., Solignac, M. and Cornuet, J.-M. (1995) Microsatellite variation in honey bee (*Apis mellifera* L.) populations: hierarchical genetic structure and test of the infinite allele and stepwise mutation models. *Genetics* 140: 679–695.

Evans, J. D. (1993) Parentage analysis in ant colonies using simple sequence repeat loci. *Molec. Ecol.* 2: 393–397.

Evans, J. D. (1995) Relatedness threshold for the production of female sexuals in colonies of a polygynous ant, *Myrmica tahoensis*, as revealed by microsatellite DNA analysis. *Proc. Natl. Acad. Sci. USA* 92: 6514–6517.

Evans, J. D. (1996a) Competition and relatedness between queens of the facultatively polygynous ant, *Myrmica tahoensis*. *Anim. Behav.* 51: 831–840.

Evans, J. D. (1996b) Queen longevity, queen adoption and posthumous indirect fitness in the facultatively polygynous ant, *Myrmica tahoensis*. *Behav. Ecol. Sociobiol.* 39: 275–285.

Fondrk, M. K., Page, R. E. and Hunt, G. J. (1993) Paternity analysis of worker honeybees using random amplified polymorphic DNA. *Naturwissenschaften* 80: 226–231.

Grosberg, R. K., Levitan, D. R. and Cameron, B. B. (1996) Characterization of genetic strucutre and genealogies using RAPD-PCR markers: a random primer for the novice and nervous. *In*: Ferraris, J. D. and Palumbi, S. R. (eds) *Molecular Zoology: Advances, Strategies and Protocols*, Wiley-Liss, New York, pp. 67–100.

Hadrys, H., Balick, M. and Schierwater, B. (1992) Applications of random amplified polymorphic DNA (RAPD) in molecular ecology. *Molec. Ecol.* 1: 55–63.

Hadrys, H., Schierwater, B., Dellaporta, S. L., DeSalle, R. and Buss, L. W. (1993) Determination of paternity in dragonflies by random amplified polymorphic DNA fingerprinting. *Molec. Ecol.* 2: 79–87.

Hamaguchi, K., Itô, Y. and Takenaka, O. (1993) GT dinucleotide repeat polymorphisms in a polygynous ant, *Leptothorax spinosior* and their use for measurement of relatedness. *Naturwissenschaften* 80: 179–181.

Hamilton, W. D. (1964) The genetical evolution of social behavior I and II. *J. Theor. Biol.* 7: 1–52.

Hooper, R. E. and Siva-Jothy, M. T. (1996) Last male sperm precedence in a damselfly demonstrated by RAPD profiling. *Molec. Ecol.* 5: 449–452.

Hughes, C. R., Queller, D. C., Strassmann, J. E. and Davis, S. K. (1993) Relatedness and altruism in *Polistes* wasps. *Behav. Ecol.* 4: 128–137.

Hunt, G. J. and Page, R. E. (1992) Patterns of inheritance with RAPD molecular markers reveal novel types of polymorphism in the honey bee. *Theor. Appl. Genet.* 85: 15–20.

Jarne, P. and Lagoda, P. J. L. (1996) Microsatellites, from molecules to populations and back. *Trends Ecol. Evol.* 11: 424–429.

Jeffreys, A. J., Wilson, V. and Thein, S. L. (1985) Individual-specific "fingerprints" of human DNA. *Nature* 316: 76–79.

Kukuk, P. F. and May, B. P. (1988) Dominance hierarchy in the primitively eusocial bee *Lasioglossum (Dialictus) zephyrum*: Is genealogical relationship important? *Anim. Behav.* 36: 1848–1849.

Kukuk, P. E., Eickwort, G. C. and May, B. (1987) Multiple maternity and multiple paternity in first generation brood from single foundress colonies of the sweat bee *Dialictus zephyrus* (Hymenoptera: Halictidae). *Insectes Soc.* 34: 131–135.

Laidlaw, H. H. and Page, R. E. (1984) Polyandry in honey bees (*Apis mellifera* L.): sperm utilization and intracolony genetic relationships. *Genetics* 108: 985–997.

LaMunyon, C. W. (1994) Paternity in naturally-occurring *Utetheisa ornatrix* (Lepidoptera: Arctiidae) as estimated using enzyme polymorthism. *Behav. Ecol. Sociobiol.* 34: 403–408.

Lester, L. J. and Selander, R. K. (1981) Genetic relatedness and the social organization of *Polistes* colonies. *Amer. Naturalist.* 117: 147–166.

Lewis, P. O. and Snow, A. A. (1992) Deterministic paternity exclusion using RAPD markers. *Molec. Ecol.* 1: 155–160.

Lynch, M. (1988) Estimates of relatedness by DNA fingerprinting. *Molec. Biol. Evol.* 5: 584–599.

Lynch, M. (1991) Analysis of population genetic structure by DNA fingerprinting. *In*: Burke, T., Dolf, G., Jeffreys, A. J. and Wolff, R. (eds) *DNA Fingerprinting: Approaches and Applications*, Birkhäuser Verlag, Basel, pp. 113–126.

Lynch, M. and Milligan, B. G. (1994) Analysis of population genetic structure with RAPD markers. *Molec. Ecol.* 3: 91–99.

Metcalf, R. A. and Whitt, G. S. (1977a) Intra-nest relatedness in the social wasp *Polistes metricus*. *Behav. Ecol. Sociobiol.* 2: 339–351.

Metcalf, R. A. and Whitt, G. S. (1977b) Relative inclusive fitness in the social wasp *Polistes metricus*. *Behav. Ecol. Sociobiol.* 2: 353–360.

Moritz, R. F. A., Kryger, P. and Allsopp, M. H. (1996) Competition for royalty in bees. *Nature* 384: 31.

Moritz, R. F. A., Kryger, P., Koeniger, G., Koeniger, N., Estoup, A. and Tingek, S. (1995) High degree of polyandry in *Apis dorsata* queens detected by DNA microsatellite variability. *Behav. Ecol. Sociobiol.* 37: 357–363.

Moritz, R. F. A., Meusel, M. S. and Haberl, M. (1991) Oligonucleotide DNA fingerprinting discriminates super- and half-sisters in honeybee colonies (*Apis mellifera* L.). *Naturwissenschaften* 78: 422–424.

Mueller, U. G., Eickwort, G. C. and Aquadro, C. F. (1994) DNA fingerprinting analysis of parent-offspring conflict in a bee. *Proc. Natl. Acad. Sci. USA* 91: 5143–5147.

Nakamura, Y., Leppert, M., O'Connell, P., Wolff, R., Holm, T., Culver, M., Martin, C., Fujimoto, E., Hoff, M., Kumlin, E. and White, R. (1987) Variable number of tandem repeat (VNTR) markers for human gene mapping. *Science* 235: 1616–1622.

Oldroyd, B. P., Smolenski, A. J., Cornuet, J.-M., Wongsiri, S., Estoup, A., Rinderer, T. E. and Crozier, R. H. (1995) Levels of polyandry and intracolonial relationships in *Apis florea*. *Behav. Ecol. Sociobiol.* 37: 329–335.

Oldroyd, B. P., Smolenski, A. J., Cornuet, J.-M., Wongsiri, S., Estoup, A., Rinderer, T. E. and Crozier, R. H. (1996) Levels of polyandry and intracolonial genetic relationships in *Apis dorsata* (Hymenoptera: Apidae). *Ann. Entomol. Soc. Amer.* 89: 276–283.

Pamilo, P. and Crozier, R. H. (1982) Measuring genetic relatedness in natural populations: Methodology. *Theor. Pop. Biol.* 21: 171–193.

Paxton, R. J., Thorén, P. A., Tengö, J. Estoup, A. and Pamilo, P. (1996) Mating structure and nestmate relatedness in a communal bee, *Andrena jacobi* (Hymenoptera, Andrenidae), using microsatellites. *Molec. Ecol.* 5: 511–519.

Peters, J. M., Queller, D. C., Strassmann, J. E. and Solí, C. R. (1995) Maternity assignment and queen replacement in a social wasp. *Proc. R. Soc. Lond. B* 260: 7–12.

Pfennig, D. W. and Reeve, H. K. (1993) Nepotism in a solitary wasp as revealed by DNA fingerprinting. *Evolution* 47: 700–704.

Queller, D. C. and Goodnight, K. F. (1989) Estimating relatedness using genetic markers. *Evolution* 43: 258–275.

Queller, D. C., Negron-Sotomayor, J. A., Strassmann, J. E. and Hughes,C.R. (1993a) Queen number and genetic relatedness in a neotropical wasp, *Polybia occidentalis. Behav. Ecol.* 4: 7–13.

Queller, D. C., Strassmann, J. E. and Hughes, C. R. (1988) Genetic relatedness in colonies of tropical wasps with multiple queens. *Science* 242: 1155–1157.

Queller, D. C., Strassmann, J. E. and Hughes, C. R. (1993b) Microsatellites and kinship. *Trends Ecol. Evol.* 8: 285–288.

Quinn, T. W. and White, B. N. (1987) Identification of restriction fragment length polymorphism in genomic DNA of the lesser snow goose (*Anser caerulescens*). *Molec. Biol. Evol.* 4: 126–143.

Reeve, H. K., Westneat, D. F., Noon, W. A. and Sherman, P. W. (1990) DNA "fingerprinting" reveals high levels of inbreeding in colonies of the eusocial naked mole-rat. *Proc. Natl. Acad. Sci. USA* 87: 2496–2500.

Reeve, H. K., Westneat, D. F. and Queller, D. C. (1992) Estimating average within-group relatedness from DNA fingerprints. *Molec. Ecol.* 1: 223–232.

Reichardt, A. K. and Wheeler, D. E. (1996) Multiple mating in the ant *Acromyrmex versicolor*: a case of female control. *Behav. Ecol. Sociobiol.* 38: 219–225.

Riedy, M. F., Hamilton, W. J. III and Aquadro, C. F. (1992) Excess of non-parental bands in offspring from known primate pedigrees assayed using RAPD PCR. *Nucl. Acid Res.* 20: 918.

Ross, K. G. (1988) Differential reproduction in multiple-queen colonies of the fire ant *Solenopsis invicta* (Hymenoptera: Formicidae). *Behav. Ecol. Sociobiol.* 23: 341–355.

Schierwater, B. and Ender, A. (1993) Different thermostable DNA polymerases may amplify different RAPD products. *Nucl. Acid Res.* 21: 4647–4648.

Schwarz, M. P. (1987) Intra-colony relatedness and sociality in the allodapine bee *Exoneura bicolor. Behav. Ecol. Sociobiol.* 21: 387–392.

Scott, M. P. and Williams, S. M. (1998) Comparative reproductive success of communally breeding burying beetles as assessed by PCR with randomly amplified polymorphic DNA. *Proc. Natl. Acad. Sci. USA* 90: 2242–2245.

Scott, M. P., Haymes, K. M. and Williams, S. M. (1992) Parentage analysis using RAPD PCR. *Nucl. Acid Res.* 20: 5493.

Scott, M. P., Lee, W.-J. and van der Reijden, E. D. (1998) The frequency and fitness consequences of communal breeding in a natural population of burying beetles. *Behav. Ecol.*; *in press*.

Seppä, P. (1992) Genetic relatedness of worker nestmates in *Myrmica ruginodis* (Hymenoptera: Formicidae) populations. *Behav. Ecol. Sociobiol.* 30: 253–260.

Seppä, P. and Walin, L. (1996) Sociogenetic organization of the red ant *Myrmica rubra*. *Behav. Ecol. Sociobiol.* 38: 207–217.

Shoemaker, D. D., Ross, K. G. and Arnold, M. L. (1994) Development of RAPD markers in two introduced fire ants, *Solenopsis invicta* and *S. richteri*, and their application to the study of a hybrid zone. *Molec. Ecol.* 3: 531–539.

Smith, J. J., Scott-Craig, J. S., Leadbetter, J. R., Bush, G. L., Roberts, D. L. and Fulbright, D. W. (1994) Characterization of random amplified polymorphic DNA (RAPD) products from *Xanthomonas campestris*: implications for the use of RAPD products in phylogenetic analysis. *Mol. Phylogenet. Evol.* 3: 135–145.

Stille, M., Stille, B. and Douwes, P. (1991) Polygyny, relatedness and nest founding in the polygynous myrmicine ant *Leptothorax acervorum* (Hymenoptera: Formicidae). *Behav. Ecol. Sociobiol.* 28: 91–96.

Strassmann, J. E., Hughes, C. R., Queller, D. C., Turillazzi, S., Cervo, R., Davis, S. K. and Goodnight, K. F. (1989) Genetic relatedness in primitively eusocial wasps. *Nature* 342: 268–270.

Strassmann, J. E., Queller, D. C., Solís, C. R. and Hughes, C. R. (1991) Relatedness and queen number in the neotropical wasp, *Parachartergus colobopterus*. *Anim. Behav.* 42: 461–470.

Strassmann, J. E., Hughes, C. R., Turillazzi, S., Solís, C. R. and Queller, D. C. (1994) Genetic relatedness and incipient eusociality in stenogastrine wasps. *Anim. Behav.* 48: 813–821.

Strassmann, J. E., Solís, C. R., Peters, J. M. and Queller, D. C. (1996a) Strategies for finding and using highly polymorphic DNA microsatellite loci for studies of genetic relatedness and pedigrees. *In*: Ferraris, J. D. and Palumbi, S. R. (eds) *Molecular Zoology: Advances, Strategies and Protocols*, Wiley-Liss, New York, pp. 163–180, 528–549.

Strassmann, J. E., Solís, C. R., Barefield, K. and Queller, D. C. (1996b) Trinucleotide microsatellite loci in a swarm-founding neotropical wasp, *Parachartergus colobopterus* and their usefulness in other social wasps. *Molec. Ecol.* 5: 459–461.

Sundström, L., Chapuisat, M. and Keller, L. (1996) Conditional manipulation of sex ratios by ant workers: a test of kin selection theory. *Science* 274: 993–995.

Visscher, P. K. (1996) Reproductive conflict in honey bees: a stalemate of worker egg-laying and policing. *Behav. Ecol. Sociobiol.* 39: 237–244.

Weatherhead, P. J. and Montgomerie, R. D. (1991) Good news and bad news about DNA fingerprinting. *Trends Ecol. Evol.* 6: 173–174.

Welsh, J. and McClelland, M. (1990) Fingerprinting genomes using PCR with arbitrary primers. *Nucl. Acid Res.* 18: 7213–7218.

Williams, J. G. K., Kubelik, A. R., Kivak, K. J., Rafalski, J. A. and Tingey, S. V. (1990) DNA polymorphisms amplified by arbitrary primers are useful as genetic markers. *Nucl. Acid Res.* 18: 6531–6535.

Zeh, J. A. and Zeh D. W. (1994) Last-male sperm precedence breaks down when females mate with three males. *Proc. R. Soc. Lond. B* 257: 287–292.

Zeh, D. W., May, C. A., Coffroth, M. A. and Bermingham, E. (1993) *MboI* and *Macrohaltica* – quality of DNA fingerprints is strongly enzyme-dependent in an insect (Coleoptera). *Molec. Ecol.*2: 61–63.

The use of microsatellites for genetic analysis of natural populations – a critical review

Christian Schlötterer[1] and Josephine Pemberton[2]

[1]Institut für Tierzucht und Genetik, Veterinärmedizinische Universität Wien, Josef-Baumann Gasse 1, A-1210 Wien, Austria
[2]Institute of Cell, Animal and Population Biology, University of Edinburgh, West Mains Road, Edinburgh, EH9 3JT, UK

Summary

Microsatellites, tandemly repeated units of 1 to 5 bp, are distributed throughout eukaryotic genomes. Length variation within microsatellites, caused by DNA slippage, can be revealed by polymerase chain reaction and used for DNA profiling. The ease of typing of microsatellites and their prevalence has promoted the application of microsatellite analysis to many research areas including molecular ecology and population genetics. This chapter provides background to microsatellite analysis and highlights examples in which microsatellite analysis has successfully extended the repertoire of molecular methods to understand dynamics and evolutionary forces acting on natural populations. Finally, the limitations of microsatellite analysis, which have emerged with increasing use, are discussed.

Introduction

The genetic study of natural populations is dependent on the availability of polymorphic, neutral markers. The introduction of isoenzyme electrophoresis revolutionized population genetics. Information about gene flow, allele frequencies and other parameters that are crucial in population biology was obtained by combining information from several loci. More recent techniques have involved exploitation of polymorphic DNA sequences for population analysis. Restriction fragment length polymorphism (RFLP) analysis was the first technique to be used. However, in general both nuclear and mitochondrial DNA RFLPs lack the degree of polymorphism needed for the study of intrapopulation structure. Molecular investigations of paternity and other family relationships for behavioral ecology studies were first made possible by multilocus DNA fingerprinting (Jeffreys et al., 1985). After treatment with a restriction enzyme and electrophoresis, genomic DNA is hybridized with a labeled fragment containing a minisatellite core sequence. This produces a pattern of many polymorphic bands. Whereas the high level of polymorphism of minisatellites is crucial for these studies, the difficulty in assigning bands to specific loci and the instability of some alleles limit this technique for certain kinds of population analysis. The invention of the polymerase chain reaction (PCR) (Saiki et al., 1988) made another class of polymorphic DNA accessible: microsatellites. In recent years microsatellites have replaced most alternative markers in ecological and evolutionary studies.

Microsatellites: a definition

Microsatellite, simple sequence repeat (SSR) or short tandem repeat (STR) are terms for stretches of repeats of short motifs ranging from mono- to pentanucleotides. Microsatellites are highly abundant in the eukaryotic genome and can normally reach a length of up to 150 bp. Because they are distributed throughout the genome, they are very useful for genome mapping (Dib et al., 1996; Dietrich et al., 1996). Polymorphisms at the microsatellite level are caused by DNA slippage (Levinson and Gutman, 1987; Tautz and Schlötterer, 1994). During replication the repeat units on the two DNA strands may anneal out of register, resulting in either an expansion or a contraction of the microsatellite on the new strand. In most cases this length difference is recognized by the DNA mismatch repair system, which removes the mismatch (Strand et al., 1993). Deficiencies in the mismatch repair system lead to increased microsatellite instability, indicating that the mismatch repair system restores the original microsatellite length. Naturally occurring microsatellite mutations are probably the result of DNA slippage events which escaped the mismatch repair system. Studies on human family panels have determined an average mutation frequency of about 0.01% for microsatellites (Weber and Wong, 1993; Weissenbach et al., 1992). This is in good agreement with rates found for microsatellites that have been introduced into yeast (Henderson and Petes, 1992). Thus it has been widely assumed that microsatellite mutation rates are constant cross species. However, recently, microsatellite mutation rates in *Drosophila melanogaster* have been estimated which are two orders of magnitude lower (Schug et al., 1997). Whether this is a singular phenomenon, or applies to other species as well must await further data.

Methods

Isolation of microsatellites and primer design

Since the first description of the use of microsatellites for DNA profiling (Litt and Luty, 1989; Tautz, 1989; Weber and May, 1989), several techniques for isolating new microsatellite loci have been published (Armour et al., 1994; Brenig and Brem, 1991; Browne and Litt, 1992; Ender et al., 1995; Ito et al., 1992; Rassmann et al., 1991). Primers allowing amplification of the microsatellite are chosen from the flanking unique-sequence DNA. For paternity testing it is ideal to select those loci likely to be most polymorphic, and some criteria are available for predicting levels of polymorphism at a new locus. Weber showed by a survey of several loci that the information content of a locus increases with the number and perfection of repeats (Weber, 1990). Good general criteria for choosing loci for primer design are (i) no alterations present in the repeat unit and (ii) more than 12 repeats present.

From a population geneticist's perspective, selection of loci most likely to be polymorphic is questionable, since the choice of loci depends on the allele distribution in the population from which the microsatellite was isolated rather than being randomly chosen. To avoid biased sampling, loci with fewer repeat units should also be included. Even microsatellite loci as short as four repeats (in the clone) are polymorphic within and among populations in some species [e.g. oaks (Steinkellner et al., 1997)].

Screening

Microsatellites are screened by PCR amplification of the locus and sizing of the PCR product(s) on a polyacrylamide gel. Most large-scale screening of microsatellites is currently carried out on automated sequencers, which detect fragments through the use of primers conjugated to fluorescent dyes and determine size by reference to in-lane standards. An alternative approach involves radioactive labeling of the PCR fragments by end-labeling of one primer or direct incorporation of radioactive nucleotides during PCR, and fragment separation on a denaturing sequencing gel. The absolute size of alleles is determined by running a standard sequencing reaction in parallel. The determination of the absolute length of the PCR product allows intergel comparison and makes the exchange of data generated in different laboratories possible. To avoid radioactive labeling, an alternative approach involves blotting PCR products onto a nylon membrane (Schlötterer, 1993; Weissenbach et al., 1992) and subsequent probing of the membrane with biotin- or digoxygenin-labeled simple sequence or PCR primer. This approach can be extraordinarily efficient if several loci with different-sized products are coamplified in the same tube (multiplex reaction). False amplification products are almost unavoidable in a multiplexing reaction, but are not visible when gels are blotted and probed with simple sequences (Schlötterer, 1993). Full-length protocols of all methods are given in Schlötterer (1997).

DNA typing with very limited amounts of DNA

By using PCR to amplify microsatellites, a wide range of sample sources becomes available. Microsatellites have been successfully amplified from bones (Hagelberg et al., 1992), plumage of museum specimens of birds (Ellegren, 1991) and single sperm cells (Hubert et al., 1992). Other DNA sources for microsatellite profiling are traces animals leave behind, such as hair roots (Morin et al., 1994), feathers or feces (Gerloff et al., 1995). Perhaps of greatest benefit for population studies, microsatellites can be screened from small organisms such as insects (Hughes and Queller, 1993).

Applications of microsatellites

Since their introduction, microsatellites have been rapidly incorporated into the technical repertoire of ecologists and evolutionary biologists. Initially, microsatellite analysis was mainly used in behavioral ecology, but there is now a steadily increasing number of applications in population genetics.

Applications in behavioral ecology

Male mating success is determined by paternity testing

Until recently, only field observations were available for estimating male mating success. The limits of this approach are obvious, and it is no surprise that genetic analysis of offspring often reveals differences from the predictions of field observation data (Amos et al., 1993; Gibbs et al., 1990). Microsatellites are a good tool for paternity analysis, as they show codominance of discrete alleles. Simple allelic subtraction allows precise determination of the paternal input, since the offspring inherits one allele from each of the parents. By analyzing several loci in this way enough information about the father can be collected to identify him. Many supporters of multilocus DNA fingerprinting might argue that for paternity testing microsatellites provide no advantage. This is true to some extent for small populations with a limited number of potential fathers. But the more males in the group, the more difficult becomes the analysis by multilocus fingerprinting, since all suspected fathers have to be run next to the mother-offspring pair. Microsatellite alleles are classified by the length of the PCR product, and the accurate length determination on a denaturing sequencing gel makes intergel comparison easy and reliable. This means that once typed an individual can be compared with any other individual just by referring to the determined lengths of the alleles. This attribute of microsatellites makes it possible to analyze populations that would otherwise not have been accessible. To demonstrate this, two very different cases of paternity analysis are described below.

Soay sheep. In a study population of Soay sheep on the island of Hirta, St. Kilda, North Atlantic, both sexes are extremely promiscuous. Field censuses show that each female has many partners per estrus, while preliminary genetic studies often eliminated all males seen consorting with an estrous female as the father of the subsequent lamb. Consequently, all males in the study area in the breeding season, of which there may be 200 or more in any particular year, have to be regarded as putative fathers for each lamb. Such a task is logistically impossible without reliable between-gel comparisons. By using 17 locus-specific polymorphisms, including 11 microsatellites, it was possible to identify the father of approximately two-thirds of lambs with 80% confidence (Pemberton et al., 1996).

Pilot whales. In contrast, our second example deals with an analysis within tightly knit social units. Pilot whales swim in groups, called pods, usually numbering 50–200 individuals of both sexes. Observation data suggested a polygamous mating system, with dominant males monopolizing females. Initial studies with multilocus fingerprints indicated that this inference might be wrong, but were inconclusive. Therefore, six microsatellite loci were isolated to study pilot whales. Fetuses of several pods were studied by comparing the fetal alleles with the alleles of the mother (allelic subtraction). For each group, the hypothesis that a single male dominated the group at mating was excluded by paternal alleles, because too many different paternal alleles were found in fetuses of the same age. Furthermore, a search for compatible males within each group revealed that the fathers of the offspring were not present in the same group. These data led to the conclusion that among pilot whales a novel mating system exists (Amos et al., 1993).

Social organization of populations

Much research in behavioral ecology is devoted to understanding the reasons why some animals live in groups, and the rules that animals use to behave altruistically or selfishly towards one another. Hamilton's concept of inclusive fitness, under which altruistic acts may be under selection if the beneficiaries are genetically related to the donor, is the most commonly cited explanation of altruism (Hamilton, 1964). Genetic techniques which can identify relatedness play a major role in distinguishing this possibility from others such as reciprocal altruism.

Pilot whales. Mammals that form groups often evidence a matrilineal structure, in which related females form the core of the group. Males either disperse or stay in the group as helpers. Microsatellites are an ideal tool for studying social groups, because the availability of well-defined alleles allows determination of kinship. For pilot whales, the existence of such a population structure has been demonstrated through the use of microsatellites (Amos et al., 1993). To demonstrate a matrilineal structure, the authors tested two basic hypotheses. First, the mother of most females should be found in the group, and second, alleles found in the older females should be the most common ones. The number of genetically compatible females (O) was compared with the expected number based on random assortment of alleles among mature females in the group (E). If the mother is absent, O-E should be equal to 0, whereas a present mother will be indicated by a value closer to 1. In fact, an estimated 95% of the young individuals were probably accompanied by their mother.

To test the second hypothesis, in principle, for every individual the frequencies of its alleles in the pod were added. This pod-specific genotype frequency index was plotted against the age of the animal. A significant positive correlation between age and frequency index was observed, indicating that the majority of – and possibly all – pod members belonged to one matrilineal family. Interestingly, males of all ages had similar values to females of the same age, suggesting that even mature males remain in their native pod. The high-quality allelic information which microsatellites provide allowed this to be further tested. Precise probability calculations confirmed the matrilineal structure and showed nicely how analysis beyond first-degree relationships is possible with microsatellites.

Social insects. Social insects have been the focus of many studies investigating the basis of sociality. The haplodiploid genetic system in Hymenoptera provides an experimental setup in which an asymmetry of relatedness is provided: workers in eusocial colonies are unequally related to the male and female reproductives they provision. Hence, in species where workers control sex allocation, a strong bias towards females is expected as long the colony contains a single, once-mated queen. If the queen is multiple mated, or several queens are present in the colony, this bias should be less extreme (as the coefficient of relatedness among workers decreases). These theoretical predictions have been tested in the polygynous ant *Myrmica tahoensis* (Evans, 1995). Using microsatellites to determine the genetic relationships of colonies, Evans showed that colonies producing male and female sexuals often had low levels of relatedness among nestmates. Colonies that produced mainly female sexuals were composed of highly related females (workers and female sexuals).

Sperm competition

The concept of sperm competition is that ejaculates from different males compete within the receiving female to fertilize her eggs. Sperm competition is likely to occur in species where females store sperm and sperm are long-lived. This phenomenon is found in many birds, insects and some mammals. Sperm competition may result in displacement of previous ejaculates by new sperm or stratification processes. In addition, the female reproductive tract may be able to influence the success of different ejaculates. Microsatellites offer, for the first time, unusual opportunities to study aspects of sperm biology. First, they can be used to show whether multiple paternity occurs within litters or broods, as for the noctule bat, *Nyctalus noctula* (Mayer et al., unpublished observation) and *Drosophila melanogaster* (Imhof et al., 1998). Second, there is the potential to study sperm composition in the female storage device by determining the origin of single sperm. Siva-Jothy and Hadrys (this volume) use a combination of microsatellite and RAPD analysis to provide insight into the outcome of sperm competition and the functions of multiple sperm storage organs in highly polygamous odonates.

Population genetic analysis

Isoenzymes provided population geneticists with discrete codominant information. These properties were essential for the development of formulae used to describe the genetic structure of a population. Hardy-Weinberg equilibrium, linkage disequilibrium, genetic drift, migration and effective population size are just a few parameters that can be determined by the use of isoenzymes, and every population genetics textbook will provide more applications. Like isoenzymes, microsatellites have discrete, codominant alleles, but they have more alleles and seem less likely to be under selection. Microsatellites can therefore be regarded as the second generation of population genetic markers.

Microsatellites and conservation genetics

Studies of endangered species are often hampered by low levels of genetic variation. To assess the severity of bottlenecks in endangered populations, a highly polymorphic class of markers is needed. Microsatellites were found to be informative in several species (e.g. cheetah, northern hairy-nosed wombat and Coelacanth), which showed almost no variation at other genetic markers (Menotti-Raymond and O'Brien, 1995; Taylor et al., 1994; Schlötterer et al. in prep.). Furthermore, the use of PCR for microsatellite analysis makes museum collections accessible for comparison with extant populations and permits the use of noninvasive sampling strategies.

Uniparentally inherited microsatellites

Most microsatellite studies focus on the use of nuclear microsatellite markers. However, uniparentally transmitted markers have been shown to be highly informative for inferring population history and to have several advantages compared with biparentally inherited markers. Multiple microsatellites on uniparentally inherited chromosomes and extranuclear DNA allow the study of haplotype genealogies uncomplicated by the problem of recombination. Combined analysis of uniparentally and biparentally transmitted microsatellites could provide additional insight into several population genetic parameters such as population structure and gene flow.

In principle, several potential sources of uniparental inherited markers are available: mitochondria, chloroplasts and Y chromosomes. However, the isolation of uniparentally inherited markers depends on the separation of the uniparentally inherited DNA regions from the remainder of the genome. Hence, most uniparentally inherited microsatellites reported so far were obtained from databanks. Uniparentally transmitted microsatellites have been described for the human Y chromosome (Freije et al., 1992), avian mitochondria (Berg et al., 1995) and several chloroplasts (Powell et al., 1995a; Powell et al., 1995b). Unfortunately, the utility of avian mitochondrial DNA (mtDNA) microsatellites in population studies is severely limited due to the high mutation rates resulting in high levels of heteroplasmy. In chloroplasts, a GenBank search for microsatellites containing a minimum of 10 repeats yielded 500 loci (Powell et al., 1996) which were polymorphic but stable enough to show no heteroplasmy. The high degree of sequence conservation of the chloroplast genome allows for cross-species amplification of microsatellites contained in the chloroplast (Powell et al., 1995b). Given that chloroplast sequences are more conserved than nuclear sequences, it is expected that cross-species amplifications will be more successful than for nuclear microsatellite loci. The mode of chloroplast inheritance varies among plant species, so in some (e.g. pines) chloroplast microsatellites could be used as paternal markers, while in others (e.g. soybeans) they could be used as maternal markers.

Microsatellites and selection

Despite several uncertainties about the exact patterns of microsatellite evolution and possible evolutionary constraints, microsatellites can be regarded to a very good approximation as neutral markers (Charlesworth et al., 1994). Based on this assumption, the use of microsatellites has been suggested to infer selection in genomic regions adjacent to the microsatellite. Directional selection on a given genomic region will reduce the observed variance in repeat number of closely linked microsatellites. Hence, significant deviations of observed variances from neutral expectations can be attributed to selection. Goldstein et al. (1996) studied five microsatellite loci in the nonrecombining portion of the Y chromosome to test whether the human Y chromosome has been exposed to a selective sweep. With the current estimates of average microsatellite mutation rate and effective population size, the authors found no evidence of a selective sweep during the last 74,000 years (Goldstein et al., 1996).

While Goldstein et al. considered all loci simultaneously, a different approach has been taken to detect selection in *D. melanogaster* populations (Schlötterer et al., 1997). Seven *D. melanogaster* populations were analyzed for 10 microsatellite loci. All loci were polymorphic in the total sample, but in some populations a significant reduction in variation was observed for a single locus (the other loci in this population were still polymorphic). Most interestingly, in different populations different loci were affected by this phenomenon. The authors conclude that this observed reduction in variation is best explained by a recent selective sweep, which may have occurred during local adaptation of the respective populations. Given the random distribution of microsatellites across the entire genome, this result could provide a microsatellite-based strategy to screen systematically for genes involved in local adaptation.

Cross-species amplification of microsatellites

A major obstacle for microsatellite analysis in the past was the need to develop PCR primers for every species. The demonstration that microsatellites and their flanking regions are conserved across species significantly promoted the widespread use of microsatellites (Schlötterer et al., 1991). In general, there is a good correlation between phylogenetic distance and the success of cross-species amplification, but several microsatellites are reported which are conserved and polymorphic over phylogenetic distances of several million years (FitzSimmons et al., 1995; Rico et al., 1996). Before isolating new microsatellite loci for a given species, we recommend that researchers screen databases and journals such as *Molecular Ecology* or *Animal Genetics*, which publish PCR primers for newly characterised microsatellites. In addition, primer sequences can often be obtained from a microsatellite newsgroup. Information about how to subscribe to this newsgroup is available from T. Chapman (Thomas_Chapman@sfu.ca).

Phylogeny reconstruction

Given the conservation of microsatellites across species, the question arises to what extent microsatellites contain useful information for reconstructing phylogenetic relationships. As for DNA sequences, reconstruction of phylogenetic relationships based on microsatellites is dependent on a thorough understanding of the mutation processes underlying the variation. Mutations in a microsatellite via DNA slippage alter the number of repeat units (Tautz and Schlötterer, 1994). This mutation process has a high probability of generating homoplasy of microsatellite alleles (identity by state/size without identity by descent). Therefore, no reliable genealogy can be directly inferred from the size of different microsatellite alleles. However, since earlier theoretical studies showed that over time the variance in repeat number remains constant and only the mean changes (Moran, 1975), genetic distance measures likely to be linearly related to time have been proposed (e.g. Goldstein et al., 1995). The model underlying these new distance measures makes some assumptions which are not fully consistent with current knowledge about microsatellite evolution. Equal mutation rates at all microsatellite loci/alleles are assumed and, more important, the possibility of limitations to microsatellite length are not considered. By assuming the most obvious microsatellite length limitation, a reduction to a single repeat unit, Nauta and Weissing showed that this constraint may lead to incorrect reconstruction of genetic distances (Nauta and Weissing, 1996). Therefore, reliable reconstruction of phylogenetic relationships of species or populations, which have been separated for long time intervals, depends on improved algorithms for genetic distances which consider more refined models of microsatellite evolution.

Discrimination between lines and species

Despite their limited use for phylogenetic reconstruction, microsatellites have been shown to be useful for identifying phenotypically almost indistinguishable groups. Sunnucks et al. used microsatellites to identify chromosomal races of *Sitobion miscanthi* which had specific genotypes despite high allelic diversity at all loci studied (Sunnucks et al., 1996). Species-specific allele distributions are often due to insertions or deletions in the genomic region flanking the microsatellites. Hence, mapped microsatellites with species-specific allele distribution could be used to study introgression of genomic regions in hybrid zones as well as for identifying genomic regions which are involved in hybrid sterility.

Limitations of microsatellites

Despite their still-increasing popularity and many successful applications, it should be borne in mind that although microsatellites are useful markers, they have a number of specific limitations.

Typing of microsatellites

Currently, microsatellite alleles are typed by determining the length of a PCR product, which contains the microsatellite. All observed length differences are attributed to a change in repeat number, as mutations in the microsatellite are much more frequent than insertion and deletions in the DNA region flanking the microsatellite. However, many microsatellite data sets show length differences of less than a single repeat unit, indicating the occurrence of insertions or deletions. If the length of a cloned microsatellite is used to infer the number of repeats in a PCR fragment, insertions and deletions in the flanking regions will lead to an increase in variance and hence inflate variance-based genetic distances. The presence of deletions or insertions in the flanking region is especially pronounced for cross-species comparisons of microsatellites. Hence, determination of the number of repeats may lead to a significant error, when based on the length of the PCR product. A more accurate typing of microsatellites would be the direct measurement of the number of repeat units with novel techniques such as DNA chips (Southern, 1996).

Null alleles

Point mutations, or insertions or deletions, in the primer-binding site may obstruct PCR amplification of microsatellites. If the mutation is not fixed in the population, only a fraction of all alleles will amplify. The nonamplifying alleles are called "null alleles" (Callen et al., 1993; Pemberton et al., 1995). If not recognized, null alleles will result in an apparent surplus of homozygous individuals in the population. In most cases, redesign of PCR primers allows successful amplification of all alleles.

Mutational biases

One important assumption in microsatellite mutation models is that mutation to lower or higher repeat unit number is equally likely to occur. Any bias, say towards gaining repeat units, would cause significant errors in genetic distance estimates. Such a bias has been suggested for human microsatellites (Rubinsztein et al., 1995) as well as for one avian microsatellite (Primmer et al., 1996). Further bias may be introduced by any correlation of microsatellite length and mutation rate. If longer microsatellites have a higher mutation rate, as suggested by Weber (1990), then this bias needs also to be considered for accurate measurements of genetic distances.

Paucity of microsatellites

With an increasing number of species from which microsatellites are cloned, it has become clear that the abundance of microsatellites differs significantly between species. While most species contain sufficient numbers of microsatellite loci for population studies, some species contain only very few microsatellites. Some lepidopteran species (C. Nussbaumer, personal communication) and some mite species (M. Navajas, personal communication) are just a few examples of species with few microsatellite loci.

PCR artifacts caused by low DNA amounts

In theory, PCR is able to amplify microsatellites from single cells. A recent study, however, demonstrated that microsatellite amplification from small amounts of template DNA is associated with a high frequency of errors (Foucault et al., 1996). Amplification of alleles with incorrect length and nonamplification of one allele in heterozygous individuals are the two types of errors which occur at very low amounts of DNA template. With increasing amounts of DNA template, a more reliable microsatellite typing was achieved. This phenomenon also occurred in studies which amplified microsatellites from feces (Gerloff et al., 1995) and shed hair (Gagneux et al., 1997). Future studies on microsatellites using low amount of template DNA will have to take precautions, such as multiple amplifications of the same locus, to avoid typing errors.

Conclusions and further perspective

Microsatellite-based DNA typing is a valuable tool for studying natural populations. For paternity assignment and assessment of genetic relatedness within social groups, microsatellites are unquestionably the marker of choice. Their usefulness in population genetics is mainly dependent on the elucidation of the exact mutational behavior of microsatellites and its incorporation into distance measurements. However, despite this limitation, microsatellites are still the best available marker to study relationships between closely related populations. New screening methods, such as parallel capillary electrophoresis, will enhance rapid throughput of large sample sizes for many loci. The high potential for full automation of microsatellite analysis will allow routine typing to occur in private or university core facilities, which will produce the required data at low cost.

Acknowledgements

The laboratory of CS is supported by the Fonds zur Förderung der wissenschaftlichen Forschung, Oesterreichische Nationalbank and the European Community biotechnology program (Biodiversity). The laboratory of JP is supported by the UK Biotechnology and Biological Sciences Research Council, the Wellcome Trust and the UK Natural Environment Research Council.

References

Amos, B., Schlötterer, C. and Tautz, D. (1993) Social structure of pilot whales revealed by analytical DNA profiling. *Science* 260: 670–672.

Armour, J. A. L., Neumann, R., Gobert, S. and Jeffreys, A. J. (1994) Isolation of human simple sequence repeat loci by hybridization selection. *Hum. Mol. Genet.* 3: 599–605.

Berg, T., Moum, T. and Johansen, S. (1995) Variable numbers of simple tandem repeats make birds of the order ciconiiformes heteroplasmic in their mitochondrial genomes. *Curr. Genet.* 27: 257–262.

Brenig, B. and Brem, G. (1991) Direct cloning of sequence tagged microsatellite sites by DNA affinity chromatography. *Nucl. Acid. Res.* 19: 5441.

Browne, D. L. and Litt, M. (1992) Characterization of $(CA)_n$ microsatellites with degenerate sequencing primers. *Nucl. Acid. Res.* 20: 141.

Callen, D. F., Thompson, A. D., Shen, Y., Phillips, H. A., Richards, R. I., Mulley, J. C. and Sutherland, G. R. (1993) Incidence and origin of "null" alleles in the $(AC)_n$ microsatellite markers. *Am. J. Hum. Genet.* 52: 922–927.

Charlesworth, B., Sniegowski, P. and Stephan, W. (1994) The evolutionary dynamics of repetitive DNA in eukaryotes. *Nature* 371: 215–220.

Dib, C., Faure, S., Fizames, C., Samson, D., Drouot, N., Vignal, A., Millasseau, P., Marc, S., Hazan, J., Seboun, E., Lathrop, M., Gyapay, G., Morissette, J. and Weissenbach, J. (1996) A comprehensive genetic map of the human genome based on 5,264 microsatellites. *Nature* 380: 152–154.

Dietrich, W. F., Miller, J., Steen, R., Merchant, M. A., Damron-Boles, D., Husain, Z., Dredge, R., Daly, M. J., Ingalls, K. A., O'Connor, T. J. et al. (1996) A comprehensive genetic map of the mouse genome. *Nature* 380: 149–152.

Ellegren, H. (1991) DNA typing of museum birds. *Nature* 354: 113.

Ender, A., Schwenk, K., Städler, T., Streit, B. and Schierwater, B. (1995) RAPD identification of microsatellites in *Daphnia*. *Molec. Ecol.* 5: 437–441.

Evans, J. D. (1995) Relatedness threshold for the production of female sexuals in colonies of a polygynous ant, *Myrmica tahoensis*, as revealed by microsatellite DNA analysis. *Proc. Natl. Acad. Sci. USA* 92: 6514–6517.

FitzSimmons, N. N., Mority, C. and Moore, S. S. (1995) Conservation and dynamics of microsatellite loci over 300 million years of marine turtle evolution. *Molec. Biol. Evol.* 12: 432–440.

Foucault, F., Praz, F., Jaulin, C. and Amor-Gueret (1996) Experimental limits of PCR analysis of $(CA)_n$ repeat alterations. *Trends Genet.* 12: 450–452.

Freije, D., Helms, C., Watson, M. S. and Donis-Keller, H. (1992) Identification of a second pseudoautosomal region near the Xq and Yq telomeres. *Science* 258: 1784–1787.

Gagneux, P., Boesch, C. and Woodruff, D. S. (1997) Microsatellite scoring errors associated with noninvasive genotyping based on nuclear DNA amplified from shed hair. *Molec. Ecol.* 6: 861–868.

Gerloff, U., Schlötterer, C., Rassmann, K., Rambold, I., Hohmann, G., Fruth, B. and Tautz, D. (1995) Amplification of hypervariable simple sequence repeats (microsatellites) from excremental DNA of wild living Bonobos (*Pan paniscus*). *Molec. Ecol.* 4: 515–518.

Gibbs, H. L., Weatherhead, P. J., Boag, P. T., White, B. N., Tabak, L. M. and Hoysak, D. (1990) Realized reproductive success of polygynous red-winged blackbirds revealed by DNA markers. *Science* 250: 1394–1397.

Goldstein, D. B., Ruiz Lineares, A., Cavalli-Sforza, L. L. and Feldman, M. W. (1995) Genetic absolute dating based on microsatellites and the origin of modern humans. *Proc. Natl. Acad. Sci. USA* 92: 6723–6727.

Goldstein, D. B., Zhivotovsky, L. A., Nayar, K., Ruiz Lineares, A., Cavalli-Sforza, L. L. and Feldman, M. W. (1996) Statistical properties of the variation at linked microsatellite loci: implications for the history of human Y chromosomes. *Molec. Biol. Evol.* 13: 1213–1218.

Hagelberg, E., Gray, I. C. and Jeffreys, A. J. (1992) Identification of the skeletal remains of a murder victim by DNA analysis. *Nature* 352: 427–429.

Hamilton, D. W. (1964) The genetical evolution of social behaviour 1&2. *J. Theoret. Biol.* 7: 1–52.

Henderson, S. T. and Petes, T. D. (1992) Instability of simple sequence DNA in *Saccharomyces cerevisiae. Mol. Cell. Biol.* 12: 2749–2757.

Hubert, R., Weber, J. L., Schmitt, K., Zhang, L. and Arnheim, N. (1992) A new source of polymorphic DNA markers for sperm typing: analysis of microsatellite repeats in single cells. *Am. J. Hum. Genet.* 51: 985–991.

Hughes, C. R. and Queller, D. C. (1993) Detection of highly polymorphic microsatellite loci in a species with little allozyme polymorphism. *Molec. Ecol.* 2: 131–137.

Imhof, M., Harr, B., Brem, G. and Schlötterer, C. (1998) Multiple mating in wild *Drosophila melanogaster*-revisited by microsatellite analysis. *Mol. Ecol.; in press.*

Ito, T., Smith, C. L. and Cantor, C. R. (1992) Sequence-specific DNA purification by triplex affinity capture. *Proc. Natl. Acad. Sci. USA* 89: 495–498.

Jeffreys, A. J., Wilson, V. and Thein, S. L. (1985) Hypervariable "minisatellite" regions in human DNA. *Nature* 314: 67–73.

Lehmann, T., Hawley, W. A. and Collins, F. H. (1996) An evaluation of evolutionary constraints on microsatellite loci using null alleles. *Genetics* 144: 1155–1163.

Levinson, G. and Gutman, G. A. (1987) Slipped-strand mispairing: a major mechanism for DNA sequence evolution. *Molec. Biol. Evol.* 4: 203–221.

Litt, M. and Luty, J. A. (1989) A hypervariable microsatellite revealed by *in vitro* amplification of a dinucleotide repeat within the cardiac muscle actin gene. *Am. J. Hum. Genet.* 44: 397–401.

Menotti-Raymond, M. and O'Brien, S. J. (1995) Evolutionary conservation of ten microsatellite loci in four species of Felidae. *J. Hered.* 86: 319–322.

Moran, P. A. P. (1975) Wandering distributions and electrophoretic profile. *Theor. Pop. Biol.* 8: 318–330.

Morin, P. A., Moore, J. J., Chakraborty, R., Jin, L., Goodall, J. and Woodruff, D. S. (1994) Kin selection, social structure, gene flow, and the evolution of chimpanzees. *Science* 265: 1193–1201.

Nauta, M. J. and Weissing, F. J. (1996) Constraints on allele size at microsatellite loci: implications for genetic differentiation. *Genetics* 143: 1021–1032.

Pemberton, J. M., Slate, J., Bancroft, D. R. and Barrett, J. A. (1995) Nonamplifying alleles at microsatellite loci: a caution for parentage and population studies. *Molec. Ecol.* 4: 249–252.

Pemberton, J. M., Smith, J. A., Coulson, T. N., Marshall, T. C., Slate, J., Paterson, S., Albon, S. D. and Clutton-Brock, T. H. (1996) The maintenance of genetic polymorphism in small island populations: large mammals in the Hebrides. *Phil. Trans. R. Soc. Lond.* 351: 745–752.

Powell, W., Morgante, M., Andre, C., McNicol, J. W., Machray, G. C., Doyle, J. J., Tingey, S. V. and Rafalski, J. A. (1995a) Hypervariable microsatellites provide a general source of polymorphic DNA markers for the chloroplast genome. *Curr. Biol.* 5: 1023–1029.

Powell, W., Morgante, M., McDevitt, R., Vendramin, G. G. and Rafalski, J. A. (1995b) Polymorphic simple sequence repeat regions in chloroplast genomes: applications to the population genetics of pines. *Proc. Natl. Acad. Sci. USA* 92: 7759–7763.

Powell, W., Machray, G. C. and Provan, J. (1996) Polymorhism revealed by simple sequence repeats. *Trends Plant Sci.* 1: 215–222.

Primmer, C. R., Ellegren, H., Saino, N. and Møller, A. P. (1996) Directional evolution in germline microsatellite mutations. *Nat. Genet.* 13: 391–393.

Rassmann, K., Schlötterer, C. and Tautz, D. (1991) Isolation of simple-sequence loci for use in polymerase chain reaction-based DNA fingerprinting. *Electrophoresis* 12: 113–118.

Rico, C., Rico, I. and Hewitt, G. (1996) 470 million years of conservation of microsatellite loci among fish species. *Proc. R. Soc. Lond. B Biol. Sci.* 263: 549–557.

Rubinsztein, D. C., Amos, W., Leggo, J., Goodburn, S., Jain, S., Li, S.-H., Margolis, R. L., Ross, C. A. and Ferguson-Smith, M. A. (1995) Microsatellite evolution-evidence for directionality and variation in rate between species. *Nat. Genet.* 10: 337–343.

Saiki, R. K., Gelfand, D. H., Stoffel, S., Scharf, S. J., Higuchi, R., Horn, G. T., Mullis, K. B. and Ehrlich, H. A. (1988) Primer-directed enzymatic amplification of DNA with a thermostable DNA polymerase. *Science* 239: 487–491.

Schlötterer, C. (1993) Non-radioactive analysis of multiplexed microsatellite reactions using a direct blotting-sequencing apparatus. *Nucl. Acid. Res.* 21: 780.

Schlötterer, C. (1998) Microsatellites *In*: Hoelzel, R. A. (ed.) *Molecular Genetic Analysis of Populations: A Practical Approach*, 2nd ed, Oxford University Press, Oxford, pp. 237–261.

Schlötterer, C., Amos, B. and Tautz, D. (1991) Conservation of polymorphic simple sequence loci in cetacean species. *Nature* 354: 63–65.

Schlötterer, C., Vogl, C. and Tautz, D. (1997) Polymorphism and locus-specific effects at microsatellite loci in natural *Drosophila melanogaster* populations. *Genetics* 146: 309–320.

Schug, M. D., Mackay, T. F. C. and Aquadro, C. F. (1997) Low mutation rates of microsatellite loci in *Drosophila melanogaster*. *Nat. Genet.* 15: 99–102.

Southern, E. M. (1996) DNA chips: analysing sequence by hybridization to oligonucleotides on a large scale. *Trends Genet.* 12: 110–115.

Steinkellner, H., Fluch, S., Lexer, C., Turetschek, E., Streiff, R. and Glössl, J. (1997) Identification and characterization of microsatellite marker in *Quercus petraea*. *Plant Mol. Biol.* 33: 1093–1096.

Strand, M., Prolla, T. A., Liskay, R. M. and Petes, T. D. (1993) Destabilization of tracts of simple repetitive DNA in yeast by mutations affecting DNA mismatch repair. *Nature* 365: 274–276.

Sunnucks, P., England, P. R., Taylor, A. C. and Hales, D. F. (1996) Microsatellite and chromosome evolution of parthenogenic Sitobion ahpids in Australia. *Genetics* 144: 747–756.

Tautz, D. (1989) Hypervariability of simple sequences as a general source for polymorphic DNA markers. *Nucl. Acid. Res.* 17: 6463–6471.

Tautz, D. and Schlötterer, C. (1994) Simple Sequences. *Curr. Opin. Genet. Develop.* 4: 832–837.

Taylor, A. C., Sherwin, W. B. and Wayne, R. K. (1994) Genetic variation of microsatellite loci in a bottlenecked species: the northern hairy-nosed wombat *Lasiorhinus krefftii*. *Molec. Ecol.* 3: 277–290.

Weber, J. L. (1990) Informativeness of human $(dC - dA)_n \cdot (dG - dT)_n$ polymorphisms. *Genomics* 7: 524–530.

Weber, J. L. and May, P. E. (1989) Abundant class of human DNA polymorphisms which can be typed using the polymerase chain reaction. *Am. J. Hum. Genet.* 44: 388–396.

Weber, J. L. and Wong, C. (1993) Mutation of human short tandem repeats. *Hum. Mol. Genet.* 2: 1123–1128.

Weissenbach, J., Gyapay, G., Dib, C., Vignal, A., Morissette, J., Millasseau, P., Vaysseix, G. and Lathrop, M. (1992) A second-generation linkage map of the human genome. *Nature* 359: 794–801.

Caution before claim: an overview of microsatellite analysis in ecology and evolutionary biology

Howard C. Rosenbaum[1,2] *and Amos S. Deinard*[3]

[1]Molecular Systematics Laboratory, American Museum of Natural History, 79th St. & CPW, New York, NY 10024-5192, USA
[2]Dept. of Biology, Yale University, PO Box 6666, New Haven, CT 06510, USA
[3]School of Veterinary Medicine, University of California, Davis, CA 95616, USA

Summary

Over the past decade, microsatellites, or short tandem repeats (STRs), have become the *en vogue* marker used in ecological and evolutionary studies for addressing questions concerning phylogenetics, population genetic structure, behavioral ecology and kinship (e.g. Queller et al., 1993a; Jarne and Lagoda, 1996; Goldstein and Pollock, 1997). High heterozygosity levels and mutation rates, a biparental mode of inheritance and their apparent widespread occurrence throughout most species' genomes make STRs an attractive marker (Tautz, 1989). In addition, the relative ease and low cost with which these markers can be typed have also contributed to their popularity in biological studies. In this chapter, we provide a general overview of STR analysis, including what is currently suspected about mutational processes and STR evolution and review the benefits and limitations, both technical and conceptual, for STR analysis. In general, we express caution in the use of STRs for resolving questions concerning phylogeny, but see greater utility for kinship and paternity studies.

Introduction

The application of STR analysis to resolve specific ecological and evolutionary questions has often occurred without a thorough understanding of the inherent weaknesses associated with these markers. These misunderstandings have, in some cases, resulted in conclusions that are unwarranted – either the significance of the data have been overstated, or the data are inappropriate for the specific biological question being examined. This does not imply that STR loci cannot provide useful biological data, but rather that the application and interpretation of STR data must be thoroughly considered in order to realize their limitations and utility. In this review,

R. DeSalle, B. Schierwater (eds) Molecular Approaches to Ecology and Evolution
©1998, Birkhäuser Verlag Basel

we provide a general overview of the use of STR markers in systematic, population and ecological research, as well as highlight specific claims and cautions with respect to the application of the data.

Definition

Although all STRs consist of repeated arrays of a specific DNA sequence (i.e. a core sequence ranging in size between 1 and 5 bp), they may be further differentiated by the specific composition of their core sequence. In other words, the repetitive unit may not always be consistent. STR arrays may be composed of perfect (or pure) repeats, in which the core repetitive unit is maintained throughout the region, or imperfect (or interrupted) repeats, in which the core repetitive unit is interrupted by nonrepeat sequences. Interrupted STRs may differ in allele size, nature and location of the core repeat interruption and more importantly are believed to express different levels of evolutionary stability and polymorphism when compared with perfect STRs (e.g. Estoup et al., 1995b). In addition, STRs may also be classified as compound STRs that consist of different types or lengths of tandemly repeated sequences adjacent to one another (Weber, 1990; Fig. 1).

To date, loci that have dinucleotide repeats are the most frequently typed STRs in biological research with approximately 8000 listings in GENBANK alone. The reason for this is that they are the most frequent repeat array in both plants (with AT/TA or GA/CT repeats being the most frequent) and animals (with CA/GT repeats being the most frequent) (Moore et al., 1991; Lagercrantz et al., 1993). More recently, however, typing of tri- and tetranucleotide repeats has gained popularity. Approximately 217 trinucleotide and 1596 tetranucleotide STRs are currently listed in GENBANK1.

PURE	CAGCAGCAGCAGCAGCAGCAGCAGCAGCAG
COMPOUND	CAGCAGCAGCTGCTGCTGCTGCTGCTGCTG
IMPERFECT	CAGCAGCAGTTCAGCAGCAGCAGTTCAGCAG
FLANKING SEQUENCE	ACTGTATTTATGCAGCAGCAGCAGCAGCAGCAGCAGCAGCAGTTACAGTTACCC

Figure 1. This is an example of the different types and compositions that make up a repetitive array of an STR. This particular array consists of 10 CAG trinucleotides. The underlined sequence is an example of the nonrepetitive DNA sequence external to or flanking the repetitive array. These regions may also contain base substitutions, insertions and deletions that influence allelic size variance at STR loci.

History

Where did the term "microsatellite" originate? By all accounts, it appears to be a misnomer attributed to a DNA analysis methodology. Prior to the discovery of restriction enzymes, DNA was analyzed through the ultracentrifugation of randomly sheared DNA in $CsCl_2$ or $CsSO_4$ gradients. DNA analyzed in this manner produced a series of diffuse bands within the gradient that reflected the aggregation of fragments of similar density. Under certain conditions, or when comparing the DNA from two different organisms, a completely different and unique band was identified that did not represent, or was not observed, in the "normal" DNA. The DNA in these narrow bands consequently became known as satellite DNA because it represented only a fraction of the total cellular DNA. The first of these "satellite" DNAs to be fully characterized were the ribosomal genes and were dubbed "alpha-satellite DNA". Further study of these alpha-satellites revealed that these DNAs consisted of a hierarchy of repeated units, ranging in size from hundreds of base pairs to kilobases, and repeated hundreds to a few thousands times (Gall and Atherton, 1974).

In the early 1980s, Jeffreys et al. (1985) identified a series of tandemly repeated regions of DNA with repeat units that were much smaller in size than the original alpha-satellites; the authors termed these newly identified loci "minisatellites". Two years later, Nakamura et al. (1987) identified loci consisting of tandemly repeated arrays, numbering anywhere from the low teens to several dozen, that were several kilobases in size. Both minisatellites and these novel loci, known as variable number of tandem repeats (VNTRs), demonstrated high levels of polymorphism (Jeffreys et al., 1985; Nakamura et al., 1987). Interestingly, although neither minisatellites nor VNTRs were characterized via the methodology that resulted in the discovery of satellite DNA, they were nonetheless grouped together – minisatellites and VNTRs were not considered "normal", but rather a form of satellite DNA.

The consequence of this misappropriation of nomenclature has been that the terms "minisatellite", and more recently (and for the same reasons) "microsatellite", are used today to describe DNA that does not truly represent satellite DNA; these loci merely represent regions of DNA that consist of tandemly repeated units. From a purely historical perspective, therefore, it is worth noting that although these terms have become incorporated into the laboratory lexicon, they do not accurately represent the true nature of the DNA being described. Today, satellite DNA has evolved to include tandem repeats of much smaller scale, different molecular nature, and possibly different origin than the alpha-satellites from which the term was originally coined.

---

[1] It is worth noting that the GENBANK estimates may be underestimated because they are not always listed by the type or number of repeats in the array. Additionally, these databases are continuously updated and will have undoubtably changed significantly by the time this chapter is published.

Methodological considerations for STR typing

To generate useful biological data from an STR, it is necessary to have a particular locus (or group of loci) available for genotyping; informative loci can be obtained by use of previously identified and characterized STRs or by identifying new loci. Identification and characterization of novel STR loci, however, remains a laborious process that involves the construction and screening of a genomic library, and subsequent subcloning of the desired DNA region (e.g. Rassman et al., 1991). Additionally, newly identified STRs may not necessarily be polymorphic and/or informative. The easier and less expensive method to identify informative STRs, therefore, remains the analysis of previously identified and characterized STR loci.

The most common source for identifying potentially useful STR loci is the numerous DNA databases such as GENBANK, GDB, EMBL. A secondary source, and one that may be more beneficial to some researchers, is published literature. Although this source of information may seem obvious, it is worth mentioning because quite often the latest references may be more beneficial than the existing databases. Not only can one identify STR markers that are known to be informative in a particular species (or a closely related species), but there may also be delays for existing data (STR markers, primer sequences and PCR conditions) to be listed in the respective databases.

There are several advantages to using known STR markers. The greatest advantage is that the marker has usually been fully characterized: the PCR primer sequences and PCR conditions are both known. Consequently, one does not have to spend time optimizing conditions. Moreover, due to the robustness of PCR, it is possible to type STR loci originally cloned in one species in other closely related species, even though sequence differences may exist in priming sites; numerous studies have successfully demonstrated and/or used the conservation of primer sites associated with STR loci for genotyping among and within closely related species (e.g. Schlötterer et al., 1991; Morin et al., 1994a; Meyer et al., 1995; Pépin et al., 1995; Valsecchi and Amos, 1997; Strassman et al., 1996; Primmer et al., 1997; Ellegren et al., 1997; Palsbøll et al., 1997a). From our own experiences, human STR primers have been used successfully in typing loci among the great apes, with few exceptions (Deinard and Kidd, 1995; Deinard, 1997). Although Primmer et al. (1997) have attempted to establish a correlation between STR utility and evolutionary distance from the species from which the loci were originally cloned, there is still no documented benchmark as to the extent to which cross-species STR analysis may be conducted.

While such cross-taxa comparisons may be an appealing approach for addressing evolutionary questions, they also represent one of the pitfalls of using STRs to understand relationships among species or populations. Specifically, the nature and dynamics of these loci may vary considerably with increasing evolutionary divergence from the species for which the loci were originally cloned (Estoup et al.,

1995b; Angers and Bernatchez, 1997; Ellegren et al., 1997; Primmer et al., 1997). Although the conservation of STR loci (or the surrounding flanking or priming sites) across taxa has been a contributing factor to their popularity, it is only recently that empirical studies have been conducted which illustrate the complexity of these loci.

Mutational processes: What do we really know?

Due to the complexity and variability of most STR loci, it is not surprising that little consensus currently exists concerning the specific mutational processes and mutational rates of STR loci. This lack of consensus is due, in part, to the observation that individual STR loci may have unique rates and/or patterns of mutation that are shared neither across loci nor across species (e.g. Weber and Wong, 1993). For example, mutational rates for STRs in mammals have been estimated to range anywhere between 10^{-5} and 10^{-2} mutations per generation (Dallas, 1992; Weber and Wong, 1993), whereas the mutation rate within maintained isofemale lines of *Drosophila melanagoster* has been estimated at 10^{-6} per locus (Schug et al., 1997). These large differences among estimated mutation rates of STR loci are just one particular element that confounds the development of a universal theory to explain STR evolution. Nevertheless, theoretical models have been formulated in an attempt to better understand the evolution of these complex loci.

The two most commonly proposed evolutionary models that attempt to explain STR evolution are the infinite allele model (IAM) and the stepwise mutation model (SMM). The central premise of both the IAM and the SMM is that the molecular mechanism (i.e. the "driving force") of STR evolution is DNA strand slippage during replication (Schlötterer and Tautz, 1992)[2]. Although the majority of data used to construct (and test) both models have been generated on human and nonhuman primates (Shriver et al., 1993; DiRienzo et al., 1994; Garza et al., 1995; Valdes et al., 1993), each model has been applied, with varying success, to explain data generated both within species and among various nonprimate species (e.g. Estoup et al., 1995b; FitzSimmons et al., 1995).

The IAM postulates that any allele generated through mutation will be novel and thus will differ from currently existing alleles – every newly generated STR allele must be a unique, heretofore unobserved allele (Kimura and Crow, 1964). A more applicable variation of this model to STR loci, known as the *k* allele model (KAM), permits newly generated alleles to occupy any *k* preexisting allelic states. However, both of these models have recently been deemed inappropriate because their as-

---

[2] In contrast, unequal crossing over or conversion-like events have been suggested as the primary mutational force in the evolution of larger repeat arrays (i.e. minisatellites) (Schlötterer and Tautz, 1992; Buard and Jeffreys, 1997; Mandal, 1997).

sumptions may be violated due to the high mutation rates of STR loci and the process by which ancestral allelic states may still be detectable after mutation (Slatkin, 1995). Unlike IAM and KAM, the SMM, which has been used to explain coalesence times between alleles or variance in allele sizes, postulates that mutation directly affects the allelic state of any given locus by increasing or decreasing the number of repeats by one (Ohta and Kimura, 1973; Goldstein et al., 1995; Slatkin, 1995). Although the SMM more accurately predicts allelic distribution for particular types of STR loci (e.g. STR with a core repeat 3–5 bp long), the model is less accurate for other STR loci (e.g. dinucleotide repeats or imperfect STRs) (Shriver et al., 1993; Valdes et al., 1993; Estoup et al., 1995b). Moreover, it has recently been argued that a two-phase model, in which the properties of the SMM are combined with infrequent, multiple-step mutations that occur with some probability, may better depict evolution of STRs than SMM alone (Valdes et al., 1993; DiRienzo et al., 1994; Garza et al., 1995). In such a model, the distribution of allele sizes in any sample is dependent upon mutation rate and historical and demographic processes (DiRienzo et al., 1994).

Although it may be possible to model observed mutational rate and mode at STR loci, such models may not take into account other evolutionary constraints that may directly affect STR evolution. The two most commonly cited forces that affect STR evolution, other than mutation rate and process, are those that influence or bias mutational directionality and those that constrain STR allele size. To confound matters, these forces may act independently or in concert (e.g. Garza et al., 1995). Unfortunately, much like modeling the STR mutation process, it is equally difficult to confirm or deny the presence of these other evolutionary forces, and even more difficult to quantify such forces if identified. For example, in a comparison of STR allele length distributions in humans and chimpanzees, Rubinsztein et al. (1995) concluded that humans have, in general, longer (i.e. larger) STR alleles than chimpanzees. The authors claimed that these data demonstrate a trend within the human lineage for longer STR alleles and that STR allele size tends to increase through evolution from ancestral to more derived states, thus indicating that directional forces act on STR evolution (Rubinsztein et al., 1995; Amos and Rubinsztein, 1996).

While the data certainly suggest that such a trend exists, the authors failed to consider all the factors that may influence the pattern of variation observed. Specifically, the authors assumed that STRs identified in humans should be equally polymorphic in chimpanzees. They failed to acknowledge that most identified STR loci are neither pursued nor published if they do not demonstrate a heterozygosity greater than 0.5 or 0.6 in an initial screening of the species from which they are isolated, because they provide little benefit for genetic linkage studies and/or little data for evolutionary studies (K. Kidd, personal communication). The consequence of comparing STR allele sizes across species, therefore, may be an inherent ascertainment bias that can drastically alter one's conclusions given a relationship between heterozygosity and allele size variance (e.g. Ellegren et al., 1995; Goldstein and

Pollock, 1997). A much more meaningful comparison to make in order to determine mutational directionality would be to examine specific STR loci within a species over time (e.g. Primmer et al., 1996) or to reciprocally compare loci selected at random from two recently diverged taxa (Ellegren et al., 1997).

Selective constraints that restrict or limit repeat number have also been suggested to occur for any given STR locus. Although examples exist of abnormally large repeat regions, most of these loci have been identified because the expanded repeat is associated with some type of pathology (e.g. fragile X syndrome, myotonic dystrophy and Huntington disease). Goldstein and Pollock (1997) echo this sentiment in the following statement: "Perhaps the most compelling evidence that the number of repeats at microsatellite loci is under some form of constraint is simply the absence of alleles of very large size. Given the high mutation rate, and the very large number of loci that have been characterized, it is clear that if the process were an *unconstrained* random walk we would expect to regularly observe loci with very large alleles" (Goldstein and Pollock, 1997, p. 337, our emphasis).

Though Goldstein and Pollock's (1997) conclusion may be true for the majority of known loci (except certain trinucleotide expansion loci), the fact that large STRs are not detected may also be an artifact resulting from the inability to characterize large repeats. Specifically, before it can be concluded that large repeats do not exist due to evolutionary constraints, one must be certain that such loci, if present in the genome, can be positively identified. Because the majority of identified STR loci have been isolated (i.e. subcloned) from genomic DNA libraries and large regions of highly repetitive DNA may be extremely difficult to clone in this manner (e.g. Rico et al., 1994), the observation that large-scale repeats do not exist may simply be the limitations of our current laboratory methods.

Recent data support this possibility. Using the novel Repeat Expansion Detection (RED) methodology[3], Sirugo et al. (1997) demonstrated that regions of long CAG/GTC repeats [i.e. $\geq (CTG)_{68}$] exist in the genomes of both humans and chimpanzees, and that significant differences in trinucleotide repeat sizes do not occur when random loci are compared. At least one other study has had similar conclusions based on comparing empirical data of homologous STR loci between different taxonomic groups (Ellegren et al., 1997). Both of these studies suggest that perceived directionality bias may merely be unrecognized ascertainment bias.

It may be impossible (or nearly so), therefore, to generate a universal model or obtain a consensus on a model that best reflects the mechanisms involved in the evo-

---

[3] The RED methodology consists of using genomic DNA as a template for the annealing and ligation of repeat-specific oligonucleotides and does not require flanking sequence determination or single-copy probes. Thus the genomic DNA acts as a "guide", and the maximum number of oligos that can be ligated is the number that can simultaneously anneal to the longest sequence, anywhere in the genome (Schalling et al. 1993; Sirugo and Kidd, 1995).

lution of STRs. STRs are complex loci that may experience a number of different mutational and/or evolutionary forces. Several different models may be required to characterize different STR loci, as well as the same STR locus at different stages of its evolution. Consequently, any data generated from an STR locus may be equally as complex as the locus itself, and thus one should be extremely cautious in making conclusions based entirely on STR-generated data. With this in mind, we can now examine the applicability and caveats of STRs for three popular aspects of ecology and evolutionary biology: phylogenetics, population biology and paternity/kinship studies. Specifically, the application of inconsistent analytical methodologies, assumptions about the homology and allelic size variance and nonamplifying alleles (null alleles) play a significant role in determining the level at which interpretations of STR data are acceptable.

Applications of STRs: To what degree can genealogical or evolutionary relationships be identified?

Phylogeny, genealogy and population structure

As we discussed above, STRs are good markers for generating useful biological data because of their high heterozygosity, Mendelian inheritance and the relative ease with which they can be typed. It is for these reasons that the use of STRs has become essential in the construction of linkage maps and for "gene hunting" (e.g. Weissenbach et al., 1992; Special Issue, *Nature Genetics*, 1994; Jorde, 1995; Dib et al., 1996; Dietrich et al., 1996). Although data derived from STRs may appear at first glance to be the "Holy Grail" of biological research, a more detailed investigation of the uses of STRs suggests that, at times, the conclusions based upon STR-derived data are often overstated.

For example, it has become increasingly common to see published reports in which STR loci (or sequences associated with these markers) have been used to reconstruct hierarchical relationships among taxa or gene flow within taxa (Bowcock et al., 1994; Meyer et al., 1995; Pépin et al., 1995; Estoup et al., 1995a; Valsecchi et al., 1997). The methodology for such comparisons often consists of calculating genetic distances between observed STR alleles, much like one would generate genetic distances from allozymes or RFLPs. These genetic distances, representing operational taxonomic units (OTUs), are then used to generate a "tree" (via a neighbor-joining algorithm or a dendogram via UPGMA) to reveal the phylogenetic relationships among samples (see Goldstein and Pollock, 1997, and references reviewed therein). Intraspecific comparisons, on the other hand, have traditionally been made by generating F statistics for the data under the assumption of an IAM (Wright, 1951). More recent analyses suggest, however, that the mutational process of STRs do not conform to an IAM, and consequently, such comparisons require the use of modified F statistics (Slatkin, 1995; Michalakis and Excoffier, 1996; Rousset, 1996; Goodman, 1997).

There are several significant problems with attempting to resolve evolutionary relationships, at either the species or population level, through STR analysis. The first problem concerns the presumption that STR data, when summarized as a distance or a statistic, are appropriate for phylogenetic or genealogical analyses; such summaries result in the reduction and/or loss of character information because they ignore any hierarchical descent relationships that may be detectable through synapomorphic characters. Thus, this methodology can only provide a relative measure of interspecific relationships. This criticism is exactly the same that has been levied at approaches using immunological and electromorph data for phylogeny reconstruction (Farris, 1981). A more recent discussion of the overall inappropriateness of using genetic distances for diagnosis of taxa has been provided by Vogler and DeSalle, (1994). With respect to intraspecific relationships, recent theoretical analyses suggest that genetic distances and modified F statistics estimated from STR data under the assumption of an SMM provide better estimates than F statistics estimates alone (Slatkin, 1995; Goldstein et al., 1995; Rousset, 1996). Such approaches, however, continually attempt to account for biases inherent to STRs (Goodman, 1997; Feldman et al., 1997), yet no objective criteria currently exists for selecting among the different and conflicting methods of data analysis (i.e. genetic distances or estimates of gene flow) (e.g. Valsecchi et al., 1997). Consequently, choosing among different genetic distances remains largely arbitrary. Other comparisons on the accuracy of genetic distances suggest that distance estimators modified specifically for the STR mutational process may not only contradict each other, but that the classical Fst statistics are more consistent with other types of genetic evidence than STR-tailored distance estimates (Calafell et al., submitted). To date, it therefore remains unclear if a distance approach can provide relevant biological information that may assist in resolving relationships between taxa or among populations.

The second problem associated with using STRs to resolve evolutionary relationships is concerned with the ability to properly identify phylogenetically informative characters; the assessment of homology is central to reconstructing phylogenetic relationships (Hennig, 1966; dePinna, 1991). In many cases (including those that employ a distance-based approach), allelic electromorph size variance, rather than DNA sequence content, is used as the only means for assigning homologous character states. Through direct sequencing of STR alleles and flanking regions, the validity of this approach has been questioned, because alleles may converge on the same size via different types of events in or surrounding the repeat array. This phenomenon has been termed "size" or "length" homoplasy (Estoup et al., 1995b; Grimaldi and Crouau-Roy, 1997; Angers and Bernatchez, 1997), but discussion of this issue is somewhat confused by a failure to distinguish primary homology from secondary homology. We hope to clarify the use of this terminology and illustrate why character state assignment based solely on STR electromorphic size variances

represents an inadequate primary homology assessment rendering them inappropriate for phylogenetic analysis.

Putative primary homology statements must be initially reasonable for any phylogenetic inference to reach conclusions that are warranted (dePinna, 1991; Brower
and Schawaroch, 1996). Specifically, it is unclear whether STR loci analyzed across
taxa or across populations meet the criteria of topographical identity and character
state identity (Brower and Schawaroch, 1996). It is assumed that the criterion of
topographical identity is satisfied given that conserved primers amplify STR loci
across taxa. However, character state identity cannot be accurately assessed just by
determining STR allelic electomorphic size variances. Variation in the size of alleles
can result from numerous processes such as (i) addition or deletion of the repeat unit
within the array, (ii) addition or deletion of another type of repeat within the array,
(iii) nonrepeated sequences or a partial repeat within the array and (iv) changes in
the sequence flanking the repetitive array. These occurrences have been documented both among alleles surveyed within a population and among distantly related
groups (Angers and Bernatchez, 1997), and can only be distinguished by direct sequencing of each STR allele. It is therefore incorrect to attribute STR alleles as homologous based on size alone if the constituent parts that comprise that size are not
identical.

These problems are illustrated in the following example. In comparing humans
and the great apes (i.e. orang-utans, gorillas, common chimpanzees and pygmy
chimpanzees) for a highly polymorphic dinucleotide STR at HOXB6, it was observed that humans and orang-utans carried (and shared in size) odd-sized alleles
(e.g. 135 bp), whereas chimpanzees and gorillas carried (and shared in size) even-
sized alleles (e.g. 138 bp) (Deinard and Kidd, 1995). Although these data could be
used (in either a genetic distance or a parsimony approach) to conclude that chimpanzees and gorillas are more closely related to each other than either is to humans
or orang-utans, one must first question the evolutionary meaning of two species
sharing STR alleles identical in size. Does the observed identity in STR size accurately reflect the number of repeats? DNA sequence data revealed that the STR consisted of two domains, a pentanucleotide (CCACA) repeat that varied across species
(chimpanzees carrying two repeats, humans and orang-utans carrying three repeats
and gorillas carrying four) and the dinucleotide (CA) repeat that was responsible for
the variation observed within each species. Even though chimpanzees and gorillas
share the same size alleles (e.g. 138 bp), based on their DNA sequence these alleles
were not truly identical characters (Deinard and Kidd, 1995; Fig. 2). In this case, the
observed pattern of allele sizes for the dinucleotide repeat initially suggests homology, but the criteria for character state identity clearly fail when the sequence composition is used. Consequently, it is invalid to argue that the identity in size (e.g.
138 bp) observed between chimpanzees and gorillas at this locus represents the same
character state.

```
Pongo          ACCCT CCCACACACA -----CCACA CCACACCACA CACACA......

Homo           ACCCT CCCACACACA -----CCACA CCACACCACA CACACA......

P. paniscus    ACCCT CCCACACACA ---------- CCACACCACA CACACA......

P. troglodytes ACCCT CCCACACACA ---------- CCACACCACA CACACA......

Gorilla        ACCCT CCCACACACA CCACACCACA ·CCACACCACA CACACA......
```

Figure 2. DNA sequences for the first domain of the HOXB6 STR. The pentanucleotide (CCACA) repeat varied across species and was responsible for variation between species, while the dinucleotide (CA) repeat was responsible for the variation observed within species. Odd-sized alleles were observed in both Gorilla *and* Pan, *while even-sized alleles were observed in* Pongo *and* Homo *(Deinard and Kidd, 1995).*

In addition, the regions immediately flanking an STR can confound homology assessment in a different sense. In certain situations, STR alleles of different size lengths that appear to be nonhomologous may actually be homologous. STR loci that differ in electromorphic size due to insertion or deletion events in the flanking sequence can have identical length and sequence composition in the repetitive region. In a study of cichlid relationships based on microsatellite flanking regions, Zardoya et al. (1996) use sequence variation outside of an STR to assess the informativeness of these characters in resolving relationships among lineages of the suborder Labroidei. Certain taxa (lineages Chalinochromis and Pseudocrenilabrus; Zardoya et al., 1996, p. 1592) which have identical length and sequence composition in the repeat unit exhibit different electromorphic sizes because of differences in the number of base pairs in the regions flanking the STR. If these repeat units were used for cladistic analysis, the states for Chalinochromis and Pseudocrenilabrus should be scored as identical, but this correct assessment would be missed if the alleles were not sequenced. A similar situation is evident for allelic size variation at an STR locus in the horseshoe crab (Ortí et al., 1997; alleles 22 and 23).

Mutational events in the flanking sequence, such as the ones described above, do offer a means to assess the value of STRs as synapomorphic characters (secondary homology) by comparing genealogies constructed via DNA sequence characters and those based upon allelic size variance. Ortí et al. (1997), upon sequencing the repetitive array and flanking region, found no correlation among the allelic states of an STR with the genealogy of the horseshoe crab reconstructed using a parsimony analysis of DNA sequence characters flanking the repeat region. In this study, attempts to conform the genealogy of STR allelic states to the topology found by DNA sequence characters resulted in significantly longer trees. The lack of a correlation between allele size and DNA sequence phylogeny does not conclusively negate the appropriateness of STR alleles as a source of character information. Further studies contrasting genealogy or phylogeny constructed from flanking sequences and other regions of the genome using simultaneous analysis of sequence characters with those obtained from STR allelic states should reveal whether STRs generally provide results congruent with other types of character information. This analysis, however, must also be considered in the context of the boundaries of cladistic analysis (Davis and Nixon, 1992; DeSalle and Vogler, 1994).

A third problem associated with using STRs to resolve evolutionary relationships is the existence of null alleles. Null alleles arise when a mutation that exists in the priming site or flanking sequence region during PCR amplification causes partial or complete loss of product (Callen et al., 1993). Reported incidences of null alleles at STR range from 30% in humans (Callen et al., 1993) to 66% in *Anopheles* (Lehmann et al., 1996). They can be difficult to detect and may only be evident when no PCR amplification product is obtained, unusually low patterns of heterozygosity or genetic incompatibilities among known pedigrees are observed or amplification using alternative primers indicates their occurrence (Paetkau and Strobeck, 1995). Consequently, null alleles may lead to a significant underestimation of allele frequency, heterozygosity and any estimates of genetic distances. Null alleles provide additional evidence why estimates based on genetic distance or distance-like approaches using STR allele size variance may be biased and must be thoroughly examined regardless of the type of question.

Several approaches, both technical and theoretical, have been suggested to detect null alleles. The most reliable method is accomplished by genotyping individuals with known familial relationships in order to assess patterns of non-Mendelian inheritance (Callen et al., 1993). Multiple independent PCR reactions of the same sample(s) is a robust way of identifying null alleles and ensures authenticity of findings (Taberlet et al., 1997). Brookfield (1996) suggests a new method to estimate null alleles, based on expected frequency of heterozygotes in Hardy-Weinberg equilibrium. This approach is limited in that all the assumptions of Hardy-Weinberg must be fulfilled in the populations analyzed. Hopefully, the performance of the method will eventually be evaluated with sufficient empirical studies to determine its accuracy rate. Once suspected, direct sequencing of STR loci can determine the origin of null alleles, and use of alternative or conserved primers may reduce their occurrence (Lehmann et al., 1996; Pemberton et al., 1995).

With respect to resolving evolutionary relationships at the species or population level through the analysis of STRs, our general conclusion is one of caution. Interpretation of results may be significantly influenced by technical problems and diverse mutational patterns associated with STRs. These occurrences not only are problematic for using STRs in a cladistic framework, but may also add to error and inconsistency among distance-based approaches. Studies of kinship and paternity within populations may in fact be less prone to these limitations.

Behavioral ecology and kinship

Of the applications described in this section, we believe that STR analysis has its greatest utility in addressing questions related to paternity assessment, behavioral ecology and kinship (e.g. kinship selection, altruism, sexual selection and inclusive fitness). Probably the greatest contribution that STR analysis has made to the field of "molecular ecology" has been with respect to paternity assessment, as it is often difficult to elucidate paternity and relatedness in some taxa; questions relating to reproductive behavior and reproductive success may now be tested through the analysis of STRs. Our improved ability to obtain DNA from alternative sources such as scat, hair, baleen and buccal cells enhances the power of utilizing STRs for this purpose because their high degree of polymorphism often translates into high degrees of informativeness (i.e. paternity exclusion). This becomes extremely important in the analysis of natural populations that may be endangered, difficult to sample non-invasively and where information on genetic relationships is lacking (Amos et al., 1993; Queller, 1993b; Höss et al., 1992; Morin and Woodruff, 1992; Rosenbaum et al., 1997).

"Genetic" tagging of individuals is another novel application of STR typing with particular relevance to ecology and demography. Multilocus genotyping of STR loci of individual animals provides a form of a permanent genetic tagging for identification. The ability to undertake these types of analyses, particularly where animals are difficult to observe, encounter, differentiate or to obtain tissues samples from, will contribute significantly to studies involving abundance estimation, range of a species or population, and behavior (Morin et al., 1994b; Reed et al., 1997; Taberlet, 1997). For example, using STR loci in this manner, Palsbøll et al. (1997b) calculated that there was 10^{-7} probability that two individuals might share the same genotype across all loci from a total 3060 skin samples of humpback whales. The authors were consequently able to conduct capture-recapture analysis for estimation of population abundance and examine the timing and movements of individual migrations of humpback whales in the North Atlantic Ocean based solely upon these genotypic (STR) data.

The problems associated with what has been termed length or size homoplasy, may not influence the resolution of the marker for this application to the same extent for analysis among closely related individuals and individuals within the same population (Primmer et al., 1997; Angers and Bernatchez, 1997). With the recent findings that size homoplasy in the repeat or flanking sequence is more widespread than originally thought, the number of loci needed to achieve to resolve kin relations accurately may be greater than originally presented (see Fig. 1, Queller et al.,

1993a). Size homoplasy among alleles would tend to cause incorrect paternity exclusion of putative fathers. This occurrence would affect diagnosis of genetic relationships if a sufficient number of STR loci were not examined. As above, thorough DNA sequencing of alleles should reveal any Mendelian incompatibilities that occur due to size homoplasy.

The amount of resolution that STRs provide for kinship studies may also vary depending upon the "openness" of the population. In closed systems that experience (essentially) no immigration or emigration, and in which all members of the population have been sampled, the probability of reconstructing genealogy is largely dependent on the heterozygosity of the STR marker. For example, in reconstructing the exact genealogical relationships among a captive breeding population of beluga whales (*Delphinapterus leucas*), Rosenbaum and Amato (in press) required only three highly heterozygous loci. In open population systems, in which immigration and emigration occur, the probability of resolving kinship relationships may be limited by individual sampling and the heterozygosity of specific STR loci. In assessing paternity in wild populations, there is a far greater probability that a particular set of males will be excluded from paternity of offspring in question and a significantly lower probability of assigning exact genealogical relationships among individuals. Depending on the heterozygosity of the marker and whether the maternal genotype is known, one (or possibly more) individuals may possess multilocus STR genotypes that are consistent with the paternally contributed alleles of the offspring. However, migration and/or incomplete sampling contribute to lack of absolute certainty that another male may possess the same genetic profile (and thus have an equal probability) for paternity as another male. These possibilities have been considered for paternity (or maternity) exclusion in taxa with diverse mating systems (Morin et al., 1994a,b; Jones and Avise, 1997; Clapham and Palsbøll, 1997; Petri et al., 1997).

Lastly, it is worth mentioning that null alleles may also confuse diagnoses of relatedness. As with the resolution of phylogenetic relationships, null alleles may reduce the informativeness of an STR locus, thus possibly compromising the ability to resolve kinship relationships among individuals. For example, in paternity testing, apparent compatibility of genotypes among nonrelated individuals can occur with null alleles. Conversely, the presence of null alleles could lead to apparent non-Mendelian inheritance and resulting paternity exclusion of males where they necessarily should not be excluded (Pemberton et al., 1995). Null alleles may be especially pervasive when using trace amounts of poorly preserved DNA (e.g. DNA extracted from hair or feces) because of overall poor PCR amplification, and thus proper controls or cautionary measures must be taken (Gagneux et al., 1997; Taberlet et al., 1997). This might involve additional screening and examining incidences of null alleles for particular STR loci among known familial relations where possible.

Conclusions

Are STRs the universal markers for ecological and evolutionary studies? Although the polymorphic nature of STRs make them extremely useful for identifying relationships among closely related individuals (e.g. Queller et al., 1993a; Jarne and Lagoda, 1996), the technical and theoretical boundaries discussed here should alert researchers to the uncertainties inherent in attempting to infer evolutionary relationships from STRs. Farris once stated "The best data are of no use unless properly analyzed" (Farris, 1981: p. 3). The statement could not be more appropriate for the analysis of STRs.

Despite attempts to justify their use in phylogenetic studies, the features and dynamics associated with STRs should prevent them from being used in either a distance or cladistic framework unless all STR alleles and flanking regions are sequenced. There are no objective criteria to select among the different estimates of gene flow, population parameters or genetic distances. Future advances with respect to understanding mutational direction and models for STRs may someday lead to more accurate methods to analyze resulting genotypic data. However, genetic distance approaches to phylogeny and genealogy specifically tailored to STRs still remain limited by the assumptions about descent relationships and allele size, if the exact structure of repeat is not determined.

Allelic size variance used in a character-based framework is clearly not appropriate because the problem of primary homology has not been properly considered in the use of STRs for phylogeny reconstruction. Assuming sequencing STRs and flanking regions reveals that they have met the criteria for primary homology, it remains to be seen whether STRs can serve as a good source of phylogenetically informative characters for any question in which cladistic analysis is appropriate. Excluding the repetitive array from the analysis and using the aligned flanking regions for informative characters is one way to bypass this problem (Zardoya et al., 1996). The types of mutational events within or surrounding the array may contribute to high levels of homoplasy, giving them limited usefulness for resolving relationships at higher levels.

STRs clearly have their greatest utility at the individual level in testing theories that relate to social dynamics, kinship and behavioral evolution. Future prospects include greater applicability to ecology and population biology. Our understanding of the complexity of relationships among individuals and how this translates further into mating behavior and social structure has wide-ranging ramifications for these fields.

Acknowledgements

We thank Richard H. Baker for extensive discussions on the topics above. Ken Kidd provided insight on the historic component of this chapter. This manuscript was greatly improved by critical reviews from Richard H. Baker, Phaedra Doukakis, Mary Egan, Marcia Firmani, Alfried Vogler and Rob DeSalle.

References

Amos, W. and Rubinsztein, D. C. (1996) Microsatellites are subject to directional evolution. *Nat. Genet.* 12: 13–14.

Amos, W., Schlötterer, C. and Tautz., D. (1993) Social structure of pilot whales revealed by analytical DNA profiling. *Science* 260: 670–672.

Angers, B. and Bernatchez, L. (1997) Complex evolution of a Salmonid microsatellite locus and its consequences in inferring allelic divergence from size information. *Molec. Biol. Evol.* 14(3) 230–238.

Bowcock, A. M., Ruiz-Linares, A., Tomfohrde, J., Minch, E., Kidd, J. R. and Cavalli-Sforza, L. L. (1994) High resolution of human evolutionary trees with polymorphic microsatellites. *Nature* 368: 455–457.

Brookfield, J. F. Y. (1996) A simple new method for estimating null allele frequency from heterozygote deficiency. *Molec. Ecol.* 5: 453–455.

Brower, A. V. Z. and Schawaroch, V. (1996) Three steps of homology assessment. *Cladistics* 12: 265–272.

Buard, J. and Jeffreys, A. (1997) Big, bad minisatellites. *Nat. Genet.* 15: 327–328.

Calafell, F., Shuster, A., Speed, W. C., Kidd, J. R. and Kidd, K. K. (1997) Short tandem repeat polymorphism evolution in humans, *in press*.

Callen, D. F., Thompson, A. D., Shen Y., Phillips, H. A., Richards, R. I., Mulley, J. C. and Sutherland G. R. (1993) Incidence and origin of "null" alleles in the (AC)n microsatellite markers. *Am. J. Hum. Gen.* 52: 922–927.

Clapham, P. J. and Palsbøll, P. J. (1997) Molecular analysis of paternity shows promiscuous mating in female humpback whales. *Proc. R. Soc. Lond. B.* 264: 95–98.

Dallas, J. F. (1992) Estimation of microsatellite mutation rates in recombinant inbred strains of mouse. *Mamm. Genome* 3: 452–456.

Davis, J. and Nixon, K. C. (1992) Populations, genetic variation and the delimitation of phylogenetic species. *Syst. Biol.* 41(4): 421–435.

Deinard, A. (1997) *The Evolutionary Genetics of the Chimpanzees*, Ph.D. dissertation, Yale University, New Haven, CT.

Deinard, A. and Kidd, K. (1995) Levels of polymorphism in extant and extinct hominoids. *In*: Brenner, S. and Hanihara, K. (eds) *The Origin and Past of Modern Humans as Viewed from DNA*, World Scientific, Singapore, pp. 149–170.

dePinna, M. C. C. (1991) Concepts and tests of homology in the cladistic paradigm. *Cladistics* 7: 367–394.

DeSalle, R. and Vogler, A. P. (1994) Phylogenetic analysis on the edge: the application of cladistic techniques at the population level. *In*: Golding, B. (ed.) *Non-Neutral Evolution, Theories and Molecular Data*, Chapman and Hall, New York, pp. 154–174.

Dib, C., Fauré, S., Fizames, C., Samson, D., Drouot, N., Vignal, A., Millasseau, P., Marc, S., Hazan, J., Seboun, E., Lathrop, M., Gyapay, G., Morissette, J. and Wessenbach, J. (1996) A comprehensive genetic map of the human genome based on 5,264 microsatellites. *Nature* 380: 152–154.

Dietrich, W. F., Miller, J., Steen, R., Merchant, M. A., Damron Boles, D., Husain, Z., Dredge, R., Daly, M. J., Ingalls, K. A., O'Connor, T. J. et al. (1996) A comprehensive genetic map of the mouse genome. *Nature* 380: 149–152.

DiRienzo, A., Peterson, A. C., Garza, J. C., Valdes, A. M., Slatkin, M. and Freimer, N. B. (1994) Mutational processes of simple-sequence repeat loci in human populations. *Proc. Natl. Acad. Sci. USA* 91: 3166–3170.

Ellegren, H., Primmer, C. R. and Sheldon, B. C. (1995) Microsatellite "evolution": directionality or bias? *Nat. Genet.* 11: 360–362.

Ellegren, H., Moore, S., Robinson, N., Byrne, K., Ward, W. and Sheldon, B. C. (1997) Microsatellite evolution-a reciprocal study of repeat lengths at homologous loci in cattle and sheep. *Molec. Biol. Evol.* 14(8): 854–860.

Estoup, A., Garnery, L., Solignac, M. and Cornuet, J.-M. (1995a) Microsatellite variation in honey bee (*Apis mellifera* L.) populations: hierarchical genetic structure and test of the infinite allele and stepwise mutation models. *Genetics* 140: 679–695.

Estoup, A., Garnery, L., Solignac, M. and Cornuet, J-M (1995b) Size homoplasy and mutational processes of interrupted microsatellites in two bee species, *Apis mellifera* and *Bombus terrestris* (Apidae). *Molec. Biol. Evol.* 12(6) 1074–1085.

Farris, J. S. (1981) Distance data in phylogenetic analysis. *In*: Funk, V. and Brooks, D. R. (eds) *Advances in Cladistics: Proceedings of the First Meeting of the Willi Hennig Society*, New York Botanical Society, Bronx, NY, pp. 3–23.

Feldman, M. W., Bergman, A., Pollock, D. D. and Goldstein, D. B. (1997) Microsatellite genetic distances with range constraints: analytic description and problems of estimation. *Genetics* 145: 207–216.

FitzSimmons, N. N., Moritz, C. and Moore S. S. (1995) Conservation and dynamics of microsatellite loci over 300 million years of marine turtle evolution. *Molec. Biol. Evol.* 12(3): 432–440.

Gagneux, P., Boesch, C. and Woodruff, D. S. (1997) Microsatellite scoring errors associated with noninvasive genotyping based on nuclear DNA amplified from shed hair. *Molec. Ecol.* 6: 861–868.

Gall, J. G. and Atherton, D. D. (1974) *J. Mol. Biol.* 85: 633–634.

Garza, J. C., Slatkin, M. and Freimer, N. B. (1995) Microsatellite allele frequencies in humans and chimpanzees, with implications for constraints on allele size. *Molec. Biol. Evol.* 12(4): 594–603.

Goldstein, D. B., Linares, A. R., Feldman, M. W. and Cavalli-Sforza, L. L. (1995) An evaluation of genetic distances for use with microsatellite loci. *Genetics* 139: 463–471.

Goldstein, D. B. and Pollock, D. D. (1997) Launching microsatellites: a review of mutation processes and methods of phylogenetic inference. *J. Hered.* 88: 335–342.

Goodman, S. (1997) Rst Calc: a collection of computer programs for calculating estimates of genetic differentiation from microsatellite data and determining their significance. *Molec. Ecol.* 6: 881–885.

Grimaldi, M.-C. and Crouau-Roy, B. (1997) Microsatellite allelic homoplasy due to variable flanking sequences. *J. Mol. Evol.* 44: 336–340.

Hennig, W. (1966) *Phylogenetic systematics*. University of Illinois Press, Chicago, 263 pp.

Höss, M., Kohn, M. and Pääbo, S. (1992) Excrement analysis by PCR. *Nature* 359: 199.

Jarne, P. and Lagoda, P. J. L. (1996) Microsatellites, from molecules to populations and back. *Trends. Ecol. Evolut.* 11: 424–429.

Jeffreys, A. J., Wilson, V. and Thein, S. L. (1985) Hypervariable "minisatellite" regions in human DNA. *Nature* 314: 67–73.

Jones, A. G. and Avise, J. C. (1997) Microsatellite analysis of maternity and the mating system in the Gulf pipefish *Syngnathus scovelli*, a species with male pregnancy and sex-role reversal. *Molec. Ecol.* 6: 203–213.

Jorde, L. (1995) Linkage disequilibrium as a gene mapping tool. *Am. J. Hum. Genet.* 56: 11–14.

Kimura, M. and Crow, J. F. (1964) The number of alleles that can be maintained in a finite population. *Genetics* 49: 725–738.

Lagercrantz, U., Ellegren, H. and Andersson, L. (1993) The abundance of various polymorphic microsatellite motifs differs between plants and vertebrates. *Nucl. Acid. Res.* 21: 1111–1115.

Lehmann, T., Hawley, W. A. and Collins, F. H. (1996) An evaluation of evolutionary constraints on microsatellite loci using null alleles. *Genetics* 144: 1155–1163.

Mandal, J. (1997) Breaking the rule of three. *Nature* 386: 767–769.

Meyer, E., Wiegand, P., Rand, S. P., Kuhlmann, D., Brack, M. and Brinkmann, B. (1995) Microsatellite polymorphisms reveal phylogenetic relationships in primates. *J. Mol. Evol.* 41: 10–14.

Michalakis, Y. and Excoffier, L. (1996) A generic estimation of population subdivision using distance between alleles with special reference for microsatellite loci. *Genetics* 142: 1061–1064.

Moore,S.S., Sargeant, L. L., King, T. J., Mattick, J. S., Georges, M. and Hetzel, D. J. S. (1991) The conservation of dinucleotide microsatellites among Mammalian Genomes allows the use of heterologous PCR primer pairs in closely related species. *Genomics* 10: 654–660.

Morin, P. A. and Woodruff, D. S. (1992) Paternity exlusing using multiple hypervariable microsatellite loci amplified from nuclear DNA of hair cells. *In*: Martin, R. D., Dixon, A. F. and Wickings, E. J. (eds) *Paternity in Primates: Genetic Tests and Theories*, Karger, Basel, pp. 63–81.

Morin, P. A., Moore, J. J., Chakraborty, R., Jin, L., Goodall, J. and Woodruff D. S. (1994a) Kin selection, social structure, gene flow, and the evolution of chimpanzees. *Science* 265: 1193–1201.

Morin, P. A., Wallis, J., Moore, J. J. and Woodruff D. S. (1994b) Paternity exclusion in a community of wild chimpanzees using hypervariable simple sequence repeats. *Molec. Ecol.* 3: 469–478.

Nakamura, Y., Leppert, M., O'Connell, P., Wolfe, R., Holm, T., Culver, M., Matin, M., Fujimoto, E., Hoff, M., Kumlin, E. and White, R. (1987) Variable number of tandem repeat (VNTR) markers for human gene mapping. *Science* 235: 1616–1622.

Ohta, T. and Kimura, M. (1973) A model of mutation appropriate to estimate the number of electrophoretically detectable alleles in a finite population. *Genet. Res.* 22: 201–204.

Ortí, G., Pearse, D. E. and Avise, J. C. (1997) Phylogenetic assessment of length variation at a microsatellite locus. *Proc. Natl. Acad. Sci. USA* 94: 10745–10749.

Paetkau, D. and Strobeck, C. (1995) The molecular basis and evolutionary history of microsatellite null allele in bears. *Molec. Ecol.* 4: 519–520.

Palsbøll, P. J., Allen, J., Bérubé, M., Clapham, P. J., Feddersen, T. P., Hammond, P. S.,

Hudson, R. R., Jørgensen, H., Katona, S., Larsen, A. H., Larsen, F., Lien, J., Mattila, D. K., Sigurjønsson, J., Sears, R., Smith, T., Sponer, R., Stevick, P. and Ølen, N. (1997a) Genetic tagging of humpback whales. *Nature* 388: 767–769.

Palsbøll, P. J., Bérubé, M., Larsen A. H. and Jørgensen, H. (1997b) Primers for the amplification of tri- and tetramer microsatellite loci in baleen whales. *Molec. Ecol.* 6: 893–895.

Pemberton, J. M., Slate, J., Bancroft, D. R. and Barrett, J. A. (1995) Non-amplifying alleles at microsatellie loci: a caution for parentage and population studies. *Molec. Ecol.* 4: 249–252.

Pépin, L., Amigues, Y., Lépingle, A., Berthier, J.-L., Bensaid, A. and Vaiman, D. (1995) Sequence conservation of microsatellites between *Bos taurus* (cattle), *Capra hircus* (goat) and related species. Examples of use in parentage testing and phylogeny analysis. *Heredity* 74: 53–61.

Petri, B., Pääbo, S., von Haeseler, A. and Tautz, D. (1997) Paternity assessment and population subdivision in a natural population of the larger mouse-eared bat *Myotis myotis*. *Molec. Ecol.* 6: 235–242.

Primmer, C. R., Ellegren, H., Saino, N. and Møller, A. P. (1996) Directional evolution in germline microsatellite mutations. *Nat. Genet.* 13: 391–393.

Primmer, C. R., Møller, A. P. and Ellegren, H. (1997) A wide-range survey of cross-species microsatellite amplification in birds. *Molec. Ecol.* 5: 365–378.

Queller, D. C., Strassman, J. E. and Hughes, C. R. (1993a) Microsatellites and kinship. *Trends. Ecol. Evolut.* 8: 285–288.

Queller, D. C., Strassman, J. E., Sol's, C. R., Hughes, C. R. and DeLoach, D. M. (1993b) A selfish strategy of social insect workers that promotes social cohesion. *Nature* 365: 639–641.

Rassman, K., Schlötterer, C. and Tautz, D. (1991) Isolation of simple-sequence loci for use in polymerase chain reaction-based DNA fingerprinting. *Electrophoresis* 12: 113–118.

Reed, J. Z., Tollit, D., Thompson, P. M. and Amos, W. (1997) Molecular scatology: the use of molecular genetics to assign species, sex and individual identity to seal faeces. *Molec. Ecol.* 6: 225–234.

Rico, C., Rico, I. and Hewitt, G. (1994) An optimized method for isolating andsequencing large (CA/GT)n (*n* > 40) microsatellites from genomic DNA. *Molec. Ecol.* 3: 181–182.

Rosenbaum, H. C. and Amato, G. A. (1998) Paternity assessment using microsatellite markers in Belugo whales (*Delphinapterus leucus*), *in press*.

Rosenbaum, H. C., Egan, M. G., Clapham, P. J., Brownell, R. L. Jr. and DeSalle, R. (1997) An effective method for isolating DNA from historical specimens of baleen. *Molec. Ecol.* 6: 677–681.

Rousset, F. (1996) Equilibrium values of measures of population subdivision for stepwise mutation processes. *Genetics* 142: 1357–1362.

Rubinsztein, D. C., Amos, W., Leggo, J., Goodburn, S., Jain, S., Li, S.-H., Margolis, R. L., Ross, C. A. and Ferguson-Smith, M. A. (1995) Microsatellite evolution-evidence for directionality and variation in rate between species. *Nat. Genet.* 10: 337–343.

Schalling, M., Hudson, T. J., Buetow, K. H. and Housman, D. E. (1993) Direct detection of novel expanded trinucleotide repeats in the human genome. *Nat. Genet.* 4: 135–139.

Schlötterer, C. and Tautz, D. (1992) Slippage synthesis of simple sequence DNA. *Nucl. Acid. Res.* 20: 211–215.

Schlötterer, C., Amos, B. and Tautz, D. (1991) Conservation of polymorphic simple sequence loci in cetacean species. *Nature* 354: 63–65.

Schug, M. D., Mackay, T. F. C. and Aquadro, C. F. (1997) Low mutation rates of microsatellite loci in *Drosophila melanogaster*. *Nat. Genet.* 15: 99–102.

Shriver, M. D., Jin, L. Chakraborty, R. and Boerwinkle E. (1993) VNTR allele frequency distribuitions under the stepwise mutation model: a computer simulation approach. *Genetics* 134: 983–993.

Sirugo, G. and Kidd, K. K. (1995) Repeat expansion detection using Ampligase thermostable DNA ligase. *Epicentre Forum* 2: 1–3.

Sirugo, G., Deinard, A., Kidd, J. and Kidd, K. (1997) Survey of maximum CTG/CAG repeat lengths in humans and non-human primates: total genome scan in populations using Repeat Expansion Detection method. *Hum. Mol. Genet.* 6: 403–408.

Slatkin, M. (1995) A measure of population subdivision based on microsatellite allele frequencies. *Genetics* 139: 457–462.

Special Issue (1994) New and improved genetic linkage maps for the human and mouse genomes. *Nat. Genet.* 7: 220–339.

Strassman, J. E., Solis, C. R., Barefield, K. and Queller, D. C. (1996) Trinucleotide microsatellite loci in a swarm-founding neotropical wasp, *Parachartergus colobopterus* and their usefulness in other social wasps. *Molec. Ecol.* 5: 459–461.

Taberlet, P., Camarra, J.-J., Griffin, S., Uhres, E., Hanotte, O., Waits, L. P., Dubois-Paganon, C., Burke, T. and Bovet, J. (1997) Noninvasive genetic tracking of the endangered Pyrenean brown bear population. *Molec. Ecol.* 6: 869–876.

Tautz, D. (1989) Hypervariability of simple sequence as a general source for polymorphic DNA markers. *Nucl. Acid. Res.* 17(16): 6463–6471.

Valdes, A. M., Slatkin, M. and Freimer, N. B. (1993) Allele frequencies at microsatellite loci: the stepwise mutation model revised. *Genetics* 133: 737–749.

Valsecchi, E. and Amos, W. (1996) Microsatellite markers for the study of cetacean populations. *Molec. Ecol.* 5: 151–156.

Valsecchi, E., Palsbøll, P. J., Hale, P., Glockner-Ferrari, D., Ferrari, M., Clapham, P., Larsen, F., Mattila, D., Sears, R., Sigurjønsson, J., Brown, M., Corkeron P. and Amos, B. (1997) Microsatellite genetic distances between oceanic populations of the humpback whale. *Molec. Biol. Evol.* 14: 355–362.

Vogler, A. and DeSalle, R. (1994) Diagnosing units of conservation management. *Conserv. Biol.* 8: 354–363.

Weber, J. L. (1990) Informativeness of human (dC-dA)n(dG-dT)n polymorphisms. *Genomics* 7: 524–530.

Weber, J. L. and Wong, C. (1993) Mutation of human short tandem repeats. *Hum. Mol. Genet.* 2: 1123–1128.

Weissenbach, J., Gyapay, G., Dib, C., Vignal, A., Morissette, J., Millasseau, P., Vaysseix, G. and Lathrop, M. (1992) A second generation linkage map of the human genome. *Nature* 359: 794–801.

Wright, S. (1951) The genetical structure of populations. *Ann. Eugenetics* 15: 323–354.

Zardoya, R., Vollmer, D. M., Craddock, C., Streelman, J. T., Karl, S. and Meyer, A. (1996) Evolutionary conseration of microsatellite flanking regionsand their use in resolving the phylogeny of cichlid fishes (Pisces: Perciformes) *Proc. R. Soc. Lond.* B 1589–1598.

Arbitrary oligonucleotides: primers for amplification and direct identification of nucleic acids, genes and organisms

Gustavo Caetano-Anollés

Department of Biology, University of Oslo, N-0316 Oslo, Norway

---

Summary

Arbitrary oligonucleotide primers can amplify a multiplicity of nucleic acid target sites and generate simple to complex patterns capable of scanning anonymous DNA or RNA templates, including genomic DNA, complementary DNA, cloned DNA fragments, and even polymerase chain reaction products. Profiles can be tailored in their complexity and polymorphic content, allowing the study of closely related organisms in a number of applications, including genetic mapping, map-based cloning, general fingerprinting, and molecular ecology, systematics and evolution. Transcript signatures produced directly from mRNA populations can be used to analyze multiple cellular states and identify differentially expressed genes. The potential of various nucleic acid-scanning techniques, novel improvements including the use of oligonucleotide arrays, and their recent application in the study of plant, fungal and bacterial genomes will be discussed.

Introduction

A wide repertoire of genetic markers can identify and catalogue distantly or closely related organisms, map useful and characteristic genetic traits, and physically arrange nucleic acid segments in genome analysis endeavors. Many of these markers are DNA-based and generally use hydrogen-bonding interactions between nucleic acid strands ("hybridization") and oligonucleotide-driven enzymatic accumulation of specific nucleic acid sequences ("amplification") to uncover polymorphism, and therefore sequence variability between individual genomes. Examples include hypervariable microsatellite markers that detect simple sequence repeat (SSR) polymorphisms by the polymerase chain reaction (PCR) (Weber and May, 1989; Litt and Luty, 1989; Tautz, 1989), minisatellite markers arising from the existence of tandemly arranged repetitive sequences of variable nature (Jeffreys et al., 1985a,b, 1988), restriction fragment length polymorphism (RFLP) analysis by Southern nucleic acid transfer (Wyman and White, 1980; Botstein et al., 1980), and genome-scanning techniques that use arbitrary oligonucleotide primers to uncover polymorphic DNA (Williams et al., 1990; Welsh and McClelland, 1990; Caetano-Anollés et al., 1991; Vos et al., 1995). These genetic markers have become of great value for the genetic analysis of eukaryotic and prokaryotic organisms. In plants, they are of special importance because they facilitate the genotypic estimation of simple and complex agronomic traits in plant-breeding programs, and allow conclusive distinction between plant cultivars or even individuals with important implications in phylogenetic analysis, parentage testing and cultivar certification. The analysis of microbes has also benefited from the use of DNA markers, providing more reliable estimators of relatedness and evolution, and facilitating their identification, for example in epidemiological studies.

One special class of nucleic acid typing does not require prior knowledge of nucleotide sequence or cloned and characterized hybridization probes, producing characteristic amplification signatures from virtually any nucleic acid (Livak et al., 1992; Bassam et al., 1995; McClelland et al., 1996). These "nucleic acid-scanning" techniques use one or more arbitrary oligonucleotides, usually quite short in sequence (typically 5–32 nt), to target a multiplicity of anonymous sites in nucleic acid templates. These sites (amplicons) are then amplified by a thermostable DNA polymerase in a PCR-like process that produces simple to complex genetic signatures (fingerprints) composed mainly of amplification products of varying length. The primer(s) drive the reaction and can be either completely arbitrary or derived (partially or totally) from known terminal, unique, repetitive or dispersed sequences. Primers hybridize to perfect or partially complementary annealing sites, and those that are closely spaced and have 3' termini facing each other are selected for initial amplification. While extensive mismatching can be tolerated, the first 5–6 nt from the 3' terminus of the primer must faithfully match those in the template (Caetano-Anollés et al., 1992). Primers as short as 5-mers can fingerprint nucleic acids (Caetano-Anollés et al., 1991, 1992). However, primers can harbor an arbitrary sequence of only 3 nt, if an extraordinarily stable minihairpin is attached at their 5' termini (Caetano-Anollés and Gresshoff, 1994).

Nucleic acid scanning is versatile and universal as demonstrated by the many applications and wide range of organisms studied and reported in thousands of publications (for selected reviews see Hadrys et al., 1992; Rafalski and Tingey, 1993; Caetano-Anollés, 1993, 1996; McClelland et al., 1995; Schierwater et al., 1997). Over 300 species have been studied in the plant kingdom alone, many of them of commercial importance (Weising et al., 1994). Similar widespread use can be found in the survey of organismal diversity and in the study of populations of bacteria, fungi and animals. For example, paternity analysis in insects (Hadrys et al., 1992, 1993; Scott and Williams, 1993) has spawned innumerable studies in animal behavioral ecology (Schierwater et al., 1997). Applications include genetic and physical mapping, plant and animal breeding, general fingerprinting, population biology, taxonomy and systematics, and map-based cloning of genes. High throughput has facilitated marker-assisted selection, trait introgression, and study of multigenic or quantitative traits (Rafalski and Tingey, 1993; Cushwa and Medrano, 1996). Finally, RNA scanning (Liang and Pardee, 1992; Welsh et al., 1992) has permitted detection and cloning of genes that are differentially expressed, without the limitations of substractive hybridization and differential screening (McClelland et al., 1995).

DNA scanning

A growing number of alternative approaches (Tab. 1) to DNA analysis have followed the inception of RAPD (Williams et al., 1990), AP-PCR (Welsh and McClelland, 1990), and DAF (Caetano-Anollés et al., 1991) analysis early in this decade. While the originally introduced multiple arbitrary amplicon profiling (MAAP) techniques differed mainly in primer length, primer-to-template ratio and profile complexity (Caetano-Anollés, 1993), the new derived approaches offered improvements in design, detection of polymorphic sequences and nature of targeted sites, increasing versatility and tailoring the distinction of nucleic acids. In many applications the nucleic acid amplifier was tailored by improving: (i) analysis of amplification products, (ii) primer design and (iii) amplification strategy.

Detection of amplification products

Amplification products can be resolved in agarose or polyacrylamide gels and labeled with radioactive isotopes or fluorophores or stained with ethidium bromide or silver. They can also be highly resolved by on-line detection with automated sequencers (Cancilla et al., 1992; Bauer et al., 1993) or capillary electrophoresis (Caetano-Anollés et al., 1995). However, more conventional alternatives presently offer higher throughput, such as the use of semiautomated miniaturized electrophoretic and staining devices (Baum et al., 1994). Other approaches increase detection of codominant or polymorphic DNA by polyacrylamide gel separation of nucleic acids according to size and base composition, such as denaturing gradient gel electrophoresis (DGGE) (He et al., 1992), temperature sweep gel electrophoresis (TSGE) (Penner and Bezte, 1994), and analysis of single-strand conformation polymorphisms (SSCP) (McClelland et al., 1994a). While all these alternatives categorize the amplification products by length using electrophoresis, one emerging strategy probes their sequence by direct hybridization to oligonucleotide arrays. The oligonucleotide arrays, also known as "DNA chips" (Southern, 1996), confine individual oligonucleotides to defined physical addresses in a solid support (nylon, glass, silicon, polypropylene, etc.) by solid-phase oligonucleotide synthesis, light-directed chemical synthesis using photolithographic masks, or accurate fluid mi-

Table 1. DNA scanning techniques[*]

Technique[†]	Primer (s)	Strategy	Reference
RAPD	Arbitrary	Standard: simple fingerprints	Williams et al. (1990)
AP-PCR	Arbitrary	Standard: complex fingerprints	Welsh and McClelland (1990)
DAF	Arbitrary	Standard: complex fingerprints	Caetano-Anollés et al. (1991)
mhpDAF	Minihairpin[‡]	Tailoring: avoidance of hairpin formation in amplified products	Caetano-Anollés and Gresshoff (1994)
D-DAF, D-RAPD	Degenerate (N, I)-arbitrary	Tailoring: increased versatility	Caetano-Anollés and Gresshoff (1994); Sakallah et al. (1995)
tecMAAP	Arbitrary	Tailoring: template endonuclease predigestion	Caetano-Anollés et al. (1993)
AFLP	Hemispecific[§]	Template restriction and ligation of adaptor cassette	Vos et al. (1995)
ASAP	Arbitrary ,mini-hairpin or 5'-degenerate microsatellite	Tailoring and motif encodance: fingerprinting fingerprints in a dual-step amplification	Caetano-Anollés and Gresshoff (1996)
NASBH	Arbitrary, mini-hairpin or 5'-degenerate microsatellite	Use of arbitrary DNA chips to probe the sequence of amplified products	Salazar and Caetano-Anollés (1996)
MP-PCR	Microsatellite	Motif encodance: SSR amplification	Meyer et al. (1993), Perring et al. (1993), Weising et al. (1995)
AMP-PCR	5'- or 3'-anchored micro-satellite	Motif encodance: inter-SSR amplification	Zietkiewicz et al. (1994)
RAMP	Arbitrary and 5'-anchored microsatellite	Motif encodance: stringent amplification with 5' anchored primers	Wu et al. (1994)
SAMPL	Hemispecific and micro-satellite	Template restriction and ligation of adaptor cassette	Morgante and Vogel (1996)
RAHM	Arbitrary or microsatellite	Motif encodance: detection of SSR loci by hybridization	Cifarelli et al. (1995)
RAMPO	Arbitrary or microsatellite	Motif encodance: detection of SSR loci by hybridization	Richardson et al. (1995)
RAMS	Arbitrary	Motif encodance: detection of SSR loci by hybridization	Ender et al. (1996)
DS-PCR	Microsatellite and arbitrary	Motif encodance: dual stringency amplification	Matioli and de Brito (1995)

crodispensing by pin transfer (nL-µL), air-jet (nL) or ink-jet (pL) technologies. In sequence-by-hybridization (SBH), arrays of short oligonucleotides, usually octamers, hybridize to overlapping complementary sequences present in the target nucleic acid molecule and can sequence short stretches of DNA of about 100 nt in length (Strezoska et al., 1991; Drmanac et al., 1993). The method can also be used in DNA fingerprinting. Gridded oligonucleotide arrays can be used as probes to detect single-nucleotide polymorphisms through multiple pairwise comparisons (Maskos and Southern, 1993; Nelson et al., 1993; Yershov et al., 1996). High-density arrays were recently used to query the sequence of the 16.6 kb human mitochondrial genome (Chee et al., 1996) and monitor the expression of numerous messenger RNA (mRNA) species in parallel (Lockhart et al., 1996). These nonelectrophoretic methods were similarly extended to the typing of amplification fragments from nucleic acid scanning reactions (Salazar and Caetano-Anollés, 1996; Beattie, 1997). In nucleic acid scanning by hybridization (NASBH), arrays of terminally degenerate oligonucleotides were used to measure genetic diversity within enterohemorrhagic isolates of *Escherichia coli* that express potent Shiga-like cytotoxins in an epidemiological study of several outbreaks of enteric disease (Salazar and Caetano-Anollés, 1996). NASBH was unprecedented in its abilities to detect types within a clonal bacterial group defined by multilocus enzyme electrophoresis (MLEE) and DAF analysis (Fig. 1).

Primer design

Arbitrary primers can be designed to increase multiplex ratio, enhance detection of polymorphic DNA or target defined nucleic acid motifs. Primer length influences fingerprint pattern (Caetano-Anollés et al., 1992). For example, DAF using very

---

Table 1. Legend

[*]Also known as multiple arbitrary amplicon profiling (MAAP), arbitrarily amplified DNA (AAD) or arbitrary primer technology (APT). Table modified from Caetano-Anollés (1996).
[†]Acronyms: AFLP, amplification fragment length polymorphism; AMP-PCR, anchored MP-PCR; AP-PCR, arbitrarily primed PCR; ASAP, arbitrary signatures from amplification profiles; DAF, DNA amplification fingerprinting; D-DAF, degenerate primer-driven DAF; D-RAPD, degenerate RAPD primer analysis; DS-PCR, double stringency PCR; mhpDAF, minihairpin primer-driven DAF; MP-PCR, microsatellite-primed PCR; NASBH, nucleic acid scanning by hybridization; RAPD, randomly amplified polymorphic DNA; tecMAAP, template endonuclease cleaved MAAP; RAHM, random amplified hybridization microsatellites; RAMP or RAMPO, random amplified microsatellite polymorphism; RAMS, randomly amplified microsatellites; SAMPL, selective amplification of microsatellite polymorphic loci; SSR, simple sequence repeat. [‡]Minihairpin primer: arbitrary primer containing a 5'-terminal minihairpin. [§]Hemispecific primer: arbitrary primer containing a 5' terminus specific for the adaptor and the restriction site.

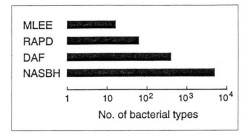

Figure 1. Detection of enteropathogenic Escherichia coli *bacterial types. The detection of different bacterial types by multi-locus enzyme electrophoresis (MLEE) can be enhanced up to 3 orders of magnitude by the use of DNA scanning techniques (RAPD and DAF), especially when coupled with DNA chips (NASBH).*

short primers (5–6 nt) produce relatively simple profiles similar to those in RAPD analysis. Similarly, primer sequence substitution with degenerate bases or inosines can increase or decrease profile complexity (Caetano-Anollés and Gresshoff, 1994; Sakallah et al., 1995), especially when substitutions are 3'-terminal. Primers can also be engineered to introduce secondary structure within certain primer domains. Minihairpin primers harboring an extraordinarily stable minihairpin and a 3' arbitrary sequence as short as 3 nt enhance detection of polymorphic DNA (Caetano-Anollés and Gresshoff, 1994, 1996; Caetano-Anollés et al., 1995). These primers generated complex "sequence signatures" from small templates such as yeast artificial chromosomes (YACs), plasmids, PCR fragments and DAF amplification products. Similarly, "gene signatures" were obtained from purified DNA fragments by low stringency amplification with a single specific primer in an hemianchored PCR (Pena et al., 1994). Scanning of small template molecules can find use in many applications, such as in the estimation of nucleotide divergence of PCR products of nuclear and cytoplasmic origin. Finally, primers can be biased to recognize sequence motifs such as those representing dispersed and repetitive DNA, or structural chromosomal domains. Primers can derive from *Alu* elements (Nelson et al., 1989), transfer RNA (tRNA) genes (Welsh and McClelland, 1991), bacterial repetitive sequences (REP, ERIC, BOX etc.; Versalovic et al., 1991), telomeres (Kolchinsky and Gresshoff, 1994), or sequence arrays in repetitive DNA, such as minisatellite core sequences (Heath et al., 1993) or microsatellite SSR (Meyer et al., 1993; Perring et al., 1993; Zietkiewicz et al., 1994; Wu et al., 1994; Weising et al., 1995; Richardson et al., 1995; Cifarelli et al., 1995; Matioli and de Brito, 1995; Ender et al., 1996; Caetano-Anollés and Gresshoff, 1996; Morgante and Vogel, 1996). The targeting of relatively conserved sequences (e.g. tRNA intergenic regions) allows the comparison of genomes at higher taxonomic levels. In contrast, the amplification of highly polymorphic sequence arrays can generate markers preponderantly codominant that express many allelic variants and can differentiate closely related individuals. Several strategies use primers complementary to microsatellite SSR loci (Tab. 1). Microsatellite primers can have 5' or 3' arbitrary anchors, can sometimes be used in combination with standard arbitrary primers and can target regions that span two closely spaced SSR loci (e.g. AMP-PCR), are terminal to one SSR (e.g. RAMP), or include the SSR locus itself (e.g. MP-PCR). Unfortunately, these techniques amplify both the SSR motif and unrelated arbitrary sequences, and generate fingerprints with relatively high multiplex ratios where codominance may be difficult to interpret (Weising et al., 1995; Vogel and Scolnik, 1997). To overcome this limitation, SSR oligonucleotides can be hybridized to arbitrarily amplified microsatellite loci (Richardson et al., 1995; Cifarelli et al., 1995; Ender et al., 1996). Alternatively, very few microsatellite loci can be amplified from DAF fingerprints with 5'-degenerate SSR primers (SSR-ASAP) and used as codominant markers (Caetano-Anollés and Gresshoff, 1996).

Amplification

Several amplifiers can enhance profile complexity and polymorphic DNA. tecMAAP uses digestion of either template or amplification products with several type II restriction endonucleases harboring 4-bp specificities (Caetano-Anollés et al., 1993). This strategy can enhance up to 100-fold the amplification of polymorphic sequences, allowing identification of near-isogenic lines and closely related plant accessions (Caetano-Anollés et al., 1993, 1995). Fungal and plant cultivars that were indistinguishable by DAF analysis were easily separated by digestion of stock DNA with up to three endonucleases prior to dilution and amplification. Moreover, near-isogenic lines of soybean (*Glycine max*) generated by ethyl methane sulfonate (EMS) mutagenesis were differentiated and markers tightly linked to the mutated locus isolated without resorting to an extensive screening of arbitrary primers (Caetano-Anollés et al., 1993). This endonuclease-linked MAAP strategy can be used to efficiently identify sequence-tagged markers linked to genes of interest, for high-resolution linkage mapping of specific genomic regions, and potentially for chromosome walking. Second and very popular is AFLP (Vos et al., 1995), a more demanding technique which ingeniously combines genome scanning and RFLP analysis. Genomic DNA is digested with one or more restriction endonucleases followed by ligation of an adaptor cassette to the ends of the generated restriction fragments. Hemispecific primers having an arbitrary 3' end and a 5' end complementary to both adaptor and restriction site are then used to amplify only a subset of restriction fragments under high-stringency amplification conditions. The number of targeted fragments appears to be a function of the length or the arbitrary 3' end of the primer (Vos et al., 1995). AFLP produces simple to highly complex fingerprints in high-resolving sequencing gels, but is subject to nonuniform targeting (i.e. clustering of markers in selected genomic regions; e.g. Keim et al., 1997) and abundant amplification of dispersed repetitive DNA. In contrast, the AFLP-derived SAMPL strategy uses a combination of hemispecific and microsatellite primers to enrich purposely for SSR loci (Morgante and Vogel, 1996). A third strategy, ASAP, generates "fingerprints of fingerprints" from arbitrarily amplified DNA (Caetano-Anollés and

Gresshoff, 1996). ASAP analysis is a dual-step molecular amplifier driven by mini-hairpin or motif-encoded primers, capable of producing highly polymorphic profiles and useful in the study of populations subjected to genetic bottlenecks (Caetano-Anollés et al., 1996). The strategy allows the combinatorial use of oligomers and therefore many fingerprinting reactions with few primers. A last though incipient strategy uses very short oligonucleotides (4–6 nt), in combination, as modular primers (G. Caetano-Anollés, unpublished observations). Each primer *per se* is incapable of appropriate amplification, but as modules they are able to initiate a "modular scanning" reaction. Modular scanning is based on the formation of strings of contiguously annealed unligated oligomers capable of priming DNA-sequencing reactions (Beskin et al., 1995). The technique provides increased versatility and a stable amplification milieu.

RNA fingerprinting

RNA can be reversed-transcribed and complementary DNA (cDNA) strands fingerprinted with arbitrary primers. Two initial studies reported this logical extension of DNA scanning. In differential display (DDRT-PCR), reverse transcription is initiated using a primer clamped at the 5' end with an oligo(dT) sequence complementary to the 3'-terminal poly(rA) tail generally present in mRNA populations (Liang and Pardee, 1992). Subsequently, an arbitrary decamer primer is used to prime DNA polymerase-mediated second-strand cDNA synthesis and drive the DNA-based amplification of 3' terminal segments of mRNA or polyadenylated heterogeneous RNA species. Alternatively, RNA arbitrarily primed PCR (RAP-PCR) uses a single arbitrary primer in a two-step procedure (Welsh et al., 1992). The primer initiates, under low-stringency conditions, first cDNA-strand reverse transcription and DNA polymerase-mediated second cDNA-strand synthesis. Ulterior high-stringency amplification produces fragments that are internal to the RNA template molecules and include open reading frames. RAP-PCR can be used on nonpolyadenylated RNA, such as that of many bacteria. Besides these two techniques, a number of RNA fingerprinting variants are possible with consequences on the normalization of RNA abundance and differential sampling of noncoding RNA domains. One important limitation of RNA fingerprinting is that profiles depend not only on primer-template mismatching but also on the relative abundance of each RNA species (Bertioli et al., 1994). Techniques such as gene expression fingerprinting (GEF; Ivanova et al., 1995) may help alleviate the inherent bias towards high copy number mRNA by providing a partly abundance-normalized cDNA sampling. Unfortunately, incomplete normalization perpetuates the underrepresentation of rare RNA. The search for rare differentially expressed targets should therefore rely on novel approaches, perhaps based on emerging strategies such as serial analysis of gene expression (SAGE; Velculescu et al., 1995), and related techniques (Kato, 1995). In SAGE, short diagnostic sequence mRNA tags are isolated by PCR amplification from type IIS restriction endonuclease-digested cDNA, concatenated, cloned and sequenced. The technique allows the simultaneous analysis of a large number of transcripts, providing characteristic gene expression patterns and a direct measure of transcript abundance.

RNA fingerprinting is a powerful tool to (i) detect and clone differentially expressed transcripts in cells that have experienced different environments or developmental programs, (ii) convey molecular phenotypes characteristic of individual cellular states, (iii) categorize transcripts and establish regulatory networks and (iv) identify convergent transcript regulatory pathways in cell lines treated with modulators of RNA abundance (McClelland et al., 1994b). These categorized transcripts can be used in genetic mapping of genes in epistatic interaction or as characters in population or phylogenetic analysis.

Arbitrary primers in population biology, taxonomy and molecular systematics

Genetic diversity has been studied with a variety of morphological, chemical, biological and molecular traits. These traits or characters are the tools of systematics, that is the analysis of samples of individuals to determine groupings within populations, delimiting species and genera, and finding phylogenetic relationships among taxa. However, a considerable number of characters must be studied to better understand genetic variation among organisms. It is obvious that only a relatively limited set of morphological, chemical and biological characters is available. Futhermore, and by way of example, isozymes and seed proteins, two kinds of molecular markers used widely in plant identification, can be limited by their potential susceptibility to environment. In contrast, DNA markers reflect the genotype and offer a wider sampling of the genome. However, to estimate genetic relationships and therefore genetic distances among individuals or populations, adequate levels of polymorphism are required. In this regard, nucleic acid-scanning techniques analyze multiple loci of moderate polymorphism, constituting an alternative to the analysis of single highly polymorphic loci (e.g. microsatellites). Discrete characters, such as MAAP markers, are useful tools only if they show variation at a particular taxonomic level (Hadrys et al., 1992). Determining their "resolving power" often requires preliminary experiments with simple and efficient sampling designs that will determine whether the character resolves at the desired level (i.e. is neither too variable nor variable enough). These experiments should consider the inclusion of individuals (populations, ecotypes, specimens from geographical locations, etc.) chosen on appropriate criteria, an appropriate sample size (unless rare specimens are being handled) and the selection of outgroups closely related to the ingroup being tested (this may be complicated in cases where little is known of overall relationships or when no suitable outgroup is available). Following preliminary studies, the character is used in conjunction with other tested characters to address the hypotheses of our research problem. In population biology the objective may be comparing individuals or populations by, for example, comparing alleles or allele frequencies at loci. In taxonomy, identification requires individual characters, while classification of taxons uses sets of individual characters of varying resolving power. In molecular systematics, the purpose may be phylogenetic reconstruction. In this case, each character is analyzed independently, and results are combined to understand population history within the evolutionary process that drives divergence and similarities of individuals and populations.

Arbitrarily amplified DNA molecules usually fall into two categories, those that are phylogenetically conserved and those that are species-, subspecies or individual-specific. This probably occurs as the result of primer target sites being randomly distributed along the genome and flanking both conserved and highly variable regions. Therefore, arbitrary primers generate a range of characters with varying resolving power that can be used to address a wide range of projects. However, characters are usually either present (1) or absent (0) in a particular individual – when comparing two genotypes, marker bands can occur in four possible configurations (1,1/1,0/0,1/0,0). Since most markers are inherited in a dominant fashion, it is not possible to distinguish heterozygotes from dominant homozygotes in diploid organisms, and assign a same character to the absent marker in haploid organisms. Furthermore, MAAP detects only a few alleles. This limitation is in part compensated for by the high number of loci amplified per fingerprint. Suitable estimators of genetic relatedness are available, despite concerns about homology and atypical inheritance. For example, Clark and Lanigan (1993) proposed a model that estimates nucleotide divergence and works best with sequence divergences that are less than 10%. Nevertheless, the model is contingent on a number of assumptions, including absence of primer-template mismatches and insertions or deletions between priming sites, Mendelian segregation and Hardy-Weinberg equilibrium. The different MAAP methods should be carefully evaluated in light of imposed theoretical limitations, to the point where comparison of divergent or unrelated taxa may justify altogether the use of DNA sequencing of PCR-amplified conserved loci (such as rDNA). As a conservative rule of thumb, most MAAP techniques should be used to distinguish organisms below the species level (cultivars, accessions, lines, clones etc.). However, closely related species have been reliably grouped with these techniques (e.g. Cerny et al., 1996). Finally, adequate optimization (e.g. Bentley and Bassam, 1996) and use of suitable DNA polymerases (e.g. Schierwater and Ender, 1993) have generated robust and transportable amplification protocols within and between laboratories that are highly reproducible, tolerate wide variations in amplification and thermal cycling parameters (cf. Schierwater et al., 1996) and minimize generation of artifactual bands.

The versatile nature of MAAP markers is suggested by how fingerprints can be tailored to be simple or complex simply by changing amplification parameters, and primer sequence, length and number (Caetano-Anollés, 1993). Screening a set of primers and choosing those that provide clear fingerprints with a useful ratio of monomorphic to polymorphic characters is sometimes easily accomplished [e.g. in *Streptococcus uberis* (Jayarao et al., 1992), root-knot nematodes (Baum et al., 1994), bermudagrass (*Cynodon*) (Caetano-Anollés et al., 1995), *Petunia* (Cerny et al., 1996) and banana (*Musa* sp.) (Kaemmer et al., 1992)]. In other cases this proves to be very difficult. For example, isolates of *Discula destructiva*, the fungal pathogen that causes anthracnose in dogwood (*Cornus*) trees, were very difficult to differentiate by DAF (Trigiano et al., 1995). In such cases, fingerprint tailoring must be used. For example, ASAP detected polymorphic DNA at 14-fold higher levels, allowing fine dissection of this highly homogeneous *D. destructiva* population (Caetano-Anollés et al., 1996). Results were consistent with the recent but separate introduction of the pathogen on the east and west coasts of North America and with a differential adaptation of the pathogen to pacific and flowering dogwoods (Fig. 2). Similarly, ASAP clearly distinguished closely related ornamental varieties of *Chrysanthemum* and geranium (*Pelargonium*) that were difficult to separate at the genetic level (reviewed in Caetano-Anollés, 1996). The identification of these varieties, often the result of spontaneous or induced somatic mutations, is impotant for

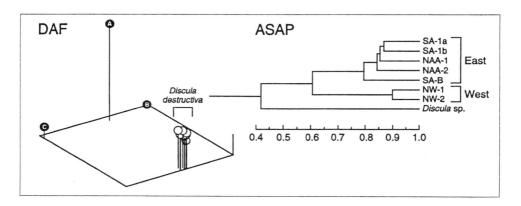

Figure 2. ASAP analysis can differentiate the highly homogenous Discula destructiva *fungal population. DAF could not distinguish fungal isolates from each other and relative to three* Discula *sp. isolates NY-326 (A), VA-17b (B) and NC-2 (C) originally defined as an outgroup and analyzed by principal coordinate analysis (PCO) (left diagram). Isolates represented the spread of the dogwood anthacnose disease in east and west coasts of the North American continent. In contrast, ASAP differentiated several subpopulations when fingerprints were subjected to cluster analysis using the UPGMA algorithm (left diagram). Results suggest different centers of origin of* D. destructiva *in west (NW) and east (SA and NAA) coasts. Genetic distances are based on Jaccard similarity coefficients. Modified from Caetano-Anollés et al. (1996).*

the floriculture industry in its efforts to support breeding and legal protection of newly developed plant cultivars.

The evaluation of molecular diversity in natural populations can be applied to population biology, molecular ecology and *in situ* conservation. Alternatively, molecular diversity can be studied in artificial or cultivated populations and used, for example, in the identification of cultivars, accessions and clones. The presumed neutrality and high allele number of microsatellites have made amplification-based SSR markers ideally suited for the study of populations and a preferred molecular method (Powell et al., 1996). However, MAAP markers have been profusely used in molecular diversity studies. Illustrative examples include conservation studies of the sea turtle *Caretta caretta* in the Eastern Mediterranean Sea, evaluation of reproductive success in odonates by kinship analysis, and establishment of taxonomical relationships of spirochete bacteria (reviewed in Schierwater et al., 1997). MAAP markers can also be used to identify accession redundancy in germplasm collections (He et al., 1995). This is crucial in *ex situ* conservation efforts and management of over a million plant accessions held in gene bank collections throughout the world.

It is clear that genome analysis has powerful new tools that promise further breakthroughs and better control of genetic and ecological resources by helping the breeder, the molecular ecologist and the population biologist better attain their goals. However, we are to envision further progress in the way genetic information is gathered, stored and analyzed. Microchip DNA analysis, holographic recording of genetic data and the use of neural networks are only few possibilities.

References

Bassam, B. J., Caetano-Anollés, G. and Gresshoff, P. M. (1995) Method for profiling nucleic acids of unknown sequence using arbitrary oligonucleotide primers, US Patent 5,413,909.

Bauer, D., Müller, H., Reich, J., Riedel, H., Ahrenkiel, V., Warthoe, P. and Strauss, M. (1993) Identification of differentially expressed mRNA species by an improved display technique (DDRT-PCR). *Nucl. Acid. Res.* 21: 4272–4280.

Baum, T. J., Gresshoff, P. M., Lewis, S. A. and Dean, R. A. (1994) Characterization and phylogenetic analysis of four root-knot nematode species using DNA amplification fingerprinting and automated polyacrylamide gel electrophoresis. *Mol. Plant-Microbe Interact.* 7: 39–47.

Beattie, K. (1997) Genomic fingerprinting using oligonucleotide arrays. *In*: Caetano-Anollés, G. and Gresshoff, P. M. (eds) *DNA Markers: Protocols, Applications and Overviews*, Wiley, New York, pp. 213–224.

Bentley, S. and Bassam, B. J. (1996) A robust DNA amplification fingerprinting system applied to analysis of genetic variation within *Fusarium oxysporum* f. sp. *cubense*. *J. Phytopathol.* 144: 207–213.

Bertioli, D. J., Schlichter, U. H. A., Adams, M. J., Burrows, P. R., Steinbiß, H.-H. and Antoniw, J. F. (1994) An analysis of differential display shows a strong bias towards high copy number mRNAs. *Nucl. Acid. Res.* 23: 4520–4523.

Beskin, A. D., Zevin-Sonkin, D., Sobolev, I. A. and Ulanovsky, L. E. (1995) On the mechanism of the modular primer effect. *Nucl. Acid. Res.* 23: 2881–2885.

Botstein, D., White, R. L., Scolnick, M. and Davis, R. W. (1980) Construction of a genetic linkage map in man using restriction fragment length polymorphisms. *Am. J. Hum. Genet.* 32: 314–331.

Caetano-Anollés, G. (1993) Amplifying DNA with arbitrary primers. *PCR Methods Applic.* 3: 85–94.

Caetano-Anollés, G. (1996) Scanning of nucleic acids by *in vitro* amplification: new developments and applications. *Nat. Biotechnol.* 14: 1668–1674.

Caetano Anollés, G. and Gresshoff, P. M. (1994) DNA amplification fingerprinting using arbitrary mini-hairpin oligonucleotide primers. *BioTechnology* 12: 619–623.

Caetano-Anollés, G. and Gresshoff, P. M. (1996) Generation of sequence signatures from DNA amplification fingerprints with mini-hairpin and microsatellite primers. *BioTechniques* 20: 1044–1056.

Caetano-Anollés, G., Bassam, B. J. and Gresshoff, P. M. (1991) DNA amplification fingerprinting using very short arbitrary oligonucleotide primers. *BioTechnology* 9: 553–557.

Caetano-Anollés, G., Bassam, B. J. and Gresshoff, P. M. (1992) Primer-template interactions during DNA amplification fingerprinting with single arbitrary oligonucleotides. *Mol. Gen. Genet.* 235: 157–165.

Caetano-Anollés, G., Bassam, B. J. and Gresshoff, P. M. (1993) Enhanced detection of polymorphic DNA by multiple arbitrary amplicon profiling of endonuclease digested DNA: identification of markers tightly linked to the supernodulation locus in soybean. *Mol. Gen. Genet.* 241: 57–64.

Caetano-Anollés, G., Callahan, L. M., Weaver, K. R., Williams, P. and Gresshoff, P. M. (1995) DNA amplification fingerprinting analysis of bermudagrass (*Cynodon*): genetic relationships between species and interspecific crosses. *Theor. Appl. Genet.* 91: 228–235.

Caetano-Anollés, G., Trigiano, R. N. and Windham, M. T. (1996) Sequence signatures from DNA amplification fingerprints reveal fine population structure of the dogwood pathogen *Discula destructiva*. *FEMS Microbiol. Lett.* 145: 377–383.

Cancilla, M. R., Powell, J. B., Hiller, A. J. and Davidson, B. E. (1992) Rapid genomic fingerprinting of *Lactococcus lactis* strains by arbitrarily primed polymerase chain reaction with ^{32}P and fluorescent labels. *Appl. Environ. Microbiol.* 58: 1772–1775.

Cerny, T. A., Caetano-Anollés, G., Trigiano R. N. and Starman, T. W. (1996) Molecular phylogeny and DNA amplification fingerprinting of *Petunia* taxa. *Theor. Appl. Genet.* 92: 1009–1016.

Chee, M. S., Huang, X., Yang, R., Hubbell, E., Berno, A. and Stern, D. (1996) Accesing genetic information with high-density DNA arrays. *Science* 274: 610–614.

Cifarelli, R. A., Gallitelli, M. and Cellini, F. (1995) Random amplified hybridization microsatellites (RAHM): isolation of a new class of microsatellite-containing DNA clones. *Nucl. Acid. Res.* 23: 3802–3803.

Clark, A. G. and Lanigan, C. M. S. (1993) Prospects of estimating nucleotide divergence with RAPDs. *Molec. Biol. Evol.* 10: 1096–1111.

Cushwa, W. T. and Medrano, J. F. (1996) Applications of the random amplified polymorphic DNA (RAPD) assay for genetic analysis of livestock species. *Animal Biotechnol.* 7: 11–31.

Drmanac, R., Drmanac, S., Strezoska, Z., Paunesku, T., Labat, I., Zeremski, M., Snoddy, J., Funkhouser, W. K., Koop, B., Hood, L. and Crkvernjakov, R. (1993) DNA sequence determination by hybridization: a strategy for efficient large-scale sequencing. *Science* 260: 1649–1652.

Ender, A., Schwenk, K., Städler, T., Streit, B. and Schierwater, B. (1996) RAPD identification of microsatellites in *Daphnia. Molec. Ecol.* 5: 437–441.

Hadrys, H., Balick, M. and Schierwater, B. (1992) Applications of random amplified polymorphic DNA (RAPD) in molecular ecology. *Molec. Ecol.* 1: 55–63.

Hadrys, H., Schierwater, B., Dellaporta, S. L., DeSalle, R. and Buss, L. W. (1993) Determination of paternities in dragonflies by random amplified polymorphic DNA fingerprinting. *Molec. Ecol.* 2: 79–87.

He, D., Ohm, H. and Mackenzie, S. (1992) Detection of DNA sequence polymorphisms among wheat varieties. *Theor. Appl. Genet.* 84: 573–578.

He, G., Prakash, C. S. and Jarret, R. L. (1995) Analysis of genetic diversity in sweetpotato (*Ipomoea batatas*) germplasm collection using DNA amplification fingerprinting. *Genome* 38: 938–945.

Heath, D. D., Iwama, G. K. and Devlin, R. H. (1993) PCR primed with VNTR core sequences yields species specific patterns and hypervariable probes. *Nucl. Acid. Res.* 21: 5782–5785.

Ivanova, N. B. and Belyavsky, A. V. (1995) Identification of differentially expressed genes by restriction endonuclease-based gene expression fingerprinting. *Nucl. Acid. Res.* 23: 2954–2958.

Jayarao, B. M., Bassam, B. J., Caetano-Anollés, G., Gresshoff, P. M. and Oliver, S. P. (1992) Subtyping *Streptococcus uberis* by DNA amplification fingerprinting. *J. Clin. Microbiol.* 30: 1347–1350.

Jeffreys, A. J., Wilson, V. and Thein, S. L. (1985a) Hypervariable "minisatellite" regions in human DNA. *Nature* 314: 67–73.

Jeffreys, A. J., Wilson, V. and Thein, S. L. (1985b) Individual-specific "fingerprints" of human DNA. *Nature* 316: 76–79.

Jeffreys, A. J., Wilson, V., Newmann, R. and Keyte, J. (1988) Amplification of human minisatellites by the polymerase chain reaction: towards fingerprinting of single cells. *Nucl. Acid. Res.* 16: 10953–10971.

Kaemmer, D., Afza, R., Weising, K., Kahl, G. and Novak, F. J. (1992) Oligonucleotide and amplification fingerprinting of wild species and cultivars of banana (*Musa* spp.). *BioTechnology* 10: 1030–1035.

Kato, K. (1996) RNA fingerprinting by molecular indexing. *Nucl. Acid. Res.* 24: 394–395.

Keim, P., Schupp, J. M., Travis, S. E., Clayton, K., Zhu, T., Shi, L., Ferreira, A. and Webb, D. M. (1997) A high-density soybean genetic map based upon AFLP markers. *Crop Sci.* 37: 537–543.

Kolchinsky, A. M. and Gresshoff, P. M. (1994) Plant telomeres as molecular markers. *In:* Gresshoff, P. M. (ed.) *Plant Genome Analysis*, CRC Press, Boca Raton, FL, pp. 113–124.

Liang, P. and Pardee, A. B. (1992) Differential display of eukaryotic messenger RNA by means of the polymerase chain reaction. *Science* 257: 967–971.

Litt, M. and Luty, J. A. (1989) A hypervariable microsatellite revealed by *in vitro* amplification of a dinucleotide repeat within the cardiac muscle actin gene. *Am. J. Hum. Genet.* 44: 397–401.

Livak, K. J., Rafalski, J. A., Tingey, S. V. and Williams, J. G. (1992) Process for detecting polymorphisms on the basis of nucleotide differences, US Patent 5,126,239.

Lockhart, D. J., Dong, H., Byrne, M. C., Follettie, M. T., Gallo, M. V., Chee, M. S., Mittman,

M., Wang, C., Kobayashi, M., Horton, H. and Brown, E. L. (1996) Expression monitoring by hybridization to high-density oligonucleotide arrays. *Nat. Biotechnol.* 14: 1675–1680.

Maskos, U. and Southern, E. M. (1993) A novel method for the analysis of multiple sequence variants by hybridisation to oligonucleotides. *Nucl. Acid. Res.* 21: 2267–2268.

Matioli, S. R. and de Brito, R. A. (1995) Obtaining genetic markers by using double-stringency PCR with microsatellites and arbitrary primers. *BioTechniques* 19: 752–758.

McClelland, M., Arensdorf, H., Cheng, R. and Welsh, J. (1994a) Arbitrarily amplified PCR fingerprints resolved on SSCP gels. *Nucl. Acid. Res.* 22: 1770–1771.

McClelland, M., Ralph, D., Cheng, R. and Welsh, J. (1994b) Interactions among regulators of RNA abundance characterized using RNA fingerprinting by arbitrarily primed PCR. *Nucl. Acid. Res.* 22: 4419–4431.

McClelland, M., Mathieu-Daude, F. and Welsh, J. (1995) RNA fingerprinting and differential display using arbitrarily primed PCR. *Trends Genet.* 11: 242–246.

McClelland, M., Welsh, J. T. and Sorge, J. A. (1996) Arbitrarily primed polymerase chain reaction method for fingerprinting genomes, US Patent 5,467,985.

Meyer, W., Mitchel, T. G., Freedman, E. Z. and Vilgalys, R. (1993) Hybridization probes for conventional DNA fingerprinting used as single primers in the polymerase chain reaction to distinguish strains of *Cryptococcus neoformans. J. Clin. Microbiol.* 31: 2274–2280.

Morgante, M. and Vogel, J. M. (1996) International Patent Publication WO96/17082.

Nelson, D. L., Ledbetter, S. A., Corbo, L., Victoria, M. F., Ramirez-Solis, R., Webster, T. D., Ledbetter, D. H. and Caskey, C. T. (1989) *Alu* polymerase chain reaction: a method for rapid isolation of human-specific sequences from complex DNA sources. *Proc. Natl. Acad. Sci. USA* 86: 6686–6690.

Nelson, S. F., McCusker, J. H., Sander, M. A., Kee, Y., Modrich, P. and Brown, P. O. (1993) Genomic mismatch scanning: a new approach to genetic linkage mapping. *Nature Genet.* 4: 11–18.

Pena, S. D. J., Barreto, G., Vago, A. R., De Marco, L., Reinach, F. C., Dias Neto, E. and Simpson, A. J. G. (1994) Sequence-specific "gene signatures" can be obtained by PCR with single specific primers at low stringency. *Proc. Natl. Acad. Sci. USA* 91: 1946–1949.

Penner, G. A. and Bezte, L. J. (1994) Increased detection of polymorphism among randomly amplified wheat DNA fragments using a modified temperature sweep gel electrophoresis (TSGE) technique. *Nucl. Acid. Res.* 22: 1780–1781.

Perring, T. M., Cooper, A. D., Rodriguez, R. J., Farrar, C. A. and Bellows, T. S. (1993) Identification of whiteflies species by genomic and behavioral studies. *Science* 259: 74–77.

Powell, W., Machray, G. C. and Provan, J. (1996) Polymorphism revealed by simple sequence repeats. *Trends Plant Sci.* 1: 215–222.

Rafalski, J. A. and Tingey, S. V. (1993) Genetic diagnostics in plant breeding: RAPDs, microsatellites and machines. *Trends Genet.* 9: 275–279.

Richardson, T., Cato, S., Ramser, J., Kahl, G. and Weising, K. (1995) Hybridization of microsatellites to RAPD: a new source of polymorphic markers. *Nucl. Acid. Res.* 23: 3798–3799.

Sakallah, S. A., Lanning, R. W. and Cooper, D. L. (1995) DNA fingerprinting of crude bacterial lysates using degenerate RAPD primers. *PCR Meth. Applic.* 4: 265–268.

Salazar, N. M. and Caetano-Anollés, G. (1996) Nucleic acid scanning-by-hybridization of enterohemorrhagic *Escherichia coli* isolates using oligonucleotide arrays. *Nucl. Acid. Res.* 24: 5056–5057.

Schierwater, B. and Ender, A. (1993) Different thermostable DNA polymerases may amplify different amplification patterns. *Nucl. Acid. Res.* 21: 4647–4648.

Schierwater, B., Metzler, D., Krüger, K. and Streit, B. (1996) The effects of nested primer binding sites on the reproducibility of PCR: mathematical modeling and computer simulation studies. *J. Comp. Biol.* 2: 235–251.

Schierwater, B., Ender, A., Schroth, W., Holzmann, H., Diez, A., Streit, B. and Hadrys, H. (1997) Arbitrarily amplified DNA in ecology and evolution. *In*: Caetano-Anollés, G. and Gresshoff, P. M. (eds) *DNA Markers: Protocols, Applications and Overviews*, Wiley, New York, pp. 313–330.

Scott, M. P. and Williams, S. M. (1993) Comparative reproductive success of communally-breeding burying beetles as assessed by PCR with randomly amplified polymorphic DNA. *Proc. Natl. Acad. Sci. USA* 90: 2242–2245.

Southern, E. M. (1996) DNA chips: analysing sequence by hybridization to oligonucleotides on a large scale. *Trends Genet.* 12: 110–115.

Strezoska, Z., Punesku, T., Radosavljevic, D., Labat, I., Drmanac, R. and Crkvenjakov, R. (1991) DNA sequencing by hybridization: 100 bases read by a non-gel-based method. *Proc. Natl. Acad. Sci. USA* 88: 10089–10093.

Tautz, D. (1989) Hypervariability of simple sequences as a general source for polymorphic DNA markers. *Nucl. Acid. Res.* 17: 6463–6471.

Trigiano, R. N., Caetano-Anollés, G., Bassam, B. J. and Windham, M. T. (1995) DNA amplification fingerprinting provides evidence that *Discula destructiva*, the cause of dogwood anthracnose in North America, is an introduced pathogen. *Mycologia* 87: 490–500.

Velculescu, V. E., Zhang, L., Vogelstein, B. and Kinzler, K. W. (1995) Serial analysis of gene expression. *Science* 270: 484–487.

Versalovic, J., Koeuth, T. and Lupski, J. R. (1991) Distribution of repetitive DNA sequences in eubacteria and applications to fingerprinting of bacterial genomes. *Nucl. Acid. Res.* 19: 6823–6831.

Vogel, J. M. and Scolnik, P. A. (1997) Direct amplification from microsatellites: detection of simple sequence repeat-based polymorphisms without cloning. *In*: Caetano-Anollés, G. and Gresshoff, P. M. (eds) *DNA Markers: Protocols, Applications and Overviews*, Wiley, New York, pp. 133–150.

Vos, P., Hogers, R., Bleeker, M., Reijans, M., van de Lee, T., Hornes, M., Frijters, A., Pot, J., Peleman, J., Kuiper, M. and Zabeau, M. (1995) AFLP: a new technique for DNA fingerprinting. *Nucl. Acid. Res.* 23: 4407–4414.

Weber, J. L. and May, P. E. (1989) Abundant class of human DNA polymorphisms which can be typed using the polymerase chain reaction. *Am. J. Hum. Genet.* 44: 388–396.

Weising, K., Nybom, H., Wolff, K. and Meyer, W. (1994) *DNA Fingerprinting in Plants and Fungi*, CRC Press, Boca Raton, FL.

Weising K., Atkinson, R. G. and Gardner, R. C. (1995) Genomic fingerprinting by microsatellite-primed PCR: a critical evaluation. *PCR Methods Applic.* 4: 249–255.

Welsh, J. and McClelland, M. (1990) Fingerprinting genomes using PCR with arbitrary primers. *Nucl. Acid. Res.* 19: 861–866.

Welsh, J. and McClelland, M. (1991) Genomic fingerprints produced by PCR with consensus tRNA gene primers. *Nucl. Acid. Res.* 19: 861–866.

Welsh, J., Chada, K., Dalal, S. S., Cheng, R., Ralph, D. and McClelland, M. (1992) Arbitrarily primed PCR fingerprinting of RNA. *Nucl. Acid. Res.* 20: 4965–4970.

Williams, J. G. K., Kubelik, A. R., Livak, K. J., Rafalski, J. A. and Tingey, S. V. (1990) DNA polymorphisms amplified by arbitrary primers are useful as genetic markers. *Nucl. Acid. Res.* 18: 6531–6535.

Wu, K., Jones, R., Danneberger, L. and Scolnik, P. A. (1994) Detection of microsatellite polymorphisms without cloning. *Nucl. Acid. Res.* 22: 3257–3258.

Wyman, A. R. and White, R. (1980) A highly polymorphic locus in human DNA. *Proc. Natl. Acad. Sci. USA* 77: 6754–6758.

Yershov, G., Barsky, V., Belgovskiy, A., Kirillov, E., Kreindlin, E., Ivanov, I., Parinov, S., Guschin, D., Drobishev, A., Dubiley, S. and Mirzabekov, A. (1996) DNA analysis and diagnostics on oligonucleotide microchips. *Proc. Natl. Acad. Sci. USA* 93: 4913–4918.

Zietkiewicz, E., Rafalski, A. and Labuda, D. (1994) Genome fingerprinting by simple sequence repeat (SSR)-anchored polymerase chain reaction amplification. *Genomics* 20: 176–183.

Part 2. Species

Rob DeSalle and Bernd Schierwater

In his now classic *The Genetic Basis of Evolutionary Change*, Lewontin (1974) summarized many aspects of evolutionary genetics that were major topics during the decade prior to publication of his book. With respect to species and the genetics of speciation Lewontin has this to say:

> "It is an irony of evolutionary genetics that, although it is a fusion of Mendelism and Darwinism, it has made no direct contribution to what Darwin obviously saw as the fundamental problem: the origin of species. (p. 159)"

He goes on to suggest that evolutionary genetics has contributed to how theories of speciation have arisen, but points out that in 1974 evolutionary genetics was no closer to "constructing a quantitative theory of speciation in terms of genotypic frequencies" (p. 160) than the shapers of the new synthesis were. To a certain extent, two decades later we are no closer to understanding the "fundamental problem" than we were in 1974, and this after extensive effort at collecting DNA sequence information from a wide variety of organisms.

But we are closer to understanding clearly the difference between what Templeton (1989) and Harrison (1990, 1991) would call the genetics of species formation and the genetics that result from species formation. The coherent discussion of speciation and species concepts by several authors in the first part of Otte and Endler's (1989) *Speciation and Its Consequences* delimits a major problem in the study of processes and patterns at the species level. This problem is tied directly to species concepts and has led some authors to suggest that no one species concept can be used in all cases (Endler, 1989; Donoghue and Baum, 1992). Part 1 of the Otte and Endler volume ("Concepts of Species") should be consulted for an excellent description of the species concepts that are currently considered useful. The reader will note that the authors of our volume adopt rather different perspectives on species concepts, and it is imporant to keep this in mind in reading the chapters.

Harrison (1991) clearly places the problem of studying species at the heart of the difference between systematics and population genetics: "It [speciation studies] falls at the interface between population biology and systematic biology and does not fit comfortably into either domain" (Harrison, 1991, p. 282). One "domain" is interested in the discovery of hierarchical relationships (systematics), while the other is interested in the characterization of the forces responsible for allele changes within populations (population genetics). The recent development of coalescent theory and its application in population genetics (Kingman, 1972; Hudson, 1990) as well as the hard empirical data collected within the so-called phylogeographic perspective (Avise et al., 1987; Avise, 1994) have allowed for a "tree-based" approach to this very difficult to examine level. Several authors have attempted to examine the problem of species by straddling the boundary and this has been a direct result of the ability of workers in the area to generate or conceptualaize gene genealogies at the level of populations (Maddison, 1995; Avise and Wollenberg, 1997).

Questions of how species divergence occurs, the genetics of speciation and the genetic consequences of species divergence are extremely important for two immediate reasons. First, the genetics of species formation has been seen as an all-important aspect of evolutionary biology for some time. Discovery of the genetic events that are responsible for "prezygotic and postzygotic barriers to gene exchange" (Harrison, 1991; p. 303) of species are to a certain extent the Holy Grail of evolutionary biology. Certainly, the discovery of the genetic loci that are involved in species divergence and isolation for as many cases as possible would be a great advance in evolutionary biology. Second, the identification of species (whether through the characterization of the genetic events responsible for species formation or through the genetics caused by speciation is used) is an important step in understanding biodiversity. The recent incoporation of systematic and population genetic approaches in conservation biology in volumes such as Karp et al. (1998), Ferraris and Palumbi (1996), Avise and Hamrick (1995), Claridge et al. (1997) and Smith and Wayne (1997) as well as articles from our earlier volume (O'Brien, 1994; Vogler, 1994) is testament to the importance of species recognition in this important applied science.

With respect to this volume the approach taken to the species boundary is very important. Four of the papers in this section (Templeton, Wakely and Hey, Amato et al. and Vogler) use explicit tree-building approaches to examine problems at the interface of populations and species. Templeton suggests that "only by straddling this interface can we actually study the process involved in the origin of a new species". By demonstrating tree-building approaches in two empirical systems (salamander mitochondrial DNA [mtDNA] and pine chloroplast DNA [cpDNA]) he establishes a "bottom-up" approach that allows for the testing of hypotheses about speciation of the organisms in these systems. The Wakely and Hey chapter, too, is concerned with testing hypotheses of species formation. Their approach is also bottom up and uses population genetic models to make predictions about the divergence of populations. A tree-building approach is used, and all of the predictions made from the models discussed by Wakely and Hey are viewed from a genealogical perspective.

The next two chapters take more of a "top-down" approach to an understanding of species and in this sense are more in line with the phylogenetic species concept. In addition, both of these chapters discuss the species problem as a problem of diagnosis, and both are therefore attempts to make a direct connection to conservation biology. Amato, Gatesy and Brazaitis describe polymerase chain reaction (PCR) methods for the rapid and efficient characterization of mtDNA haplotypes in the *Caiman crocodilus* complex. Their top-down approach uses a phylogenetic species concept and the population aggregation analysis approach (PAA; Davis and Nixon, 1992) to diagnose "units" in their study. Vogler uses *Cicindela dorsalis* of North America as a study group to examine extinction of lineages in the context of the phylogenetic species concept. Vogler's top-down character-based approach examines the effects of extinction of "linker" populations (populations that might connect otherwise diagnosable units) on cladogenesis in these important insects.

Several papers from our previous volume make inroads to problems of the genetic differences between species (Kreitman and Wayne, 1994; Streit et al., 1994; Rand, 1994; Hartl and Lozovskya, 1994). Along with these papers the last two papers in this section examine aspects of the genetics of species differences. Routman and Cheverud demonstrate the utility of quantitative trait linkage (QTL) analysis in understanding the effects of epistasis at the population level. As these authors point out, the methodology they describe has been important in the study of some systems where drastic morphological differences and reproductive isolation have evolved. Ochman and Groisman's chapter seeks the genetic basis of differences between *Escherichia coli* and *Salmonella enterica*. Although these two species of bacteria are not prticularly closely related (divergence time of 100 MY) the techniques outlined by these authors for the identification and recovery of species-specific DNA sequences as well as the evaluation of the biological role of species-specific sequencs ("functional genomics") are important approaches to linking the genetic changes produced by speciation with function of those changes.

Future work at this difficult level where population genetics and systematics meet, using molecular techniques, will most likely elaborate on three areas of research. First, molecular information will be an important component of the examination of the "bridge" between population genetics and systematics. Whether or not research in this area will come to a definite conclusion that the "straddling" of the two fields is fruitful, will depend on how well the bottom-up approach deals with the emergence of hierarchy from reticulating systems. Second, diagnosis of units using molecular techniques will more than likely increase in importance in conservation issues. Molecular techniques are needed that can quickly and accurately assess attributes for large numbers of individuals in studies of conservation units, and these are well within our grasp. Limits of the utility of diagnosis using phylogenetic methods (see Frost and Kluge, 1994, for example) and molecular data should also be a major area of exploration. Finally, approaches such as QTL studies and functional

genomics will be increasingly important in the search for genetic differences involved in speciation – the fundamental problem of species formation.

References

Avise, J. C. (1994) *Molecular Markers, Natural History and Evolution,* Chapman and Hall,
New York.

Avise, J. C. and Hamrick, J. L. (1995) *Conservation Genetics: Case Histories from Nature,*
Chapman and Hall, New York.

Avise, J and Wollenberg, K. (1997) Phylogeneticsand the origin of species. *Proc. Natl. Acad.
Sci. USA* 94: 7748–7755.

Avise, J. C., Arnold, J., Ball, R. M., Bermingham, E., Lamb, T., Neigel, J. E., Reeb, C. A. and
Saunders, N. C. (1987) Intraspecific phylogeography: The mitochondrial bridge between
population genetics and systematics. *Annu. Rev. Ecol. Syst.* 18: 489–522.

Baum, D. and Shaw, K. (1994) The genealogical species concept. *In:* Hoch, P. and
Stephenson, A. G. (eds) *Experimental and Molecular Approaches to Plant Biosystematics,*
Missouri Botanical Garden, St. Louis, MO, pp. 289–303.

Claridge, M. F., Dawah, H. A. and Wilson, M. R. (1997) *Species: The Units of Biodiversity,*
Chapman and Hall, New York.

Davis, J. and Nixon, K. C. (1992) Populations, genetic variation and the delimitation of phylogenetic species. *Syst. Biol.* 41(4): 421–435.

Endler, J. A. (1989) Conceptual ond other problems in speciation. *In:* Otte., D. and Endler,
J. A. (eds) *Speciation and ist Consequences,* Sinauer Associates, Sunderland, MA, pp.
625–648.

Ferraris, J. D. and Palumbi, S. R. (eds) (1996) *Molecular Approaches to Zoology and
Evolutions,* Wiley-Liss, New York.

Frost, D. and Kluge, A. (1994) A consideration of epistemology in *Syst. Biol.,* with special
reference to species. *Cladistics* 10: 259–294.

Harrison, R. G. (1990) Hybrid zones: Windows on evolutionary process. *Oxf. Surv. Evol.
Biol.* 69–128.

Harrison, R. G. (1991) Molecular changes at speciation. *Annu. Rev. Ecol. Syst.* 281–308.

Hartl, G. B. and Lozovskya (1994) Genome evolution: between the nucleosome and the chromosome. *In:* Schierwater, B., Streit, B., Wagner, G, P. and DeSalle, R. (eds) *Molecular
Ecology and Evolution: Approaches and Applications,* Birkhäuser, Basel, pp. 579–592.

Hudson, R. R. (1990) Gene genealogies and the coalescent process. *Oxf. Surv. Evol. Biol.* 7:
1–44.

Karp, A., Isaac, P. G. and Ingram, D. S. (eds) (1998) *Molecular Tools for Screening
Biodiversity,* Chapman and Hall, London.

Kingman, J. F. C. (1982) The coalescent. *Stochast. Proc. Appl.* 13: 235–248.

Kreitman, M. and Wayne M. L. (1994) Organization of genetic variation at the molecular level: lessons from *Drosophila. In:* Schierwater, B., Streit, B., Wagner, G, P. and DeSalle, R.
(eds) *Molecular Ecology and Evolution: Approaches and Applications,* Birkhäuser, Basel,
pp. 299–321.

Lewontin, R. (1974) *The Genetic Base of Evolutionary Change,* Columbia University Press,
New York.

Maddison, D. L. (1995) Phylogenetic histories within and among species. *In*: Hoch, P. and Stepehenson, A. G. (eds) *Experimental and Molecular Approaches to Plant Biosystematics*, Missouri Botanical Garden, St. Louis, MO, pp. 273–287.

Mishler, B. and Donoghue (1982) Species concepts: a case for pluralism. *Syst. Zool.* 31: 493–503.

O'Brien, S. J. (1994) Perspectives on conservation genetics. *In*: Schierwater, B., Streit, B., Wagner, G, P. and DeSalle, R. (eds) *Molecular Ecology and Evolution: Approaches and Applications*, Birkhäuser, Basel, pp. 275–280.

Otte, D. and Endler, J. A. (1989) *Speciation and its Consequences*, Sinauer Associates, Sunderland, MA.

Rand, D. (1994) Concerted evolution and RAPping in mitochondrial VNTRs and the molecular geography of cricket populations. *In*: Schierwater, B., Streit, B., Wagner, G, P. and DeSalle, R. (eds) *Molecular Ecology and Evolution: Approaches and Applications*, Birkhäuser, Basel, pp. 227–246.

Smith, T. B. and Wayne, R. (1997) *Molecular genetic approaches in conservation*. Oxford University Press, Oxford.

Streit, B., Städler, T., Kuhn, K., Loew, M., Brauer, M. and Schierwater, B. (1994) Molecular markers and evolutionary processes in hermaphroditic freshwater snails. *In*: Schierwater, B., Streit, B., Wagner, G, P. and DeSalle, R. (eds) *Molecular Ecology and Evolution: Approaches and Applications*, Birkhäuser, Basel, pp. 247–260.

Templeton, A. R. (1989) The meaning of species and speciation: A genetic perspective. *In*: Otte, D. and Endler, J. A. (eds) *Speciation and its Consequences*, Sinauer Associates, Sunderland, MA, pp. 3–27.

Vogler, A. P. (1994) Extinction and the formation of phylogenetic lineages: Diagnosing units of conservation management in the tiger beetle *Cicindela dorsalis*. *In*: Schierwater, B., Streit, B., Wagner, G, P. and DeSalle, R. (eds) *Molecular Ecology and Evolution: Approaches and Applications*, Birkhäuser, Basel, pp. 261–274.

The role of molecular genetics in speciation studies

Alan R. Templeton

Department of Biology, Washington University, St. Louis, MO 63130-4899, USA

Summary

Systematists and population geneticists can both use molecular data sets to construct evolutionary trees (species and gene trees, respectively), and then use the resulting historical
framework to test a variety of hypotheses. The greatest prospect for future advances in
our understanding of speciation is to extend these historical approaches to the
species/population interface, for only by straddling this interface can we actually study the
processes involved in the origin of a new species. This chapter illustrates how the bottom-
up historical approaches used in population genetics can be extended upward to this critical interface in order to separate the effects of population structure from population history, to rigorously test the species status of a group and to test hypotheses about the
process of speciation by using gene trees to define a nested, statistical analysis of biogeographic and other types of data.

Introduction

Molecular genetics has long been applied to speciation studies, with some of the earliest protein electrophoretic studies focusing on problems of taxonomy and systematics (Hubby and Throckmorton, 1965; Johnson et al., 1966). This tradition of applying molecular genetics to problems related to species and speciation has continued and expanded, as is evident from the review article by Harrison (1991). This
chapter will focus on some issues not covered in detail in Harrison's review and upon newer techniques and approaches that have not yet had a major impact in the
written literature of this area but that are likely to play an important role in future
studies on species and speciation.

Harrison (1991) points out that the study of speciation falls at the interface between population and systematic biology and currently does not fit comfortably into either domain. Presently, much of the work in molecular systematics centers upon documenting character state distributions from either DNA sequence or restric-

tion site data to infer nonanastomosing lineages. Once these lineages have been identified and their evolutionary relationships estimated, the phylogenies can be used to test a variety of hypotheses about the pattern of macroevolution and speciation. This historical approach has proven to be an extremely powerful analytical technique in testing evolutionary hypotheses when species – the units of analyses in such studies – are well defined entities (Harvey and Pagel, 1991).

Molecular population geneticists have primarily focused upon the patterns of genetic variation found within and among subpopulations of interbreeding organisms in order to study microevolutionary forces such as gene flow, genetic drift and natural selection. Traditionally, these inferences were based upon the number of alleles (or haplotypes), their frequencies and their geographical distributions. However, as more and more of the genetic surveys have come to use DNA sequence or restriction site data, it has also been possible to estimate the genealogical structure of the alleles as well. With the rapid development of coalescent theory over the last decade (Hudson, 1990; Kingman, 1982a, b; Ewens, 1990), an increasingly rich theoretical framework is developing within population genetics for dealing with allele genealogies and allele frequency distributions in an integrated fashion. This microevolutionary historical approach has already proven to be a powerful tool for studying the relationship of genotype to phenotype (Templeton et al., 1987, 1988, 1992; Templeton and Sing, 1993; Templeton, 1995; Markham et al., 1996), natural selection (Antonarakis et al., 1984; Golding, 1987; Golding and Felsenstein, 1990; Hartl and Sawyer, 1991; O'Brien, 1991; McNearney et al., 1995) and population structure (Avise et al., 1988; Hudson et al., 1992; Slatkin, 1989; Slatkin and Maddison, 1989, 1990; Templeton, 1993; Templeton et al., 1995; Templeton and Georgiadis, 1996).

This common use of historical approaches by both molecular systematists and molecular population genetics offers the greatest hope for the synthesis of population genetic and systematic approaches to the study of speciation (Harrison, 1991). What is needed is for the historical approaches of the systematist to be extended downward to the species/population interface, and for the historical approaches of the population geneticist to be extended upward to this interface. Such extensions will not be easy. The molecular systematic approach works best when the species are well differentiated genetically, thereby facilitating phylogenetic resolution. In this regard, the *genetics of species differences* (i.e. the study of how genetic variation is partitioned among species regardless of the role, if any, that that variation played in the speciation process) must be clearly distinguished from the *genetics of speciation* (the study of genetic differences that directly contribute to the traits that allow new species to evolve) (Templeton, 1981). The more distant from the actual speciation process, the greater the expected genetic differences among the species, but the more difficult it becomes to infer what genetic differences were involved in speciation vs. what genetic differences were consequences of speciation (Templeton, 1981). The closer to the speciation process, the greater the ability to focus upon the genetics of speciation; but at the same time, traditional phylogenetic resolution breaks down.

From a population genetics perspective, this species/population interface is diffi-cult to deal with because of the need to distinguish in the allele genealogies between the effects of tokogenetic (birth) relationships among individuals within populations and phylogenetic relationships among populations. Part of this difficulty arises from the need to determine what patterns in the gene genealogy are due to population structure (recurrent, tokogenetic events, such as gene flow) vs. population history (nonrecurrent events that affect whole populations of individuals simultaneously, such as colonization and fragmentation events) (Templeton et al., 1995). A second source of difficulty arises from the sharing of molecular polymorphisms across species, the sorting of ancestral polymorphisms among species and interspecific in-trogression – all of which can lead to discordance between gene trees and species trees (Harrison, 1991). Thus, when studying a system close to or in the process of speciation, well-resolved molecular phylogenies of populations are not to be ex-pected.

Despite these difficulties, this interface still offers the greatest prospect for the use of molecular genetics for the problem of speciation (Harrison, 1991). This is not to say that molecular genetic studies that focus upon the genetics of species differ-ences are unimportant. As shown in Harrison's (1991) review, such studies have been and will undoubtedly continue to be a powerful tool in inferring species sta-tus, in estimating the evolutionary relationships among species and in testing many hypotheses about the speciation process well after it is completed. Because this use of molecular genetics was reviewed by Harrison (1991), this chapter will focus up-on the extension of historical approaches to the species/population interface. Furthermore, the only extension that will be discussed is the bottom-up extension of molecular population genetics to this interface, because some progress in making this extension has already been achieved.

Distinguishing population structure from population history

One common use of genetic surveys in population genetics is to study and quantify how genetic variation is distributed over geographic space within what is assumed to be a single species. When the genetic variation is also organized into an allele or haplotype genealogy, the resulting analysis of how geography overlays allele ge-nealogy has been called "intraspecific phylogeography" (Avise, 1989). Such analy-ses commonly find a strong association between the geographical location of hap-lotypes and their evolutionary position within a haplotype tree, but the demonstra-tion of such an association *per se* tells one very little about the causes of associa-tions. For example, does the population really constitute a single species with gene flow connecting all geographical regions, but with the gene flow sufficiently re-stricted to cause isolation by distance and hence geographical association among evolutionarily related haplotypes? Alternatively, did some subset of the population recently expand its range or undergo a colonization event, bringing along with it only a subset of its genetic variation into this new geographic area, thereby creating a geographical association (Larson, 1984)? Or was the population split into different subpopulations by fragmentation events in the past such that the subpopulations subsequently behaved as separate evolutionary lineages for a sufficient length of time to create strong current geographical associations (Larson, 1984)? A further complication is that these alternative explanations for geographical association are not mutually exclusive: all could be operating in the group being studied. Nevertheless, the implications of these different causes of geographical association for inferences about species status and speciation are quite different, so it is important to be able to discriminate among them and identify where they have occurred.

Fortunately, much more information than mere association can be gathered from a geographical overlay upon haplotype trees – different causes of geographical association can yield qualitatively different patterns that can be assessed through rigorous statistical testing (Templeton et al., 1995; Templeton and Georgiadis, 1996; Templeton, 1997a). This testing is achieved by first converting the haplotype tree into a nested series of clades (branches) by using the nesting rules given in Templeton et al. (1987) and Templeton and Sing (1993). The geographical data are then quantified in two main fashions: the clade distance, D_c, which measures the geographical range of a particular clade; and the nested clade distance, D_n, which measures how a particular clade is geographically distributed relative to its closest evolutionary sister clades (i.e. clades in the same higher-level nesting category). In particular, the clade distance measures the average distance that an individual bearing a haplotype from the clade of interest lies from the geographical center of all individuals bearing haplotypes from the same clade. The nested clade distance measures the average distance that an individual bearing a haplotype from the clade of interest lies from the geographical center of all individuals bearing haplotypes from the next higher-level nesting clade that contains the clade of interest. Contrasts in these distance measures between tip clades (clades that are not interior nodes in the haplotype tree) and the clades immediately interior to them in the haplotype network are important in discriminating the potential causes of geographical structuring of the genetic variation (Templeton et al., 1995), as will be discussed later. The statistical significance of the different distance measures and the interior-tip contrasts are determined by random permutation testing that simulates the null hypothesis of a random geographical distribution for all clades within a nesting category given the marginal clade frequencies and sample sizes per locality.

If statistically significant patterns are detected, they next need to be interpreted biologically. Templeton et al. (1995) consider three major biological factors that can cause a significant spatial or temporal association of haplotype variation. The first factor is restricted gene flow, particularly gene flow restricted by isolation by distance. Because restricted gene flow implies only limited movement by individuals during any given generation, it takes time for a newly arisen haplotype to spread geographically. Obviously, when a mutation first occurs, the resulting new haplotype is found only in its area of origin. With each passing generation, a haplotype lineage that persists has a greater and greater chance of spreading to additional locations via restricted gene flow. Hence, the clade distances should increase with time under a model of restricted gene flow. If an outgroup can be successfully used to root the haplotype tree, any series of nested clades can be polarized temporally in an unambiguous fashion. However, often intraspecific haplotype trees cannot be rooted reliably by the outgroup method or other standard rooting procedures (Templeton, 1993; Castelloe and Templeton, 1994). Fortunately, in a nested series of clades, a nesting clade has to be as old or older than all the lower level clades nested within it. Hence, as nesting level increases, there is a nondecreasing age series even when the root is not known. Accordingly, the clade distances are expected to increase with increasing nesting level. This expected increase will continue until either the highest nesting level is reached or, if the gene flow is sufficiently high relative to the coalescent time of the haplotype tree, a nesting level will be reached in which the clades are uniformly distributed over the entire sampled geographical range, and all higher nesting levels will replicate that pattern. Another aspect of the expected patterns under restricted gene flow is that when a mutation occurs to create a new haplotype, that new haplotype obviously resides initially within the range of its ancestral haplotype. Since the ancestral haplotype is older than its mutational offshoot, it should have a wider geographical distribution. Therefore, when the new haplotype starts spreading via gene flow, it will often remain within the geographical range of its ancestor for many generations, particularly under an isolation-by-distance model. Because there is a strong tendency for the ancestral haplotypes to be immediately interior to the derived haplotypes in terms of the topology of the haplotype network (Castelloe and Templeton, 1994), this means that there will be a strong tendency under restricted gene flow for tip clades to have a geographical range smaller and often nested within the range of the clades that are immediately interior to them. Moreover, because the ancestral haplotype is expected to be most frequent near its site of geographical origin, most mutational derivatives of the ancestral haplotype will also occur near the ancestral site of geographical origin. This means that the geographical centers of all the clades nested together should be close; hence, the clade distances and nested-clade distances should show similar patterns under restricted gene flow.

An example of the pattern expected from recurrent but restricted gene flow is illustrated by human mitochondrial DNA (mtDNA) at the intercontinental level, as pictorially summarized in Figure 1 (for a more detailed analysis, see Templeton, 1993). As can be seen, the same pattern of more restricted geographical patterns on the tips relative to the interiors occurs again and again at many clade levels. This fractal-like geographical pattern in the gene tree is strong evidence for a recurrent evolutionary force, such as gene flow.

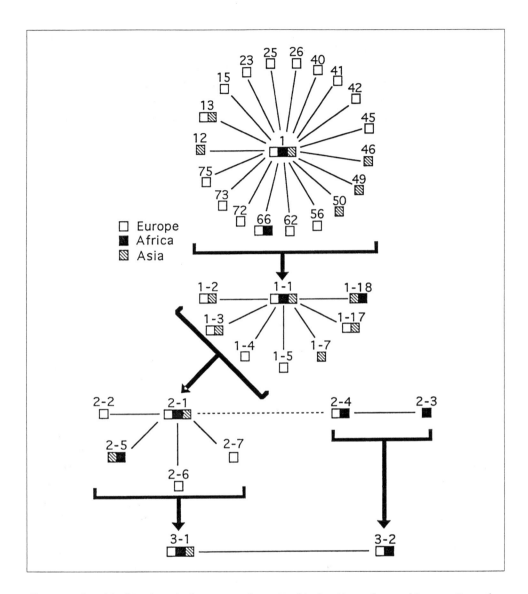

Figure 1. Graphical portrayal of a recurrent geographical pattern observed in a portion of the human mtDNA cladogram from Old World sample locations. Arabic numbers refer to haplotypes, 1-# refers to sets of haplotypes nested together into one-step clades, 2-# refers to one-step clades nested together into two-step clades, and 3-# refers to two-step clades nested together into three-step clades. All of the clades portrayed gave statistically signifi-cant deviations from the null hypothesis of no geographical association. Straight lines indi-cate a single restriction-site change in the mtDNA cladogram. Brackets with an arrow indi-cate the sets of lower-level clades that are pooled together into a single, higher-level clade. Details are given in Templeton (1993).

Range expansion (including colonization) is the second factor that can create a geographical association with the haplotype network. When range expansion occurs, those haplotypes found in the ancestral population(s) that were the source of the range expansion will become widespread geographically (large clade distances), and the distinction between the relative geographical ranges of tip vs. interior clades expected under restricted gene flow breaks down or can even be reversed. Moreover, some of the haplotypes found in the expanding populations can become quite distant from some of the older haplotypes that are confined to the ancestral, preexpansion area (large nested-clade distances), particularly when long-distance colonization is involved. As mutations first start to accumulate in the colonizing population, they will be tips with large nested-clade distances because the interior haplotypes from which they mutated will also be found in the ancestral range. Examples of this type of pattern yielding statistically significant inferences of range expansion and colonization through the nested-clade analysis are given in Templeton (1993, 1997a).

The third factor that can create associations between haplotype trees and geography is past fragmentation of a population into two or more subpopulations displaying little or no subsequent gene flow. When the nesting level reaches the temporal period at which the fragmentation event occurred, the clade distance cannot increase beyond the geographical ranges of the fragmented subpopulations, but the nested-clade distances will generally show a marked increase when the fragmented clades are allopatric, as is typically the case. If the fragmentation event is an old one relative to the rate at which mutations accumulate, the branch lengths between the clades displaying large nested-clade distances but plateaued clade distances will tend to be longer than the average branch length in the tree (due to the accumulation of mutations that differentiate the fragmented subpopulations). Examples of this type of pattern yielding statistically significant inferences of fragmentation through the nested-clade analysis are given in Templeton (1998a).

There is nothing about the evolutionary factors of restricted gene flow, range expansion events or fragmentation events that make them mutually exclusive alternatives. One of the great strengths of the nested-clade statistical test of Templeton et al. (1995) is that it explicitly searches for the combination of factors that best explains the current distribution of genetic variation and does not make *a priori* assumptions that certain factors should be excluded or be regarded as unlikely. Moreover, by using the temporal polarity inherent in a nested design (or by outgroups when available), the various factors influencing current distributions of genetic variation are reconstructed as a dynamic process through time. For example, consider the nested analysis of the tiger salamander (*Ambystoma tigrinum*) mtDNA data given in Routman (1993). In these salamanders, the lower clade levels primarily show the pattern expected under isolation by distance – a broadening of geographical distribution as one moves from tips to interior, as is illustrated in Figure 2A. When these geographical data were quantified and subjected to a nested

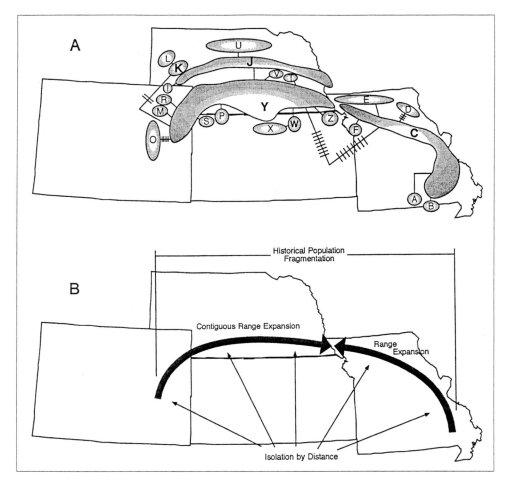

Figure 2. (A) The mitochondrial DNA cladogram for A. tigrinum *overlaid upon the sampling locations. Haplotypes are indicated by letters (A to Z) in this case, as described in Routman (1993). Lines with no tick marks indicate that the two haplotypes so joined differ by a single mutational change. For haplotypes connected by lines with tick marks, the number of tick marks indicates the minimum number of mutational changes. Most sample locations were polymorphic for more than one haplotype, which could not be easily portrayed, so the stippled areas only indicate a rough approximation to geographic extent and location of each haplotype. (B) A summary of the statistically significant inferences made with the A.* tigrinum *mtDNA cladogram.*

analysis, they yielded a statistically significant inference of restricted gene flow (Templeton et al., 1995). This observation is compatible with the fact that these salamanders are pond breeders and may display pond fidelity, which would result in restricted gene flow (Routman, 1993). However, a few clades showed a significant, nonrecurrent reversal of this pattern (Templeton et al., 1995). These clades contain the tip haplotypes with the elongated geographical distributions shown in Figure 2A. In these cases, the nested analysis and inference key identified two different significant range expansions, one in populations of the subspecies *A. tigrinum mavortium* and the other in populations of the subspecies *A. t. tigrinum*. In both cases, the inferred expanding clades are now found in areas uninhabitable during the Pleistocene, so range expansions had to have occurred over the last 18,000 years. Finally, at the four-step clade level, further broadening of geographical distribution abruptly ceases; yet, the geographical centers of the two clades at this level are distant from one another (Fig. 2A). The two clades defined at this level differ minimally by 14 steps (Fig. 2A), whereas most other steps in the mtDNA cladogram differ by only one restriction site change. These two clades identify a statistically significant fragmentation event (Templeton *et. al.*, 1995) and correspond to a western group and an eastern group with only a narrow zone of overlap in northwestern Missouri (Fig. 2A). These two clades also correspond to two named subspecies (*A. t. mavortium* in the west, and *A. t. tigrinum* in the east) that were most likely separated during the last glaciation (Routman, 1993). Hence, the current geographical distribution of mtDNA haplotypes in this salamander reflects the joint action of recurrent but restricted gene flow and the historical events of Pleistocene fragmentation followed by postglacial range expansions (Templeton et al., 1995). These statistically significant inferences are summarized in Figure 2B.

The results reported here and elsewhere (Templeton et al., 1995; Templeton and Georgiadis, 1996; Templeton, 1998a) indicate that population structure can be distinguished from population history with molecular data. This is critical for the study of speciation, because many proposed speciation mechanisms require population range expansion, colonization or fragmentation; whereas evidence of recurrent gene flow would argue against speciation having occurred.

Species inference through testing null hypotheses

The first issue that needs to be addressed when studying speciation is deciding what is or is not a species. Until one has identified the taxa that constitute "species", studies on speciation are impossible. The dominant species concept in practice has been and continues to be the morphological species concept simply because morphological data are the most abundant. However, this practical dominance does not obviate the need for a theoretical species concept as a tool for understanding evolutionary processes, inferring generalities, serving as a guide for research programs and generating hypotheses. However, for a theoretical species concept to adequately serve these purposes, it must also be capable of practical implementation. Molecular data can be used in a powerful way to achieve such an implementation.

All theoretical species definitions seek some biological universal that is not tied to particular cases. Two major universals have been used most extensively: the idea of a reproductive community and the idea of an evolutionary lineage. The universal of a reproductive community is used by the biological species concept (Mayr, 1992), in which the boundaries of that community are defined by reproductive isolating mechanisms, and by the recognition concept (Paterson, 1985), in which the boundaries are defined by shared fertilization systems. The universal of an evolutionary lineage, which is more of a true biological universal than is a reproductive community (Templeton, 1989), is used by the evolutionary species concept (Wiley, 1981), the phylogenetic species concept (Cracraft, 1989) and the cohesion species concept (Templeton, 1989). Because modern molecular data (particularly DNA sequence and restriction site data) contain much phylogenetic information, the greatest impact of molecular genetics will undoubtedly be upon those species concepts that use the universal of an evolutionary lineage. This potential impact will be illustrated by a closer examination of the cohesion species concept.

A cohesion species is an evolutionary lineage that serves as the arena of action of basic microevolutionary forces, such as gene flow (when applicable), genetic drift and natural selection (Templeton, 1989). Under the cohesion species concept, the boundaries of an evolutionary lineage are defined by the mechanisms that limit the action of gene flow, genetic drift and natural selection. These lineage-defining mechanisms in turn fall into two major categories. The existence of a lineage by definition requires reproduction, and the two classes of cohesion mechanisms focus respectively upon the genetic and demographic attributes of reproduction in the lineage. The first are those mechanisms that determine the boundaries of *genetic exchangeability* (reproductive isolating mechanisms or fertilization mechanisms). Genetic exchangeability directly determines the boundaries for gene flow and can have a powerful, sometimes the dominant, impact on the limits of action of drift and selection. The second class of mechanisms is concerned with the basic *demography* or *ecology* of reproduction. If organisms are in some sense equivalent or interchangeable in their demographic or ecological reproductive attributes, their descendants or genes can either replace (through drift) or displace (through selection) the descendants or genes of other individuals in the lineage even if the lineage is not reproducing sexually. Hence, the limits of drift and selection as evolutionary defining mechanisms can be influenced by (and, in genetically closed systems of reproduction, completely determined by) the demographic attributes of reproduction. Therefore, the derived adaptations and ecological attributes that are shared by the members of the lineage and that determine or constrain the demography of reproduction define the boundaries of demographic interchangeability.

The cohesion species is therefore a population of organisms that constitute a distinct evolutionary lineage that also represents a reproductive community in either a genetic or adaptational/ecological sense. To implement this concept, it can be rephrased as a set of testable null hypotheses [for more details, see Templeton

(1998b)]. The first null hypothesis to be tested is that all organisms under investigation constitute a single evolutionary lineage. If this null hypothesis is not rejected, there is no evidence for more than one cohesion species and the implementation procedure is terminated. If this null hypothesis is rejected, the population constitutes two or more diagnosable evolutionary lineages.

Molecular data have had and will continue to have a major impact on testing this null hypothesis of only one lineage. For example, Routman (1993) concluded that the two subspecies of tiger salamander in his study were distinct evolutionary lineages because they defined geographically contiguous, monophyletic and genetically well differentiated groupings. The statistical analysis of Templeton et al. (1995) supports the conclusion of two lineages defined by a past fragmentation event (Fig. 2). This nested-clade statistical test therefore provides a rigorous and objective manner of identifying evolutionary lineages. Whenever the inference key given in Templeton et al. (1995) leads to the conclusion of fragmentation from a statistically significant pattern of spatial distance measures, this first null hypothesis in inferring species has been rejected. This testing procedure has the property of being able to infer fragmentation events even when the fragmented populations are polyphyletic or paraphyletic to one another with respect to the gene tree due to recent shared ancestry. For example, Templeton (1998a) analyzed the human mtDNA sequence data given in Torroni et al. (1993a, b) on Siberians and Amerindians. The nested analyses of spatial distances of Templeton et al. (1995) led to several statistically significant results, and the inference key given as an appendix to Templeton et al. (1995) led to the conclusions of one or two colonization events from Siberia to North America, followed by contiguous range expansion throughout the Americas and fragmentation of the Amerindian populations from their Asian ancestors (Templeton, 1998a). Two non-nested clades inferred the American colonization event, implying either that more than one colonization event occurred or that the colonizing population was large and carried over much ancestral polymorphism. In either event, Amerindians are definitely polyphyletic with respect to Asians in the mtDNA haplotype tree. It is the nested nature of the statistical test of Templeton et al. (1995) that allows it to successfuly deal with fragmentation events that are still polyphyletic or paraphyletic due to ancestral polymorphisms. This is a critical property of any statistical test being used to study species and speciation, because these cases of discordancy between gene trees and population-level fragmentation events are expected to be common in those cases that are closest to the speciation process and hence potentially the most valuable in understanding the nature of that process.

Another complication in inferring more than one lineage occurs when there is sporadic hybridization among the lineages. Once again, when dealing with populations that are actively undergoing speciation, incomplete reproductive isolation will frequently be encountered. Fortunately, a nested analysis of the gene tree can deal with this complication as well. An example of such a case is provided by the work of Matos (1992) on Mexican pine trees. Matos (1992) studied restriction-site vari-

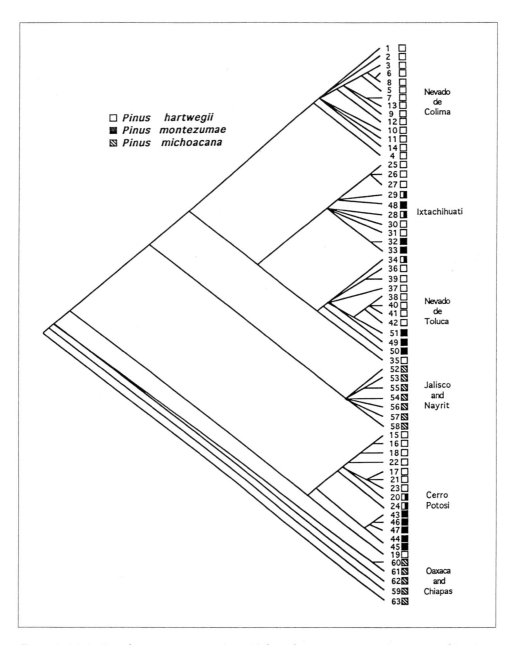

Figure 3. Majority rule consensus tree (>50%) for P. hartwegii, P. montezumae *and* P. michocana *chloroplast DNA as surveyed with restriction mapping. The cladogram was obtained using the program PAUP using a heuristic search with multiple replications of stepwise random addition, restriction site gains vs. losses weighted 1:1.3, no outgroup, and midpoint rooting. See Matos (1992) for more details.*

ation in chloroplast DNA (cpDNA) among Mexican pine trees. Three morphologically distinct prior categories (*Pinus hartwegii*, *P. montezumae* and *P. michocana*) are considered, all of which are interfertile. One taxon is allopatric (*P. michocana*), and the other two are sympatric in the sense that most individual trees live within pollen dispersal distance of individuals of the other category (cpDNA is inherited through the pollen in pine trees). Figure 3 shows the majority rule consensus tree using a standard phylogenetic analysis based on parsimony using the program PAUP. As can be seen, the species do not sort out in a simple fashion, and some haplotypes are even shared between the two sympatric taxa. However, another way of estimating the cladogram that simultaneously evaluates statistical confidence is given in Templeton et al. (1992). This method was specifically developed for gene trees and generates a plausible set of cladograms (that is, it includes all linkages among haplotype pairs until the cumulative probability of all the linkages is greater than 0.95). Figure 4 gives the plausible set of cladograms estimated by Matos (1992) using the algorithm of Templeton et al. (1992).

To test the null hypothesis that the prior taxonomic categories (defined by morphological criteria) have no phylogenetic associations, Matos (1992) converted the plausible set of cladograms into a nested statistical design going up to five-step clades. The design incorporates cladogram uncertainty and allows one to perform a nested categorical data analysis using the methods given in Templeton and Sing (1993). Highly significant associations were found at all clade levels except for the four-step level. As can be seen from inspecting Figure 4, tip clades are predominantly of a single categorical type, and even the heterogeneity of taxonomic categories found within interior clades is nonrandomly distributed. Only at the four-step clade level does one get random distributions of categorical types across the members of the higher-order clades. The randomness at this clade level involves only the sympatric populations, which were sampled on three isolated mountaintops. When the analysis is extended to five-step clades, significant associations again reappear.

The lack of a significant effect at the four-step level for the sympatric taxa indicates that introgression among taxa has been important in this group of pine trees. That this is introgression and not sorting of ancestral polymorphisms is indicated by the fact that a separate coalescence to a shared pollen ancestor occurs for each of the three isolated mountaintops; that is, interspecific coalescence and sympatry are completely associated. Nevertheless, this analysis also indicates that such interspecific introgression is not behaving as a recurrent, lineage-defining force. As can be seen from Figure 4, there is evidence for only a single interspecific introgression on each mountaintop, following which there can be some shared polymorphism among taxa but with all subsequent mutational and coalescent events being strictly confined to the named taxa. Hence, on the time scale measured by mutational change in cpDNA, interspecific introgression behaves as an evolutionarily sporadic factor that is at least two orders of magnitude less common than the recurrent forces of mutation and coalescence operating within each of these taxa. Therefore, the nest-

143

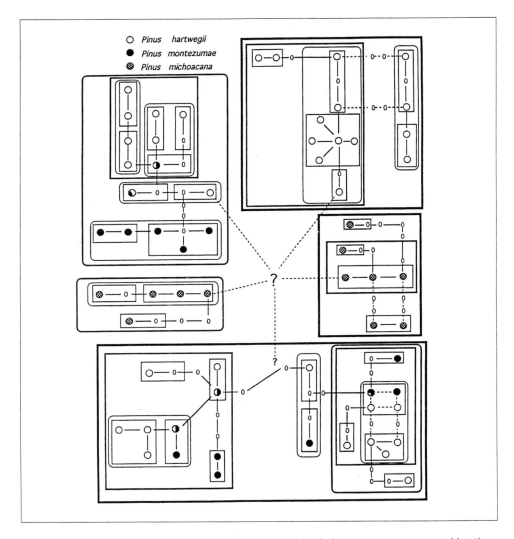

Figure 4. The unrooted chloroplast DNA 95% plausible cladogram set as estimated by the procedure of Templeton et al. (1992). Each solid line represents a single restriction-site change. A "0" represents an inferred intermediate haplotype that was not found in the sample. Dashed lines represent multiple restriction-site changes. The question mark indicates extreme ambiguity in how these dashed lines connect to one another. The taxa in which each haplotype is found is indicated by an appropriately shaded box. Boxes indicate the nested design structure used in the statistical analyses associated with these pine trees. Rectangular, narrow-lined boxes indicate one-step clades; narrow-lined boxes with rounded corners represent two-step clades, double-thick lined rectangles represent three-step clades, double-thick lined boxes with rounded corners represent four-step clades and triple-thick lined rectangles represent five-step clades. Degenerate clades are not indicated. See Matos (1992) for details.

ed analyses indicate that these three taxonomic categories of pine trees are indeed behaving as three different evolutionary lineages despite occasional hybridization and introgression among the sympatric lineages.

After testing for the existence of evolutionary lineages, the question still remains as to which lineages or sets of lineages should be elevated to species status. Being an evolutionary lineage is a necessary, but not sufficient, condition to establish a population as a species under the cohesion concept. Given that two or more evolutionary lineages have been inferred, one or both of two additional null hypotheses must be tested to infer a cohesion species: (i) the evolutionary lineages are genetically exchangeable (if sexual outcrossers), and/or (ii) the evolutionary lineages are demographically interchangeable. An evolutionary lineage is elevated to cohesion species status only when one or both of these additional null hypotheses is/are rejected. An example of testing cohesion species status of lineages through genetic exchangeability is given in Templeton (1998b), so I will limit this presentation to demographic interchangeability.

As mentioned above, demographic interchangeability arises from the shared fundamental adaptations of a population and can be reflected by the fundamental niche or selective regime (*sensu* Baum and Larson, 1991) that the population occupies. It is essential that the evaluation of these traits occur in a historical context (Baum and Larson, 1991), and once again molecular data can play a critical role by defining a nested, historical design. The null hypothesis of demographic interchangeability will be tested by overlaying adaptations or selective regimes upon the gene trees using the same type of statistical analyses used to study gene tree associations with geographical location or taxonomic categories.

As an example, consider the *A. tigrinum* case in which the cladistic geographical analysis indicated the existence of two evolutionary lineages. In terms of demographic interchangeability, there are several major adaptive traits that are variable within these salamanders (Collins et al., 1993), one of which is facultative neoteny. In their normal lifehistory, tiger salamanders have an aquatic larval phase followed by a metamorphosis into a sexually mature terrestrial phase. However, some adults can become sexually mature while remaining in the aquatic phase if they are inhabiting a relatively permanent pond – a phenomenon known as facultative neoteny. Neoteny is an extensively studied phenomenon in salamanders and has profound adaptive and ecological consequences (Gould, 1977). When the salamanders were collected for the genetic studies, the number of neotenic individuals was also recorded (Routman, 1993). Almost all salamanders were collected in cattle or large ponds, so there is little if any environmental heterogeneity in the stability of the aquatic environment for this collection (which does not imply that the overall stabilities of the aquatic environments are equivalent for these two "subspecies"). Furthermore, crosses with the neotenic *A. mexicanum* to both of these subspecies revealed that the difference in facultative neoteny is genetic and consistent with a single locus of major effect (Shaffer and Voss, 1996).

The ecological variable of interest for the salamanders is therefore the numbers of neotenic and metamorphosed adults in the sample. These data can also be subjected to a cladistic nested-contingency analysis as described in Templeton and Sing (1993). The results were very clear: every neotenic salamander collected was in the western lineage, leading to a highly significant rejection of the null hypothesis of demographic interchangeability across these two lineages. Therefore, these two tiger salamander lineages are different species under the cohesion concept.

For the Mexican pine trees, some data on selective regimes exists in the form of the elevations of the collection sites. Elevation is potentially a quantitative variable, but in this case, each mountaintop was sampled in a categorical fashion (Matos, 1992). Matos (1992) therefore performed the nested permutational contingency analysis as described in Templeton and Sing (1993). She detected highly significant associations between the cpDNA clades and selective regime that were in turn associated with taxonomic category. In particular, *P. hartwegii* clades are found at high elevations, while *P. montezumae* clades are significantly associated with low elevations. Hence, the prior categories are not only significant evolutionary lineages, but they have significant ecological distinctness as well despite an evolutionary history of sporadic hybridization between taxa and fragmentation across mountaintops. Because these elevational categories are within pollen dispersal distance on each mountaintop and the same ecological pattern is observed in each of the three isolated mountaintops, these differences are most likely genetically determined. Therefore, the prior taxonomic categories in this case are inferred to be statistically significant cohesion species on the basis of a significant lack of demographic interchangeability.

Note that not all diagnosable lineages (*sensu* Cracraft, 1989) were elevated to species status in the pine tree example (see Fig. 4). Moreover, because of the complications of interspecific introgression in this case, the lineages grouped together as species do not define monophyletic groups relative to the gene tree. It is precisely this robustness to discrepancies between gene and species trees that is essential when working at the population/species interface.

Although the above discussion was limited to the cohesion concept, other species concepts can also be implemented as a set of testable hypotheses (e.g. Cracraft, 1989). There are many advantages to this approach. First, phrasing a species concept as a set of testable hypotheses provides an explicit research program for future studies on species and speciation. By examining which hypotheses have been tested and which have not, and the quantitative results of the test statistics, an investigator can quickly focus upon those aspects of the data or the analyses that are the weakest. This provides a guideline for additional data to be gathered. Moreover, as additional data are gathered, they can be integrated into the same inference scheme as the previous data sets. These additional data are then subjected to rigorous quantitative analyses, so subjective assessments as to whether particular data weaken or strengthen the previous inferences are avoided.

Another advantage is that the criteria, data, methods of analyses and degree of support are made completely explicit. This allows the inferences to be easily evaluated by others, even those who advocate other species concepts. It also insures that the practical implementation of the species concept is consistent with the concept's theoretical basis. For example, Mayr (1992) investigated the applicability of the biological species concept to higher plants. He presented two criteria for biological species, "interbreeding communities" and "reproductively isolated from other such groups", and concluded that the biological species concept worked well in 93.6% of the 838 species native to the town of Concord, New Hampshire. Although he excluded apomictic plants as good examples of biological species, he did include autogamous species which do not define "interbreeding communities" (Templeton, 1983, 1989). On the other hand, Mayr (1992) also included as good biological species taxa that are members of well-studied syngameons even though they are not "reproductively isolated from other such groups". Autogamous species are included because there were no "difficulties in species recognition in a local area" (Mayr, 1992, p. 236), and syngameon members are included as long as hybridization does not obliterate the morphological "integrity of the parent species" (p. 233). The remaining species in this study were judged to fit the biological species concept solely on the grounds that "they raise no problems of species identity", often because they are the only representative of their genus in the local area. Thus, in practice, the only use of the biological species concept by Mayr (1992) was to exclude apomictic species. All species supposedly identified as being good "biological species" were actually judged using a strictly morphological species concept (Whittemore, 1993), even to the extent that morphological criteria overrode the stated criterion of "interbreeding" and "isolation" whenever a potential conflict was noted (autogamous and syngameon species). By rephrasing a species concept as a set of testable hypotheses, this type of discrepancy between theoretical and functional definitions of species can be avoided. Moreover, by subjecting the stated criteria to explicit quantitative testing with the available data, the subjective assessments of what is "significant" and what is not can also be avoided. All species concepts would benefit from such a rephrasing, and molecular data will play an increasingly important role in implementing species concepts.

Testing hypothesis about the speciation process

After species have been defined, one can address the issue of how species originated (Templeton, 1989). Modes of speciation have been classified by either geographical (Bush, 1975) or evolutionary-mechanistic (Templeton, 1981) criteria. The geographical and mechanistic classifications are not completely independent, as some of the mechanisms have geographical implications. As a consequence, it is sometimes possible to use geographical data in a historical analysis to reject certain mechanisms of speciation, although not necessarily to prove a particular mechanism (as will be illustrated later in this section with studies on Hawaiian swordtail crickets in the genus *Laupala*).

Although geographical and mechanistic modes are not independent, it is important to emphasize that the geographical and mechanistic classifications of speciation modes cannot be equated. For example, Lynch (1989) presents a method for testing among three different geographical modes of speciation: vicariant (allopatric type I in Bush's 1975 terminology, involving a widespread ancestral species fragmented into geographically isolated daughter species, each of large size), peripheral isolates (or peripatric, corresponding to allopatric type II in Bush's 1975 terminology, involving a widespread parental species with daughter species on the periphery of the parental range, with the daughter species having a much more limited geographical range) and sympatric speciation, in which speciation occurred without geographical isolation. Although the geographical mode of peripatric speciation is commonly equated to the mechanistic mode of founder-induced speciation (genetic transilience in the terminology of Templeton, 1981) (Mayr, 1982, 1954), these two classifications of modes of speciation are not the same. For example, one of the more compelling examples of genetic transilience is found within the Hawaiian *Drosophila* (Carson and Templeton, 1984). The founding events involve interisland dispersal followed by a large increase in population size and range. Indeed, this increase in population size is a necessary component of the mechanism of genetic transilience (Carson and Templeton, 1984; Templeton, 1980). As a consequence, the Hawaiian *Drosophila* speciation events that arise from interisland founder events result in well-separated, allopatric species in which the daughter species tends to be as widespread as the parental species (or in many cases even more widespread, because the newer islands tend to be larger than the older islands). This pattern clearly satisfies the criteria of vicariant speciation and not peripatric speciation in Lynch's methodology. As this example shows, geographical criteria alone cannot be used to infer the evolutionary mechanisms responsible for the speciation process, and Lynch (1989) wisely and correctly limited his conclusions strictly to the geographical modes.

The Hawaiian *Drosophila* reveal a more serious flaw in Lynch's (1989) methodology: in these flies the ancestral species undergoes no change in geographical range, and the daughter species is established by dispersal. Hence, the geographical pattern that Lynch (1989) describes for vicariant speciation (vicariant speciation is defined by Lynch as that involving a widespread ancestral species that is fragmented into daughter species) clearly can be derived from the mode of speciation that he describes as peripatric speciation. The Lynch method therefore cannot be used even to infer geographic modes of speciation or their frequencies.

Lynch's method and other vicariant biogeographical methods fail as a tool for inferring modes of speciation for two primary reasons. First, such methods use only species phylogenies and work best when the species are well resolved, thereby avoiding the species/population interface that Harrison (1991) described as so critical for understanding the mechanisms of speciation. Such macroevolutionary, top-down approaches can be quite useful for testing ideas about factors that influence the rate or pattern of speciation (e.g. Mitter et al., 1988, 1991; Farrell and Mitter, 1990; Page, 1990), but usually offer little insight into the microevolutionary details of how speciation is achieved.

The second problem with Lynch's method and vicariance biogeoraphy in general is the static way in which it treats geographical data. Geographical range is treated as a static species character that rarely changes. Lynch (1989) can infer vicariant speciation from a geographical pattern that, as we have just seen, can arise from peripatric speciation because he *assumes* a static geographical range (Lynch, 1989, p. 547), and not upon data analysis. Yet other historical studies indicate that species ranges can change dramatically (e.g. Hillis, 1985; Templeton, 1998a). Moreover, the very geographical pattern that he uses to infer vicariant speciation is itself often an artefact of this assumption. Crandall (1993) points out that by reconstructing ancestral species ranges as the union of the ranges of the descendant species, there is an automatic bias in favor of vicariant speciation.

What is needed to avoid these problems is a microevolutionary, bottom-up approach that extends through the population/species interface. Such a bottom-up approach is not a replacement for the top-down, macroevolutionary approach, but rather is a complement to it. The methodology for such a bottom-up approach has already been outlined in the previous section and is based upon obtaining haplotype trees from extensive sampling both below and above the species level. To see the utility of this approach, the tiger salamander example of the previous section will be reexamined. Under the cohesion or phylogenetic (Cracraft, 1989) species concept, there are two species that correspond to two named "subspecies" at this time, *mavortium* (in the west) and *tigrinum* (in the east). An analysis such as Lynch's would simply conclude that an ancestral species exists that covered the entire east-west range and that it was fragmented into eastern and western daughter species. Note, however, that no evidence that can discriminate between fragmentation and dispersal is actually presented or used; dispersal is simply assumed not be the explanation for the current pattern (Lynch, 1989, p. 547). In contrast, the nested geographical distance analysis based upon the mtDNA haplotype tree treats geography as a dynamic entity and attempts to explain the current patterns of haplotype biogeography by reconstructing this dynamic geographical history. In this case, fragmentation of the eastern from the western populations was also inferred, but in a manner that could discriminate among the alternatives. Hence, the hypothesis of fragmentation is tested, not assumed.

Moreover, within both subspecies of salamander, the analysis indicated that geographical range was not static, but very dynamic after the fragmentation event, with the eastern population undergoing a range expansion along a south-north axis, and with the western population undergoing a range expansion along a west-east arc bending towards the north. Furthermore, within each population there is also considerable isolation by distance. Note that far more insight and detail into the geographical components of these salamanders' recent evolutionary history can be achieved in this way, that biogeography is treated as an ever-changing dynamic entity that can be altered by many forces or events and most important, that alternative explanations for the current geographical pattern are tested, not merely eliminated by assumption. By performing this nested geographical distance analysis, the fundamental types of events that are important in the speciation process can be identified and the strength of data support quantified, including such factors as fragmentation, contiguous range expansion and long-distance dispersal or colonization events (Templeton et al., 1995; Templeton, 1998a). Often, this information alone can identify the mode of speciation, or at least eliminate many possibilities. For example, the nested geographical distance analysis indicates with strong statistical support that the east-west tiger salamander speciation event falls into the vicariant speciation mode of Lynch (1989). In terms of the mechanistic classification of speciation, this biogeographic pattern eliminates all but adaptive divergence and genetic transilience (because we do not know what the inbreeding and variance effective sizes of the fragmented isolates were) or a mixture of these two mechanisms (note, unlike the geographical modes of speciation, the mechanistic modes are not necessarily mutually exclusive; more than one can combine to influence a single process of speciation just as more than one microevolutionary force, such as selection and drift, can combine to influence a single anagenetic process – see Templeton, 1981).

Another advantage of this bottom-up approach is that it is not limited just to geographical data. The same haplotype cladogram can be used as a statistical design to study morphological and ecological data and to test patterns of concordance or discordance among these different data sets. As pointed out by Crandall (1993), more resolution into the mechanisms of speciation is often possible by using these nongeographical data sets along with geography than is possible using any one data set alone. For example, Shaw (1993) constructed an mtDNA sequence cladogram that spanned the population/species interface in the genus *Laupala*, an endemic group of Hawaiian swordtail crickets. Using the exact nested-contingency analysis of Templeton and Sing (1993), Shaw (1993) detected significant associations with the prior taxonomic categories developed by Otte (1989) on the basis of male calling songs. As with the pine tree work of Matos (1992), many of the prior species categories were not monophyletic in the estimated cladogram set, but in this case there was no sharing of haplotypes across species designations in areas of sympatry, and the lack of monophyly was due to paraphyly and/or polyphyly among allopatric populations from different species categories. Hence, unlike the pine tree example, this pattern does not seem to involve any hybridization among the prior species categories, but rather is indicative of lineage sorting of ancestral polymorphisms in these recently evolved categories. Besides this evidence for ancestral mtDNA polymorphisms being carried across speciation events, the geographical pattern of speciation in this group (using the prior categories of Otte, 1989) was that almost all speciation occurred within islands. This pattern is in great contrast to the Hawaiian *Drosophila*, in which mtDNA polymorphisms are not carried across speciation events and in which most speciation is associated with interisland transfer events (Carson and Templeton, 1984; DeSalle and Templeton, 1988). The *Laupala* pattern is therefore inconsistent with the transilient mechanisms of speciation (Templeton, 1981), in great contrast to the Hawaiian *Drosophila*.

The *Laupala* therefore most likely speciated through a divergence mechanism (Templeton, 1981), and Otte (1989) has further argued that reproductive character displacement of the pulse period of the male calling song has played an important role in the speciation process of this group. Shaw (1993) tested that possibility by laying the male pulse period phenotype over her mtDNA cladogram. She then identified comparisons for species in sympatry vs. species in allopatry and reconstructed the change in pulse period from the ancestral condition using both linear parsimony and squared-change parsimony. Mann-Whitney U-tests were then performed on the amount of evolutionary change in this phenotype as estimated by these two parsimony procedures in order to test the effects of sympatry versus allopatry. The results are shown in Table 1. Both methods of estimating the amount of phenotypic change gave similar results, with sympatric populations showing more evolutionary change than allopatric ones, at the 0.077 significance level for linear parsimony and the 0.042 significance level for squared-change parsimony. Hence, both methods of phenotypic change estimation yield results that hover around the 5% significance level, giving borderline support to the hypothesis of reproductive character displacement.

Table 1. Median male calling song pulse rate ancestor-descendant differences for Laupala species in sympatry vs. species in allopatry under linear parsimony and squared-change parsimony character state estimates. Details are in Shaw (1993).

Geographical setting	Sample size	Median	Average rank	Sig. level[*]
Linear parsimony				
Sympatry	17	0.1520	14.5	
Allopatry	8	0.0260	9.9	
Overall	25	-	13.0	0.077
Squared-change parsimony				
Sympatry	17	0.1970	14.8	
Allopatry	8	0.0575	9.2	
Overall	25	-	13.0	0.042

[*]The statistical significance is determined by a Mann-Whitney U-Test.

The above example illustrates that more insight into the process of speciation can be obtained when the historical analyses treat geographical range as a dynamic entity subject to a variety of alterations, and when more than just geographical data are included in the analysis. Molecular data sets are critical in these analyses, because molecular data are one of the few types of data that can be used to estimate a historical framework for analytical design that straddles the species/population interface. Only by straddling this interface is it possible to study the details of the modes of speciation.

Conclusions

Harrison (1991) pointed out that progress in applying molecular genetics to the problem of speciation requires studies at the population/species interface. In this chapter, this interface was approached from below, examining how allele or haplotype trees can be partially influenced by within-species, tokogenetic factors, and population-level, phylogenetic events. A statistical methodology originally developed for intraspecific studies can be extended upward to make inferences from haplotype trees at this interface from a variety of data types, including but not limited to biogeographic data. The results outlined in this chapter indicate that much information exists in molecular data about the meaning of species and the process of speciation, and that some of this information can be extracted by exploring the population/species interface with bottom-up statistical tools. As more attention is paid to this interface, a greater variety of tools will undoubtedly be developed to investigate this critical interface. In particular, one great need is to extend downward the top-down historical approaches traditionally used to study problems above the species level (Harvey and Pagel, 1991) to this population/species interface, just as this chapter shows how the bottom-up, intraspecific historical approaches could be extended up to this interface. The refinement of analytical tools at this difficult level of biological organization will make the future of molecular studies about the speciation process a most exciting one.

Acknowledgements

This work was supported by NIH grant R01 GM31571. Many of the analyses reported in this chapter were performed while the author was a visiting fellow at Merton College, University of Oxford, and the support of Merton College is gratefully acknowledged. This work was greatly strengthened by comments upon an earlier draft by Eric Routman, Christopher Phillips, Nicholas Georgiadis and Anne Gerber, I wish to thank all of them for their excellent suggestions.

References

Antonarakis, S. E., Boehm, C. D., Serjeant, G. R., Theisen, C. E., Dover, G. J. and Kazazian, J. (1984) Origin of the β^S-globin gene in Blacks: the contribution of recurrent mutation or gene conversion or both. *Proc. Natl. Acad. Sci. USA* 81: 853–856.

Avise, J. C. (1989) Gene trees and organismal histories: a phylogenetic approach to population biology. *Evolution* 43: 1192–1208.

Avise, J. C., Ball, R. M. and Arnold, J. (1988) Current versus historical population sizes in vertebrate species with high gene flow: a comparison based on mitochondrial DNA lineages and inbreeding theory for neutral mutations. *Molec. Biol. Evol.* 5: 331–344.

Baum, D. A. and Larson, A. (1991) Adaptation reviewed: a phylogenetic methodology for studying character macroevolution. *Syst. Zool.* 40: 1–18.

Bush, G. L. (1975) Modes of speciation. *Annu. Rev. Ecol. Syst.* 6: 339–364.

Carson, H. L. and Templeton, A. R. (1984) Genetic revolutions in relation to speciation phenomena: the founding of new populations. *Annu. Rev. Ecol. Syst.* 15: 97–131.

Castelloe, J. and Templeton, A. R. (1994) Root probabilities for intraspecific gene trees under neutral coalescent theory. *Mol. Phylogenet. Evol.* 3: 102–113.

Collins, J. P., Zerba, K. E. and Sredl, M. J. (1993) Shaping intraspecific variation: development, ecology and the evolution of morphology and life history variation in tiger salamanders. *Genetica* 89: 167–183.

Cracraft, J. (1989) Species as entities of biological theory. *In*: Ruse, M. (ed.) *What the Philosophy of Biology is*, Kluwer, Dordrecht, Netherlands, pp. 31–52.

Crandall, K. A. (1993) *A phylogenetic analysis of speciation in the crayfish subgenus Procericambarus (Decapoda: Cambaridae)*, Ph.D. thesis, Washington University, St. Louis, Missouri.

DeSalle, R. and Templeton, A. R. (1988) Founder effects and the rate of mitochondrial DNA evolution in Hawaiian *Drosophila*. *Evolution* 42: 1076–1084.

Ewens, W. J. (1990) Population genetics theory – the past and the future. *In*: Lessard, S. (ed.) *Mathematical and Statistical Developments of Evolutionary Theory*, Kluwer, Dordrecht, Netherlands, pp. 177–227.

Farrell, B. and Mitter, C. (1990) Phylogenesis of insect/plant interactions: Have Phyllobrotica leaf beetles (Chrysomelidae) and the Lamiales diversified in parallel? *Evolution* 44: 1389–1403.

Golding, G. B. (1987) The detection of deleterious selection using ancestors inferred from a phylogenetic history. *Genet. Res.* 49: 71–82.

Golding, B. and Felsenstein, J. (1990) A maximum likelihood approach to the detection of selection from a phylogeny. *J. Mol. Evol.* 31: 511–523.

Gould, S. J. (1977) *Ontogeny and Phylogeny*. Belknap Press, Cambridge, MA.

Harrison, R. G. (1991) Molecular changes at speciation. *Annu. Rev. Ecol. Syst.* 22: 281–308.

Hartl, D. L. and Sawyer, S. A. (1991) Inference of selection and recombination from nucleotide sequence data. *J. Evol. Biol.* 4: 519–532.

Harvey, P. H. and Pagel, M. D. (1991) *The Comparative Method in Evoluionary Biology*, Oxford University Press, Oxford.

Hillis, D. M. (1985) Evolutionary genetics of the Andean lizard genus *Pholidobolus* (Sauria: Gymnophthalmidae): phylogeny, biogeography and a comparison of tree construction techniques. *Syst. Zool.* 34: 109–126.

Hubby, J. L. and Throckmorton, L. H. (1965) Protein differences in *Drosophila*. II.

Comparative species genetics and evolutionary problems. *Genetics* 52: 203–215.

Hudson, R. R. (1990) Gene genealogies and the coalescent process. *Oxf. Surv. Evol. Biol.* 7: 1–44.

Hudson, R. R., Slatkin, M. and Maddison, W. P. (1992) Estimation of levels of gene flow from DNA-sequence data. *Genetics* 132: 583–589.

Johnson, F. M., Kanapi, C. G., Richardson, R. H., Wheeler, M. R. and Stone, W. S. (1966) An operational classification of *Drosophila* esterases for species comparisons. *In*: Wheeler, M. R. (ed.) *Studies in Genetics. III. Morgan Centennial Issue*, University of Texas Press, Austin, pp. 517–532.

Kingman, J. F. C. (1982a) The coalescent. *Stochast. Proc. Appl.* 13: 235–248.

Kingman, J. F. C. (1982b) On the genealogy of large populations. *J. Appl. Prob.* 19A: 27–43.

Larson, A. (1984) Neontological inferences of evolutionary pattern and process in the salamander family Plethodontidae. *Evol. Biol.* 17: 119–217.

Lynch, J. D. (1989) The gauge of speciation: on the frequencies of modes of speciation. *In*: Otte., D. and Endler, J. A. (eds) *Speciation and ist Consequences*, Sinauer Associates, Sunderland, MA, pp. 527–553.

Markham, R. B., Schwartz, D. H., Templeton, A., Margolick, J. B., Farzadegan, H., Vlahov, D. and Yu, X. (1996) Selective transmission of human immunodeficiency virus type 1 variants to SCID mice reconstituted with human peripheral blood mononuclear cells, *J. Virol.* 70: 6947–6954.

Matos, J. (1992) *Evolution within the Pinus montezumae Complex of Mexico: Population Subdivision, Hybridization and Taxonomy*, Ph.D. thesis, Washington University, St. Louis, MO.

Mayr, E. (1954) Change of genetic environment and evolution. *In*: Huxley, J., Hardy, A. C. and Ford, E. B. (eds) *Evolution as a Process*, Princeton University Press, Princeton, NJ, pp. 157–180.

Mayr, E. (1982) Processes of speciation in animals. *In*: Barigozzi, C. (ed.) *Mechanisms of Speciation*, Alan R. Liss, New York, pp. 1–19.

Mayr, E. (1992) A local flora and the biological species concept. *Amer. J. Bot.* 79: 222–238.

McNearney, T., Hornickova, Z., Templeton, A., Birdwell, A., Arens, M., Markham, R., Saah, A. and Ratner, L. (1995) Nef and LTR sequence variation from sequentially derived human immunodeficiency virus type 1 isolates. *Virology* 208: 388–398.

Mitter, C., Farrell, B. and Wiegmann, B. (1988) The phylogenetic study of adaptive zones: has phytophagy promoted insect diversification? *Amer. Naturalist.* 132: 107–128.

Mitter, C., Farrell, B. and Futuyma, D. J. (1991) Phylogenetic studies of insect-plant interactions: insights into the genesis of diversity. *Trends Ecol. Evol.* 6: 290–293.

O'Brien, S. J. (1991) Ghetto legacy. *Curr. Biol.* 1: 209–211.

Otte, D. (1989) Speciation in Hawaiian crickets. *In*: Otte., D. and Endler, J. A. (eds) *Speciation and ist Consequences*, Sinauer Associates, Sunderland, MA, pp. 482–526.

Page, R. D. M. (1990) Temporal congruence and cladistic analysis of biogeography and cospeciation. *Syst. Zool.* 39: 205–226.

Paterson, H. (1985) The recognition concept of species. *In*: Vrba, E. (ed.) *Species and Speciation*, Transvaal Museum, Pretoria, South Africa, pp. 21–29.

Routman, E. (1993) Population structure and genetic diversity of metamorphic and paedomorphic populations of the tiger salamander, *Ambystoma tigrinum*. *J. Evol. Biol.*, 6: 329–357.

Shaffer, H. B. and Voss, S. R. (1996) Phylogenetic and mechanistic analysis of a developmentally integrated character complex: alternate life history modes in Ambystomatid salamanders. *Amer. Zool.* 36: 24–35.

Shaw, K. L. (1993) *The evolution of song groups in the Hawaiian cricket genus* Laupala, Ph.D. thesis, Washington University, St. Louis, MO.

Slatkin, M. (1989) Detecting small amounts of gene flow from phylogenies of alleles. *Genetics* 121: 609–612.

Slatkin, M. and Maddison, W. P. (1989) A cladistic measure of gene flow inferred from the phylogenies of alleles. *Genetics* 123: 603–613.

Slatkin, M. and Maddison, W. P. (1990) Detecting isolation by distance using phylogenies of genes. *Genetics* 126: 249–260.

Templeton, A. R. (1980) The theory of speciation via the founder principle. *Genetics* 94: 1011–1038.

Templeton, A. R. (1981) Mechanisms of speciation – a population genetic approach. *Annu. Rev. Ecol. Syst.* 12: 23–48.

Templeton, A. R. (1983) Natural and experimental parthenogenesis. *In*: Ashburner, M., Carson, H. L. and Thompson, J. N. (eds) *The Genetics and Biology of Drosophila*, Academic Press, London, pp. 343–398.

Templeton, A. R. (1989) The meaning of species and speciation: A genetic perspective. *In*: Otte, D. and Endler, J. A. (eds) *Speciation and ist Consequences*, Sinauer Associates, Sunderland, MA, pp. 3–27.

Templeton, A. R. (1993) The "Eve" hypothesis: a genetic critique and reanalysis. *Amer. Anthropol.* 95: 51–72.

Templeton, A. R. (1995) A cladistic analysis of phenotypic associations with haplotypes inferred from restriction endonuclease mapping or DNA sequencing. V. Analysis of case/control sampling designs: Alzheimer's disease and the Apoprotein E locus. *Genetics* 140: 403–409.

Templeton, A. R. (1998a) Nested clade analyses of phylogeographic data: testing hypotheses about gene flow and population history. *Molec. Ecol.* 7: 381–397.

Templeton, A. R. (1998b) Species and speciation: geography, population structure, ecology, and gene trees. *In*: Howard, D. J. and Berlocher, S. H. (eds) *Endless Forms: Species and Speciation*, Oxford University Press, Oxford, *in press*.

Templeton, A. R. and Georgiadis, N. J. (1996) A landscape approach to conservation genetics: conserving evolutionary processes in the African Bovidae. *In*: Avise, J. C. and Hamrick, J. L. (1995) *Conservation Genetics: Case Histories from Nature*, Chapman and Hall, New York, pp. 398–430.

Templeton, A. R. and Sing, C. F. (1993) A cladistic analysis of phenotypic associations with haplotypes inferred from restriction endonuclease mapping. IV. Nested analyses with cladogram uncertainty and recombination. *Genetics* 134: 659–669.

Templeton, A. R., Boerwinkle, E. and Sing, C. F. (1987) A cladistic analysis of phenotypic associations with haplotypes inferred from restriction endonuclease mapping. I. Basic theory and an analysis of alcohol dehydrogenase activity in *Drosophila*. *Genetics* 117: 343–351.

Templeton, A. R., Crandall, K. A. and Sing, C. F. (1992) A cladistic analysis of phenotypic associations with haplotypes inferred from restriction endonuclease mapping and DNA sequence data. III. Cladogram estimation. *Genetics* 132: 619–633.

Templeton, A. R., Routman, E. and Phillips, C. (1995) Separating population structure from population history: a cladistic analysis of the geographical distribution of mitochondrial

155

DNA haplotypes in the Tiger Salamander, *Ambystoma tigrinum. Genetics* 140: 767–782.

Templeton, A. R., Sing, C. F., Kessling, A. and Humphries, S. (1988) A cladistic analysis of phenotypic associations with haplotypes inferred from restriction endonuclease mapping. II. The analysis of natural populations. *Genetics* 120: 1145–1154.

Torroni, A., Schurr, T. G., Cabell, M. F., Brown, M. D., Neel, J. V., Larsen, M., Smith, D. G., Vullo, C. M. and Wallace, D. C. (1993a) Asian affinities and continental radiation of the four founding native American mtDNAs. *Amer. J. Hum. Genet.* 53: 563–590.

Torroni, A., Sukernik, R. I., Schurr, T. G., Starikovskaya, Y. B., Cabell, M. F., Crawford, M. H., Comuzzie, A. G. and Wallace, D. C. (1993b) mtDNA variation of aboriginal Siberians reveals distinct genetic affinities with Native Americans. *Amer. J. Hum. Genet.* 53: 591–608.

Whittemore, A. T. (1993) Species concepts: a reply to Ernst Mayr. *Taxon* 42: 573–583.

Wiley, E. O. (1981) *Phylogenetics: The Theory and Practice of Phylogenetic Systematics.* Wiley, New York.

Testing speciation models with DNA sequence data

John Wakeley[1] and Jody Hey

Department of Ecology, Evolution, and Natural Resources, Nelson Biological Labs, Rutgers University, P.O. Box 1059, Piscataway, NJ 08855, USA
[1]Present address:, Department of Organismic and Evolutionary Biology, Harvard University, Cambridge, MA 02138, USA

Summary

This chapter reviews an approach to the study of speciation that is based on patterns of genetic variation within and between closely related species. Historically, research on the genetic mechanisms of speciation, and of species divergence, is very difficult – suffering from both practical difficulties in data collection and from theoretical problems. The method outlined in this paper is based on genealogical models of population divergence. We describe a hierarchy of models, and show how these fit into a hypothesis-testing framework that overcomes some of the theoretical problems of studying speciation. The method also advances the empirical study of speciation. Since testing of the models relies only on comparative DNA sequence data from closely related species, it can be applied to existing species regardless of whether it is practical or possible to generate hybrids.

Introduction

Research on the mechanisms of speciation is difficult, and speciation studies have traditionally suffered from two related theoretical uncertainties. One is the "species problem", which is the long-standing debate over the meaning of the word *species* and about the best means of identifying them. Disagreement over the nature of species has contributed to the second problem, which is the present lack of a hypothesis-testing framework for studying speciation.

We do not propose to solve the first of these two problems; rather, we sidestep it in two ways. First, we focus on mathematical models of the genetic divergence of populations. These are simple extensions of well-known population genetic models, and can be applied equally to populations and to species. For example, two separate populations (or species) that were once one will diverge over time unless there is gene flow between them. Under virtually any model that includes mutation, the two will accumulate differences. So long as our focus is on the level of genetic variation within and between populations, we can model divergence without regard to the delineation of the populations into named species. Second, in discussing speciation, we focus mainly on sexually reproducing organisms. Thus, we are able to concentrate on a point of relative agreement among workers in this field. Reproductive isolation, or the inability to interbreed with outsiders, is for most the hallmark of sexual species. In this context, the development of reproductive isolation is synonymous with speciation. The mathematical models that we explore also fall into a hierarchy of complexity, and this hierarchy helps to overcome the second of the two traditional theoretical difficulties in studying speciation (i.e. the lack of a hypothesis-testing framework).

The study of speciation is also difficult from an empirical standpoint, and most studies face one of two common obstacles. The first difficulty occurs for studies of organisms from closely related populations that may be incipient species. These studies face the uncertainty that such present-day examples of population differentiation may not be representative of speciation events in general. A second difficulty can arise when genetic approaches are applied to clearly distinct species. Some methods require crosses and hybrid formation, and these cannot usually be done on clearly distinct species that have separated long ago, for the simple reason that crosses often do not yield fertile progeny. An alternative kind of data, including comparative DNA sequences from multiple loci, can overcome these difficulties. If data are collected from within each of two closely related (but clearly distinct) species, for multiple loci, then the patterns of intraspecific and interspecific polymorphism can be interpreted in light of models of speciation (Hey and Kliman, 1993; Hey, 1994; Hilton et al., 1994).

In this chapter, we describe a conceptual framework for mathematical models of speciation. We begin with a simple model of variation within a single population, then we go on to describe several divergence models and show how they vary in their predictions about patterns of DNA sequence variation within and among populations. Some of these models are simpler than others, and our basic premise is that the hierarchy of model complexity permits a statistical approach to compare models of divergence. The statistical framework that we develop starts with the simplest possible extensions of single-population models to the case of two species. When a simple model can be rejected, another, more complicated one can be proposed and tested. In this way, data can point to successively more precise descriptions of divergence. If two models are both consistent with some data, then we are inclined to work with the simpler of the two until contradictory data appear.

All of the models are genealogical, and they hold a close correspondence to comparative DNA sequence data sets collected from within and between closely related species. The models help to inform our intuition about speciation, and their correspondence to the data permits the actual testing of speciation models.

The standard neutral model

We begin by briefly reviewing a model that makes predictions about levels and patterns of DNA sequence variation within a single population. This model is widely used and forms the basis of most of the statistical tests developed to detect historical patterns of natural selection in DNA sequence data (Hudson et al., 1987; Tajima, 1989b; Fu and Li, 1993). We then show how this model can be extended to study divergence between populations and how these models of divergence are related to theories of speciation.

A single population

The standard, single-population, neutral model is a combination of the well-known Wright-Fisher model (Fisher, 1930; Wright, 1931; Ewens, 1979; Hudson, 1990) and the assumption that mutations have negligible consequences on fitness (Kimura, 1983. In brief, mating is random, the population has had a constant size for a very long period of time, and mutations are neutral and occur in such a way that individual base positions are segregating at most only a single mutation at any point in time.

Under this model, variation is lost through genetic drift at a rate that is proportional to the inverse of the population size (i.e. $1/N$), and variation is input to the population at a rate that is proportional to the mutation rate, u. In practice, the compound parameter $\theta = 4Nu$ has proven useful for describing the amount of genetic variation that is expected under the model. Typically, estimating θ for a population or species is the first step in a population genetic study.

Assume that we have taken a sample of homologous DNA sequences from a diploid population which conforms to this standard, neutral model. If the mutation model is at least approximately true, then most base positions will not be variable in the sample, and it is not difficult to align the DNA sequences. When the sequences are aligned in rows, for example, some base positions may be revealed as polymorphic, and these appear simply as columns in which not all of the base values are identical. If we take a sample of n DNA sequences, then under our simple model the expected number of these polymorphic, or segregating, sites is

$$E(S) = \theta \sum_{i=1}^{n-1} \frac{1}{i} . \tag{1}$$

Suppose that we have a sample of two sequences ($n = 2$). Then it is easy to see from (1) that the expected number of polymorphic sites is equal to θ. In fact, the average number of polymorphic sites among pairs of sequences in a larger sample, commonly called pairwise differences, has this same expectation. If we add a third sequence, then expression (1) means that we expect to add half again as many polymorphic sites as were observed with just two sequences. As more sequences are added, the expected number of polymorphic sites in the entire sample rises but at a slower and slower rate.

If we count up all of the polymorphic sites S, we can use (1) to estimate θ:

$$\hat{\theta} = \frac{S}{a_n},$$
(2)

$$\text{where} \quad a_n = \sum_{i-1}^{n-1} \frac{1}{i}.$$

Expression (2) is commonly referred to as Watterson's estimator of θ. Watterson (1975) also derived the variance of S that will arise under the combination of Fisher-Wright and neutral mutation models. An understanding of the variance can be especially useful when data have been collected from multiple loci from the same population. Even though the data are from the same organisms, if they are from independently segregating loci, then the variance among them for S is expected to follow Watterson's expressions. We will return to this idea of using variation among loci when we discuss ways to evaluate models of population divergence.

Expression (1) is one way to connect the model parameter θ to an observable quantity, in this case S. However, it does not make use of all of the information regarding polymorphisms that is available from a set of aligned DNA sequences. In particular, each polymorphic site has an associated frequency, because each site divides the sample into two groups. There are, for instance, sites at which a single sequence is different from all the others, and sites where two sequences bear one nucleotide and the other $n - 2$ bear another, and so on. Under our simple model, the expected numbers of polymorphic sites in each frequency class are known. Thus, if we could distinguish the ancestral from the mutant base at a particular site, and if we let ξ_i represents the number of sites at which the mutant nucleotide has frequency i/n in the sample, then

$$E(\xi_i) = \frac{\theta}{i}$$
(3)

(Tajima, 1989b; Fu and Li, 1993; Fu, 1995). Without an outgroup sequence, it is impossible to distinguish mutants in frequency i/n from those in frequency $(n - i)/n$. Thus, we are limited to measuring $\eta_1 = (\xi_i + \xi_{n-i})/\delta$. In this expression δ is just an adjustment for the special case when n is even and $i = n/2$. Thus δ is equal to 1 if $i \neq n - i$ and equal to 2 if $i = n - i$. Then η_i has expectation

$$E(\eta_i) = \left(\frac{\theta}{i} + \frac{\theta}{n-i} \right) / \delta \qquad (4)$$

(Fu, 1995) also gives the variances and covariances of site frequencies.

The genealogies of nucleotide sites

In the absence of recombination, a sample of homologous DNA sequences will have a single gene tree history, or genealogy. The single-population model described above predicts a particular probability distribution of genealogies. A typical one of these is pictured in Figure 1. Under the assumption typically made that the number of sequences sampled is much smaller than the total number of individuals in the population ($n \ll N$), the genealogy of a sample is a random bifurcating tree and is expected to have characteristic branch lengths; namely, the times between successive common ancestor events (the nodes in the genealogy) are expected to be shorter when there are many ancestral lineages and longer when there are fewer. Thus, for the genealogy in Figure 1, we expect that $t_6 < t_5 < t_4 < t_3 < t_2$. More specifically, under the Wright-Fisher model, t_i is approximately exponentially distributed with parameter $i(i-1)/(4N)$, so that the expectation of t_i is equal to $4N/[i(i-1)]$.

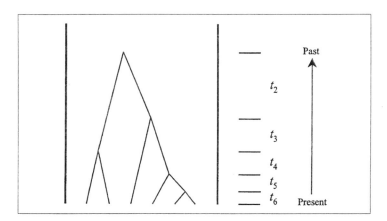

Figure 1. One possible genealogy of a sample of six sequences from a Wright-Fisher population. The thick lines represent the population's boundaries, emphasizing its finite size. The thin lines trace the ancestral lineages of the sample back (up) in time. The times, t_i, are the periods during which there were i ancestral lineages of the sequences in the sample.

The quantities S and η_i (which we can calculate from data) contain information about the genealogy of a sample, because the mutations that cause the variation observable in a sample must occur on the ancestral lineages like those depicted in Figure 1. For example, η_i is the number of mutations which occurred on lineages that left i or $n - i$ descendants in the sample. The total number of polymorphic sites, S, is simply the total number of mutations that occurred on the entire genealogy. Deviations from the standard model can be understood easily if we keep these relations between observable quantities and underlying genealogies in mind. For example, η_1 will increase as t_6 increases in Figure 1 because the greater t_6 is, the longer are the lineages leading to single descendent sequences. Thus, a larger than expected value of η_1 might indicate the failure of the standard model and suggest a different history that would cause this kind of genealogy.

In general, we cannot directly observe the genealogy of a sample, but even if we could, the particular one we saw would be just a single point in the universe of all possible genealogies. Looking at Figure 1, it is easy to see that this universe is incredibly large. Not only are there a very great number of different possible patterns of branching, but for each one of these there are an infinite number of possible t_i.

Multiple loci and recombination

When a number of unlinked loci are sampled, but within each of them no recombination occurs, then each one represents a single, independent draw from the sample space of genealogies. Thus, equations (1) and (4) would apply to each locus separately, but they would also apply to the total S and η_i for all the loci combined. The variances of S and η_i among the loci should follow the expressions given by Watterson (1975) and Fu (1995). Suppose that the data from several independent small loci are pooled, and S and η_i calculated for all the loci together, as if they constituted one large composite locus. In this case the variances of the total S and η_i would not follow those same expressions. Rather the variances for the composite locus would be considerably reduced. The reason is that the composite locus would be made up of several segments, each providing information independent from the others, and would not be just a single realization of the genealogical process.

Recombination is another component which can be included in the model. Like mutation, it is important in shaping the pattern of genetic variation, but in a very different way. The effect of recombination within a locus is similar to that of sampling multiple loci. The expected values of S and of η_i in expressions (1) and (4) are unaffected by recombination, but the variances are smaller. When recombination occurs in the history of some sequences, its effect is to break the sequence up into pieces that segregate more or less independently of each other. Thus, different segments within a locus that undergoes recombination may have different genealogies, just as different loci may. This means that observed single-locus values of S and η_i will tend to be closer to their expectations when the recombination rate is high than when it is low. Sampling multiple loci and the occurrence of recombination within loci both act to increase the number of independent observations, thus decreasing the variance. In general, lower variances permit more accurate estimates of parameters like θ, and will also give us more power to test the other assumptions of the model.

It is important to note that, because of recombination and independent segregation, some loci or even small regions within a single locus may be subject to selection and thus differ markedly from the standard neutral model, while most loci or most sites within a single locus conform well to the model. Sites that are under selection will influence the histories of adjacent regions, but the magnitude of this effect will decrease with the recombinational distance (Hudson and Kaplan, 1988). Genetic "hitchhiking", the phenomenon that neutral or deleterious mutations can be swept to fixation if they are tightly linked to a positively selected variant (Hudson et al., 1987}, is one well-known example of this. Thus, we need to draw a distinction between processes which act on single loci or small regions of the genome and ones which affect all loci identically.

For instance, if we use Tajima's (1989b) or Fu and Li's (1993) tests on a single locus, and find that the data are not compatible with the standard neutral model, without sampling more loci, we cannot know whether the process we have uncovered is locus-specific or occurs at the population level. A rapid growth in population size and a selective sweep will have nearly identical effects on the character of genetic variation (Tajima, 1989a; Slatkin and Hudson, 1991), as will population subdivision and balancing selection (Kaplan et al., 1988; Simonsen et al., 1995). However, selection acts on specific, limited regions, whereas population growth and subdivision act on the entire genome. This represents an advantage rather than a problem. By sampling many different loci, we can potentially distinguish between these two kinds of processes.

Neutral two-population models of divergence

The simple model of a single population can be extended to include two (or more) populations. To do so, we must introduce new model parameters that describe the historical relationship between the two populations. We must also consider the data, which will have a component that is not apparent when just one population has been sampled, i.e. divergence between the two populations. When data come from just one population, all polymorphisms appear as variations within that population, but with two populations we must also consider DNA sequence variation that distinguishes the two populations.

In the following sections, we review some simple extensions of single-species models to models of population divergence. Three kinds of models are discussed: isolation without gene flow; limited migration; and mixtures of migration and isolation.

Isolation

In the isolation model, a single ancestral species splits into two descendants. The split is assumed to happen instantaneously at some time in the past. After that time, the two descendent populations are assumed to be completely isolated from each other – there is no genetic exchange between them at any time thereafter. All three populations, the ancestor and the two descendants, are Wright-Fisher populations (see section "A single population"). Figure 2a gives a graphic depiction of the isolation model.

The model has three Wright-Fisher populations. Each may have its own unique effective size, N, so that there may be three characteristic parameters, θ. We propose that this general version of the isolation model is a good starting point for the study of speciation. It is fairly simple, and it corresponds roughly to the case of population divergence under complete allopatry. Other isolation models impose restrictions on the relative sizes of the three populations: Takahata and Nei (1985) and others assumed that all three populations or species are of the same size, and Hudson et al. (1987) assumed that the ancestral population is equal in size to the average of the two descendant population sizes. The salient feature of the isolation model is the complete absence of genetic exchange. In designing statistical tests, the more general version of the isolation model is preferable; we would not want to reject the model simply because of differences in population size among species.

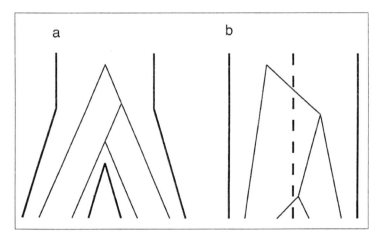

Figure 2. The strict isolation model, (a), and the equilibrium migration model, (b). As in Figure 1, the thick lines represent population boundaries, now with dashed lines to indicate that migration can occur, and thin lines trace the genealogy of each sample back (up) in time.

Wakeley and Hey (1997) considered this general isolation model with two descendant and one ancestral population size: N_1, N_2, and N_A. Four parameters then characterize the model, and these are θ_1, θ_2, θ_A, and T, where T is the time of separation measured in units of $2N_1$ generations. Correspondingly, four categories of segregating sites characterize variation within and between the two species. The first two of these comprise sites that are polymorphic in one of the species but monomorphic in the other. The numbers of these are called S_{x1} and S_{x2}, for the counts in species 1 and 2, respectively. Next are sites which show fixed differences, that is, which are monomorphic in both species but with different nucleotides. These are called S_f and were previously studied by Hey (1991). Last, there are sites at which the same polymorphism segregates in both species. The number of these shared polymorphisms is referred to as S_s. Wakeley and Hey (1997) derived the expectations of each of these four mutually exclusive categories of segregating sites in a sample and used them to estimate the four parameters of the isolation model. The expectations depend on all four parameters, θ_1, θ_2, θ_A and T, so the parameters of the model are estimated by solving numerically for values that most closely equate the expectations to observations. Extensive simulations showed that the numbers of exclusive, fixed and shared polymorphisms do contain the information necessary to estimate ancestral population parameters. Wakeley and Hey (1997) also derived the expectations of the joint site frequencies in a two-species sample.

Figure 3 shows two examples of plots of the four types of polymorphisms as a function of the time of isolation. The only difference in the parameters for Figures 3a and 3b is that in 3a the common ancestral population was equal to one of the descendants, whereas in 3b, the ancestor was much larger than either descendant. Note the large difference in the shapes of the curves, especially for times less than $T = 1$. Clearly the size of the common ancestor can have a very large impact on the levels of variation in each of the four classes.

Migration

Restricted, but nonzero genetic exchange between two populations is another possible cause of differentiation. Migration models stand in contrast to isolation models which permit no gene flow, but in practice the two give similar predictions about many aspects of genetic variation (Slatkin and Maddison, 1989; Takahata and Slatkin, 1990).

The most-studied migration model assumes that both populations are of the same size and have been exchanging migrants at a constant rate for an essentially infinite length of time into the past. This simple equilibrium migration model, pictured in Figure 2b, has just two parameters, the population migration rate, M, and a single θ. It is easily compared with the simple isolation model studied by Takahata

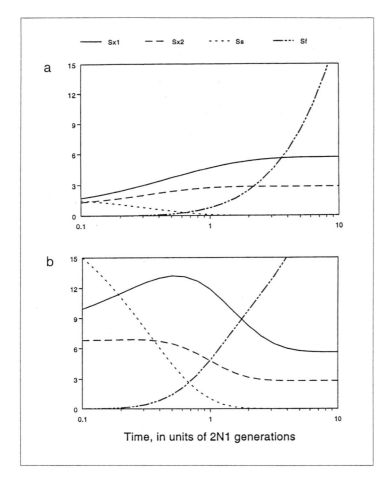

Figure 3. Expected values for S_{x1}, S_{x2}, S_s and S_f under the general isolation model as a function of T (time in units of $2N_1$ generations). For both figures, $n_1 = n_2 = 10$, $\theta_1 = 20$ and $\theta_2 = 1.0$ (a), $\theta_A = 1.0$ (b) $\theta_A = 10$.

and Nei (1985) and others, which is characterized by the time of separation, T, measured in units of $2N$ generations, and a single θ.

In order to compare these two models, we need a measure of divergence for which the expectation and the variance is known under both. At present, the only measure that is sufficiently well understood in both contexts is the average number of pairwise differences. For two populations, there are three measures of pairwise difference: the averages within each population, d_1 and d_2, and the average between populations, d_{12}. These are easy to calculate from sequence data. For instance, d_{12} is calculated simply by comparing each sequence from one population with each from the other, determining the number of differences between the two sequences, and taking the average. It is a well-known result that when $T = 1/(2 M)$ these two models give identical predictions about the expected values of d_1, d_2 and d_{11} (Li, 1976; Gillespie and Langley, 1979). However, the two models make different predictions about the variances. Wakeley (1996a) derived expressions for the variances of pairwise differences in the two-population equilibrium migration model and compared them with those found under isolation by Takahata and Nei (1985). The results showed that, when the expectations of the average numbers of pairwise differences are the same in both models, the variances are larger under migration than under isolation. This can, again, be understood by considering the genealogy of a sample. Figure 2 compares the isolation model and the migration model, and shows the genealogy of a sample of two gene copies from each population. Looking first at the variance of between-population pairwise differences, it is easy to see that under isolation the occurrence of interpopulation common ancestors is restricted in time to the ancestral population. Under migration, however, there may be both very recent and very ancient interpopulation events. The genealogies shown in Figure 2 illustrate this. Because interpopulation common ancestor events occur over a broader range under migration than under isolation, the variance is larger. The variance of intrapopulation pairwise differences is inflated also, but does not depend on there being more than one interpopulation common ancestor.

Wakeley (1996b) used this result to devise a test of the simple isolation model. The test is formulated such that the isolation model is rejected, for given values of average pairwise difference, when the variances are too large. The usefulness of the variance of pairwise differences in testing the simple isolation model suggests a test of the more general isolation model studied by Wakeley and Hey (1997) that could also detect migration. Other things being equal, the variances of the numbers of fixed, shared and exclusive polymorphisms among loci will be higher in models that include migration than in ones that assume strict isolation. Thus, a test could be made that rejects the four-parameter isolation model in favor of some sort of limited migration. We expect that such a test will not be as sensitive to changes in population size as Wakeley's (1996b) test, because fewer restrictions on population sizes are imposed.

Mixed models

Consider a situation in which the isolation model has been fitted to a data set and the fit is so poor that the model is rejected. This would occur if we found variances among loci that were considerably larger than expected under the isolation model. The next step is to consider realistic alternatives that included limited migration. If we are studying species, the equilibrium migration model, described above, is not appropriate, because it predicts a standing level of differentiation between the two populations which does not increase over time. It is a model of equilibrium divergence but not speciation. Realistic models of speciation must include some element of isolation, such that gene flow is ultimately prevented. Of course, natural selection may be a part of this, and we address this below. Focusing for the moment on neutral models of divergence, the alternatives to a strict isolation model are mixed models that include some migration and some isolation.

We consider two mixed isolation-migration models that appear to be reasonable alternatives and that may aid in our attempts to understand the divergence of species. These two models are pictured in Figure 4, and while no theoretical work has been done on either of them, we can predict some of their characteristics from what is known about strict isolation and equilibrium migration. The first, shown in Figure 4a is a hybrid of the isolation model and the model considered by Wakeley (1996c). Originally there was a single Wright-Fisher population, then a period of migration between two nascent species and, finally, complete isolation. The second model, shown in Figure 4b, is one in which the original Wright-Fisher population gradually splits into two which exchange migrants. The migration rate is initially a very high value (i.e. as if there were just one panmictic population) and then decays until isolation is complete. By adjusting the parameters of these two models, we can

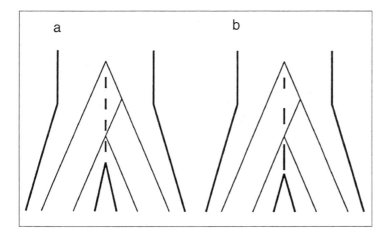

Figure 4. The two mixed neutral models discussed in the text: (a) the hybrid of migration and isolation, and (b) the decay-of-gene-flow model. The dashed lines in (b) are drawn to indicate that gene flow occurs readily at first, then becomes less likely as time passes.

mimic a great number of different scenarios for speciation, ranging from strict isolation to very recent divergence with a long history of migration.

We can expect that both of these mixed models will have a greater variance among loci than a strict isolation model. They will be able to explain more phenomena than strict isolation, but at the price of greater complexity, i.e. more parameters. The hybrid model requires at least two more parameters: a time of onset for migration, and a rate of migration. The decay-of-gene-flow model will require one extra parameter: a rate of decay in migration rate. These two models may be extremely similar in their predictions about genetic variation and divergence, and so may be very difficult to distinguish. All other things being equal, the decay-of-gene-flow model may be preferable, since it requires fewer parameters.

Adding natural selection

The models discussed so far have all assumed that natural selection has not shaped the pattern of variation. If speciation could occur simply as the by-product of divergence via genetic drift between populations – with no natural selection – then the isolation model and the mixed models could be considered neutral models of speciation. However, it is well known that natural selection can have a large impact, and most theories of speciation include selection. As in the traditional within-species neutralist-selectionist debate, we can adopt the neutral models of divergence as null models of speciation, and ask whether observed patterns of genetic variation require that natural selection be invoked in addition to, or together with, the processes of migration and isolation.

Variation among loci is even more important here than in the context of single populations. Some loci may be subject to selection, and this may contribute to species differences, while other, unlinked loci may conform perfectly to the neutral isolation model or to one of the mixed models. The neutral models of section 3 do not specify the causes of isolation and migration. Under a neutral isolation or mixed model, we generally imagine some sort of geographic barrier. However, when natural selection enters, the possibility exists that selection itself is the cause of isolation and divergence.

Divergence models and speciation theories

As mentioned above, the isolation model of divergence corresponds roughly to (neutral) allopatric speciation. One of the most interesting ways in which natural selection might contribute to divergence under allopatry is if one population is very small and isolated from a larger primary population. Such an isolated population may be in new circumstances, both environmental and genetical. For instance, if the population is very small, individuals may become quite inbred. A variety of scenarios have been envisioned whereby such a founder population might undergo considerable and novel adaptations (Mayr, 1963; Carson, 1978; Templeton, 1981), and these could increase the rate of divergence for a period of time just after the formation of the population. These adaptations may collectively have a pleiotropic effect such that the new population is reproductively isolated from the primary population, should they again come into contact.

These kinds of allopatric speciation models that require a small founder population yield specific predictions of patterns of variation. Small founder population models predict that data should show evidence of a population bottleneck, i.e. reduced variation for a period of time after the bottleneck. Even if the population rebounds in size and variation accumulates, Tajima's (1989b) or Fu and Li's (1993) tests could detect residual effects on the genealogies of nucleotide sites. Also, a population bottleneck will affect all loci in the same way, so every locus sampled should give similar evidence.

When natural selection is considered within the context of a model that includes migration, a fairly general situation emerges in which natural selection contributes to speciation. In brief, natural selection acting to reduce migration is tantamount to selection for reproductive isolation, and thus will further the process of speciation. This situation can arise whenever there are two groups of organisms that can exchange genes, but in which the hybrids have lower fitness than the parents. There seem to be at least two categories of natural selection that can arise in this context. One kind acts on loci that are the sites of differential adaptation in the two populations or that are epistatic with loci associated with differential adaptation. Individuals that are heterozygous at these loci, for one allele from each population, may have poor fitness. Natural selection is then manifest as poor hybrid fitness. The second kind of locus is one that can contribute to mate choice and where alleles that lead to preferential mating within a population (i.e. avoidance of hybrid mating) are favored by natural selection. This kind of adaptation could arise if some individuals are involved in relatively unsuccessful hybrid matings, and again, the source of selection is poor hybrid fitness.

Models that include these types of selection have traditionally been divided into several geographic categories: sympatric speciation wherein incipient species exist at least partly within overlapping geographic ranges; parapatric speciation where the populations abut one another; and models in which divergence begins to accrue under allopatry, but reproductive isolation is not complete before the populations come back into contact.

All these situations have the potential to lead to the evolution of premating barriers to gene flow and thus to speciation. Whether or not speciation occurs depends on the details of available genetic variation, levels of hybridization and gene flow, and the magnitude of selection coefficients against hybridization. Hereafter we will consider these models collectively. We may call them gene flow-selection models, as they all have in common the feature that gene flow can occur during speciation, but that it is most restricted for those loci that are the cause of low hybrid fitness.

Gene flow, selection and hypothesis testing

The gene flow-selection models could be quite difficult to evaluate directly, as has been done for the neutral models. Formal models would need additional, possibly many more, parameters than the neutral models. However, the gene flow-selection models exhibit two basic differences from neutral models: gene flow-selection models have different predictions, particularly regarding variation among loci; and they are more complex. Thus gene flow-selection models are ideal alternative hypotheses, to be contrasted with simpler null speciation models in statistical tests. While it may be difficult to specify the best gene flow-selection model for a particular data set, it may be possible to reject neutral models and show that, as a class, gene flow-selection models must be considered.

For some models, and some patterns of variation, these tests may be difficult, and it may be difficult to distinguish between a gene flow-selection model and a neutral model. For example, a model of sympatric or parapatric speciation with a period of migration before complete reproductive isolation would be impossible to distinguish from the neutral hybrid model of the section "Mixed models" if the loci under study are not linked to selected loci. A gene flow-selection model might be indicated if the variation found among loci is too great to be explained by neutral mixed models. This will occur only when some of the loci sampled are either under selection or are linked to selected loci and others are not. Loci that are not affected by selection for divergence or speciation will reflect the underlying, population-level process of migration. Despite the inherently large variances expected under migration, selected loci may appear as outliers.

This framework – isolation or mixed models as null and gene flow-selection models as alternative – can only be informative to the extent that the two classes of models differ in the predictions about divergence. At the level of genealogies, gene flow-selection models have a fundamental difference from neutral isolation or mixed models. Under neutral models, all loci are subject to common factors of shared effective population size and a presence or absence of gene flow between populations. Under gene flow-selection models, different loci may be subject to different levels of selection against gene flow. Thus, tests which examine variation among loci will be very useful. However, tests which focus on the pattern of variation at a single locus in the way Tajima's (1989b) and Fu and Li's (1993) do in the context of single populations may also be helpful in rejecting mixed neutral models in favor of speciation via selection.

Consider a comparative DNA data set collected from two divergent populations, and suppose the data include information on variation within and between the populations from each of several loci. The isolation and mixed models predict that all loci will have experienced common effective population sizes, and that divergence began at a single time point for all loci. In contrast, the gene flow-selection models predict that some portions of the genome may have relatively high levels of gene flow and little divergence, while loci linked to sites under selection (because of low hybrid fitness) will experience low or zero gene flow. If the data set includes loci of both types (or with sufficient linkage to both types), then some loci may show very little divergence, while others may show considerable divergence. This variation among loci may be much higher than is expected under the isolation or mixed models.

When neutral models are rejected in favor of gene flow-selection models, the process will also necessarily generate hypotheses of natural selection that may be amenable to additional tests. For example, a locus that does not reveal evidence of gene flow, when contrasted with others that do, may be a candidate for linkage to a site that is under selection against gene flow. If either an isolation or mixed model of divergence were true in this case, then the locus is just an outlier of the neutral model distribution. Thus if the locus could be subjected to an independent test of gene flow, or natural selection, in a controlled setting, it might become possible to further distinguish the gene flow-selection models from the neutral models.

Examples from *Drosophila*

Table 1 shows a three-locus data set collected from two species of the *D. melanogaster* species complex. The values shown are the numbers of S_{x1}, S_{x2}, S_f and S_s (see section "Isolation") observed for *D. mauritiana* and *D. simulans*. These numbers lead to the following parameter estimates: $\theta_1 = 30.2$, $\theta_2 = 23.0$, $\theta_A = 28.6$ and $T = 0.6$. Thus the isolation model fit indicates a common ancestor population intermediate in size to both descendants. It turns out that a statistical test of a specific isolation model that assumes the ancestor had intermediate population size has been in wide use for some years. The HKA test, though primarily used to test for the effect of natural selection on patterns of variation, is also a test of this specific isolation model (Hudson et al., 1987). When the data in Table 1 were put to this test, the fit between the data and model expectations was very good (Hey and Kliman, 1993).

Table 2 shows another three-locus data set, this time from *D. pseudoobscura* and *D. persimilis*. Note that the data in Table 1 show some variation among loci, particularly in S_s, but that the variation among loci is much greater in the data in Table 2. Indeed, one locus (Adh) shows no fixed differences and a very large number of shared polymorphisms. When the general isolation model was fit to these data, the model parameter estimates were: $\theta_1 = 28.7$, $\theta_2 = 24.9$, $\theta_A = 102.9$ and $T = 0.48$. Thus, taken together the data suggest a model in which the ancestral population size was far larger than that of either descendant. When a statistical test of the quality of fit between the data and this isolation model was made (via computer simulation) the fit was quite poor, and the model was rejected (Wang et al., 1997). In the paper describing these findings, we conclude that an isolation speciation

Table 1. Segregating sites in D. simulans *and* D. mauritiana

	n_1	n_1	S_{x1}	S_{x2}	S_s	S_f
Period	6	6	43	37	11	3
Zeste	6	6	18	9	0	1
Yp2	6	6	3	4	0	2

Note. Species 1 is *D. simulans* and species 2 is *D. mauritiana*; n_1 and n_2 are the number of sequences. These data are from Kliman and Hey (1993) and Hey and Kliman (1993).

Table 2. Segregating sites in D. pseudoobscura *and* D. persimilis

	n_1	n_1	S_{x1}	S_{x2}	S_s	S_f
Period	11	11	42	30	6	2
Hsp82	11	11	33	9	1	8
Adh	99	6	333	27	67	0

Note. Species 1 is *D. pseudoobscura* and species 2 is *D. persimilis*; n_1 and n_2 are the number of sequences. These data are from Wang et al. (1997), Wang and Hey (1996) and Schaeffer and Miller (1991, 1992).

model does not fit the *D. pseudoobscura/D. persimilis* data, and other models that include migration must be considered.

Conclusion

Traditional approaches to the study of speciation have faced practical difficulties (i.e. it has not always been clear what kind of data should be collected) and epistemological uncertainties (due to the species problem and a lack of a hypothesis-testing framework). This chapter has outlined an approach for the study of speciation that overcomes some of these shortcomings. We have described several formal population genetic models of speciation that generate specific predictions of patterns of genetic variation. These predictions bear a very close correspondence to the kinds of observations that are made using multilocus comparative DNA sequence data sets, so it is possible to fit the speciation models to data. Finally, we show how some speciation models are more complex than others, and how this complexity permits a hypothesis testing hierarchy.

Acknowledgements

This work was supported by NIH grant GM 17745 to JW and NSF grant DEB-9306625 to JH.

References

Carson, H. L. (1978) Speciation and sexual selection in Hawaiian *Drosophila*. *In*: Brussard, P. F. (ed.) *Ecological Genetics: The Interface*, Springer-Verlag, New York, pp. 93–107.

Ewens, W. J. (1979) *Mathematical Population Genetics*, Springer-Verlag, Berlin.

Fisher, R. A. (1930) *The Genetical Theory of Natural Selection*, Clarendon, Oxford.

Fu, X.-Y. (1995) Statistical properties of segregating sites. *Theoret. Pop. Biol.* 48: 172–197.

Fu, X.-Y. and Li, W.-H. (1993) Statistical tests of neutrality of mutations. *Genetics* 133: 693–709.

Gillespie, J. H. and C. H. Langley (1979) Are evolutionary rates really variable? *J. Mol. Evol.* 13: 27–34.

Hey, J. (1991) The structure of genealogies and the distribution of fixed differences between DNA sequences from natural populations. *Genetics* 128 831–840.

Hey, J. (1994) Bridging phylogenetics and population genetics with gene tree models. *In*: Schierwater, B., Streit, B., Wagner, G, P. and DeSalle, R. (eds) *Molecular Ecology and Evolution: Approaches and Applications*, Birkhäuser, Basel, pp. 435–449.

Hey, J. and Kliman, R. M. (1993) Population genetics and phylogenetics of DNA sequence variation at multiple loci within the *Drosophila melanogaster* species complex. *Molec. Biol. Evol.* 10: 804–822.

Hilton, H., Kliman, R. M. and Hey, J. (1994) Using hitchhiking genes to study adaptation and divergence during speciation within the *Drosophila melanogaster* complex. *Evolution* 48: 1900–1913.

Hudson, R. R. (1990) Gene genealogies and the coalescent process. *In*: Futuyma, D. J. and Antonovics, J. (eds) Oxford Surveys in *Evolutionary Biology*, vol. 7, Oxford University Press, Oxford, pp 1–44.

Hudson, R. R. and Kaplan, N. L. (1988) The coalescent process in models with selection and recombination. *Genetics* 120: 831–840.

Hudson, R. R., Kreitman, M. and Aguade, M. (1987) A test of neutral molecular evolution based on nucleotide data. *Genetics* 116: 153–159.

Kaplan, N. L., Darden, T. and Hudson, R. R. (1988) Coalescent process in models with selection. *Genetics* 120: 819–829.

Kimura, M. (1983) *The Neutral Theory of Molecular Evolution*, Cambridge University Press, Cambridge.

Kliman, R. M. and Hey, J. (1993) DNA sequence variation at the *period* locus within and among species of the *Drosophila melanogaster* complex. *Genetics* 133: 375–387.

Li, W.-H. (1976) Distribution of nucleotide difference between two randomly chosen cistrons in a subdivided population: the finite island model. *Theor. Pop. Biol.* 10: 303–308.

Mayr, E. (1963) *Animal Species and Evolution*, Belknap Press, Cambridge, MA.

Schaeffer, S. W. and Miller, E. L. (1991) Nuceotide sequence analysis of *adh* genes estimates the time of geographic isolation of the B ogota population of *Drosophila pseudoobscura*.

Proc. Natl. Acad. Sci. USA 88: 6097–6101.

Schaeffer, S. W. and Miller, E. L. (1992) Estimates of gene flow in *Drosophila pseudoobscura* determined from nucleotide sequence analysis of the alcohol dehydrogenase region. *Genetics* 132: 471–480.

Simonsen, K. L., Churchill, G. A. and Aquadro, C. F. (1995) Properties of statistical tests of neutrality for DNA polymorphism data. *Genetics* 141: 413–429.

Slatkin, M. and Hudson, R. R. (1991) Pairwise comparisons of mitochondrial DNA sequences in stable and exponentially growing populations. *Genetics* 129: 555–562.

Slatkin, M. and Maddison, W. P. (1989) A cladistic measure of gene flow inferred from the phylogenies of alleles. *Genetics* 123: 603–613.

Tajima, F. (1989a) DNA polymorphism in a subdivided population: the expected number of segregating sites in the two-population model. *Genetics* 123: 229–240.

Tajima, F. (1989b) Statistical method for testing the neutral mutation hypothesis by DNA polymorphism. *Genetics* 123: 585–595.

Takahata, N. and Nei, M. (1985) Gene genealogy and variance of interpopulational nucleotide differences. *Genetics* 110: 325–344.

Takahata, N. and M. Slatkin (1990) Genealogy of neutral genes in two partially isolated populations. *Theor. Pop. Biol.* 38: 331–350.

Templeton, A. R. (1981) Mechanisms of speciation – a population genetic approach. *Annu. Rev. Ecol. Syst.* 12: 23–48.

Wakeley, J. (1996a) The variance of pairwise nucleotide differences in two populations with migration. *Theor. Pop. Biol.* 49: 39–57.

Wakeley, J. (1996b) Distinguishing migration from isolation using the variance of pairwise differences. *Theor. Pop. Biol.* 49: 369–386.

Wakeley, J. (1996c) Pairwise differences under a general model of population subdivision. *J. Genetics* 75: 81–89.

Wakeley, J. and Hey, J. (1997) Estimating ancestral population sizes. *Genetics* 145: 847–855.

Wang, R.-L and Hey, J. (1996) Speciation history of *Drosophila pseudoobscura* and close relatives: inferences from DNA sequence variation at the period locus. *Genetics* 114: 1113–1126.

Wang, R.-L., Wakeley, J. and Hey, J. (1997) Gene flow and natural selection in the origin of *Drosophila pseudoobscura* and close relatives. *Genetics* 147: 1091–1106.

Watterson, G. A. (1975) On the number of segregating sites in genetical models without recombination. *Theor. Pop. Biol.* 7: 256–276.

Wright, S. (1931) Evolution in Mendelian populations. *Genetics* 16: 97–159.

PCR assays of variable nucleotide sites for identification of conservation units: n example from *Caiman*

George Amato[1], John Gatesy[2] and Peter Brazaitis[1]

[1]Science Resource Center, Wildlife Conservation Society, 2300 Southern Blvd., Bronx, New York, NY 10460, USA
[2]Department of Ecology and Evolution, University of Arizona, Tucson, AZ 85721, USA

Summary

A number of authors have suggested that the best approach for identifying units of conservation is to follow a systematics model of character analysis (Amato, 1991; Cracraft, 1991; Vogler and DeSalle, 1994). This approach requires the use of an operational, character-based, diagnostic species concept. The use of the phylogenetic species concept has the utility and philosophical logic appropriate for this task (Barrowclough and Flesness, 1993; Vogler and DeSalle, 1994; Cracraft, 1997). Additionally, there is a large body of literature that uses this framework, along with a parsimony-based character analysis to identify patterns of phylogeny (Cracraft, 1983; Nelson and Platnick, 1981; Nixon and Wheeler, 1990).

While we advocate this approach, we recognize that one of its limiting factors is sample size. We propose that by selective direct sequencing plus rapid sampling of variable target characters by polymerase chain reaction (PCR) assays of specific sites, sufficiently large numbers of individuals can be accurately, inexpensively and quickly surveyed for diagnostic characters. This procedure is demonstrated by a survey of variable nucleotide sites in the *Caiman crocodilus* complex.

Introduction

Both conservation biologists and systematists attempt to characterize biological diversity. Conservation biologists have primarily been concerned with identifying "natural" units which form the basis of both *in situ* and *ex situ* conservation programs (Amato, 1994). While there may be many parochial reasons to conserve specific local populations, most conservation biologists and managers recognize the value of a generalized, scientifically based, objective methodology for determining units of conservation across taxonomic and geographic boundaries. Despite a consensus on the importance of such an approach (O'Brien and Mayr, 1991; Amato et al., 1995), there are broad disagreements about which scientific disciplines provide the best intellectual framework for this endeavor (Avise and Ball, 1990; Vogler and DeSalle, 1994). A failure to recognize that it is the competing ontologies of population genetics and systematics which are largely responsible for this controversy has led to applications of each discipline's strengths at inappropriate hierarchical levels. In this paper we propose using an operational, character-based diagnostic approach which equates phylogenetic species with objective units of conservation. By confining our explorations to a level that reveals historically distinct entities, we can employ this logically defensible methodology without trespassing into the morass of intraspecific tokagenetic patterns (Hennig, 1966; Avise and Acquadro, 1982; Avise and Ball, 1990). Using phylogenetic species as our terminal taxonomic units ensures that we will be examining the finest level of resolution within this objective framework.

In 1985 a special meeting sponsored by the American Association of Zoological Parks and Aquariums (AAZPA) was held in Philadelphia to address the "subspecies dilemma" for endangered species. Subspecies designations have historically been applied to everything from undifferentiated local populations to biological species. At this meeting the term *evolutionarily significant unit* (ESU) was introduced to describe the "natural" unit that should be the focus of conservation efforts. Since this time it has been suggested that the ESU corresponds to phylogenetic species (Cracraft, 1991; Barrowclough and Flessness, 1993; Vogler and DeSalle, 1994) and biological species (O'Brien and Mayr, 1991; Avise, 1996) and even "reciprocally monophyletic populations" (Moritz, 1994). While introduced as a specific term, ESU now has so many definitions as to render it useless. We agree with Cracraft (1997) that the term should be abandoned, and once again propose that phylogenetic species be the unit of conservation. Phylogenetic species are defined as "an irreducible cluster of organisms, diagnosably distinct from other such clusters, and within which there is a parental pattern of ancestry and descent" (Cracraft, 1989). Operationally, phylogenetic species are diagnosably distinct populations (or groups of populations) where every individual shares the diagnostic character (or suite of characters). It is noteworthy that other lower-level, systematics-based approaches have been proposed. Vrana and Wheeler (1992) have proposed using only individuals as terminal taxa. While this approach is more assumption-free than defining populations for a phylogenetic species analysis, it has less application to conservation where populations are the unit of management.

Systematics analyses at this level have been problematic due to a paucity of discrete morphological characters and incomplete geographic sampling. However, the advent of molecular systematics has provided useful new data sets, especially DNA sequences, for use in lower-level systematics studies (Hillis and Moritz, 1995; Avise, 1996). PCR technology has been the driving force in the greatly accelerated rate of data collection. While all types of characters (morphological, behavioral, karyotypic and genetic) are useful and important, easily obtained DNA sequence data provides enormous numbers of genetic characters useful for a systematics approach to identifying conservation units. Along with this improved ability to generate large numbers of characters has been an explosion of ideas and algorithms for analyzing molecular character data for phylogenetic study (Swofford, 1990; Felsenstein, 1990; Farris, 1989). The large number of available molecular phylogenetic studies has also provided useful information about character evolution for conservation. While identifying a phylogenetic species rests only on demonstrating that a population has diagnostic characters (which are shared by all members but are not found in other groups), the use of higher-level phylogeny reconstruction may be important in identifying regions of DNA that are useful for characterization (Cracraft, 1997; Amato et al., 1993).

PCR has also allowed the amplification of target DNA sequences from nontraditional biological samples. Before PCR, biochemical techniques required careful preparation of large quantities of fresh blood or organ tissues (for vertebrates) or entire organisms (for smaller invertebrates). This proved especially problematic for conservation research because samples were often needed from animals that were handled infrequently, existed only in small, isolated populations and might only be handled by field researchers who had difficulty in obtaining and preserving biological samples. PCR advances allowed the use of such samples as hair, small skin biopsies, shed feathers, dried blood, museum specimens and others (Amato et al., 1993; Walsh et al., 1991; Garza and Woodruff, 1992). Now with the collection of materials made easier, as well as the generation of large numbers of molecular genetic characters, we are better prepared than ever to tackle questions concerning units of conservation.

While DNA characters offer a wealth of discrete character information for such lower-level systematics problems (e.g. Vogler and DeSalle, 1993), the expense and time involved in direct sequencing has often resulted in the sampling of small numbers of specimens. In contrast, many morphological characters can be scored rapidly and inexpensively. Although PCR technology has accelerated the rate of data collection, most DNA sequencing studies employ only a few exemplars that are used to represent a particular population or subspecies. The use of direct sequencing for broad character surveys is not generally cost-effective given the number of phylogenetically relevant characters that are discovered per dollar and hour.

PCR based sampling methods

One option is to use polymerase chain reaction assays of specific sites (PCRASS) to increase the scope of a lower-level systematic analysis. The simplest approach to this assay is the identification of potentially informative (polymorphic) sites by direct sequencing of a subset of available samples. Sites can then be assayed by designing PCR primers that match the polymorphic sites at the 3' end of the primer. Sommer et al. (1992) describe various methods of assuring specificity of assaying for single base changes with any primer:template mismatch. This methodology has been termed PCR amplification of specific alleles (PASA), allele-specific amplification

(ASA), allele-specific PCR (ASP) and amplification refractory mutation system (ARMS) (Sommer et al., 1992; Newton et al., 1989; Nichols et al., 1989; Okayama et al., 1989; Wu et al., 1989). In all cases amplification takes place if there is a perfect match between the primer with the correct sequence to complement the polymorphic site and no amplification if there is a mismatch at the variable site. Varying magnesium concentrations, enzyme concentrations and cycling conditions allows for optimization of this assay (Sommer et al., 1992). Both primers can be used to provide a positive and negative control, excluding amplification problems from confusing the results. Additionally, any samples that were assayed as ambiguous can be completely sequenced.

Other techniques are designed to identify nucleotide sequence changes (down to the level of a single base insertion/deletion/substitution) without completely sequencing all samples. Polziehn et al. (1995) used a variation of this technique that they term "primer-generated RFLPs" (PG-RFLPs) to survey populations of North American bison (*Bison bison*). In this study, Strobeck and Polziehn use allele specific PCR primers which also contain restriction-site sequences with the variable site as part of the restriction site. The purpose was to enhance their ability to discriminate between absence of PCR products, apparently due to problems of amplification specificity. In addition to PG-RFLPs, they surveyed for polymorphic sites by simple restriction-enzyme assays where direct sequencing had revealed a polymorphism that fell within a restriction-site sequence. However, this does not unambiguously identify a specific site.

Lessa and Applebaum (1993) reviewed related techniques for identifying allelic variation. Single-strand conformational polymorphism (SSCP) (Orita et al., 1989) is based on the change in mobility of single-stranded DNA fragments that vary by base substitutions, insertions and deletions. PCR products are denatured by heating in the presence of formamide, and are separated on acrylamide gels. Strands migrating differentially can then be sequenced to identify base changes.

Two other techniques, denaturing gradient gel electrophoresis (DGGE) (Myers et al., 1986, 1989a, b) and temperature gradient gel electrophoresis (TGGE) (Whartell et al., 1990) rely on the separation of denatured or "melted" double-stranded DNA. The physical properties of the DNA fragment when it reaches its melting point in a gradient gel allow for separation based on a single base change. However, direct sequencing is necessary to identify the position of the change. Carefully controlling conditions should allow for an inference of base-change identity for identically migrating fragments. Another related method for assessing allelic variation is coupled amplification and sequencing (CAS) (Ruano and Kidd, 1991a, b). This method allows for the simultaneous direct sequencing of the two complementary strands of a PCR product identified as a different allele by a DGGE gel.

One particularly novel use of this technology related to conservation was the use of PCRASS to assess which species of sturgeon were represented in commercially available containers of caviar. DeSalle and Burstein (1996) isolated DNA from a large sample of caviar eggs and then compared diagnostic DNA characters from sturgeon reference sequences by PCR amplification with species-specific primers.

The *Caiman crocodilus* complex

There are five currently recognized species of caiman – *C. latirostris*, *C. crocodilus*, *Melanosuchus niger*, *Paleosuchus trigonatus* and *P. palpebrosus*. The most widespread species is *C. crocodilus*, which ranges from Central America through northern South America including the Amazon and Orinoco river systems south and west to to Brazil, Peru, Bolivia, Paraguay and Argentina. Taxonomic designations for the "*C. crocodilus* complex" have historically been problematic (Brazaitis, 1973; Frair and Behler, 1983). At the present time the complex is most frequently described as three subspecies – *C. c. crocodilus* (northern South America, Amazon and Orinoco Rivers), *C. c. fuscus* (Central America and northern Columbia and Venezuela), and *C. c. yacare* (southern and western regions of Brazil bordering Bolivia, Argentina and Paraguay including Rio Paraguay and Rio Pilcomayo) (King and Rocca, 1987; Medem, 1983). At various times *C. c. yacare* has been designated a full species (Daudin, 1802; Medem and Marx, 1955; Carvalho, 1955; Medem, 1960, 1983; King and Burk, 1989). Full species designations were based on a variety of morphological analyses of distinctive skull morphology, color, pattern and scalation correlated with geographic distribution. The number of taxonomic units has been further confused by a leather industry booklet naming two additional subspecies (Fuchs, 1974). These designations were introduced into the literature by Wermuth and Mertens (1977) in spite of the fact that they were based on unreferenced, incomplete tannery skins from unknown collectors and localities, and without deposited voucher specimens (Frair and Behler, 1983). All of this complicates the fact that the subspecies have had different levels of protection under the Convention for International Trade in Endangered Species (CITES).

Our interest in applying a molecular systematics approach to identifying phylogenetic species/conservation units for the *C. crocodilus* complex is in part a response to the confusion in the literature. Accurately identifying these units has important implications for designing *in situ* and *ex situ* conservation strategies for this group. Additionally, identifying numbers and distributions of caiman taxa will impact the commercial trade in skins and potentially provide important forensic tools for monitoring the trade.

Methods

Samples used in this study are a subset of samples collected by a number of independent field studies (Brazaitis et al., 1990, 1992). These field surveys sampled all major populations and important river systems in the range countries. A variety of protocols were employed to preserve blood samples including desiccating whole blood on sterile cotton surgical sponges with room temperature storage in sealed plastic bags and preserving whole blood in an equal volume of buffer containing 100 mM EDTA, 100 mM Tris, 2% SDS. All samples were obtained without harm to the study animals, which were immediately released after blood samples, morphological measurements and photographs were taken. These sampling procedures

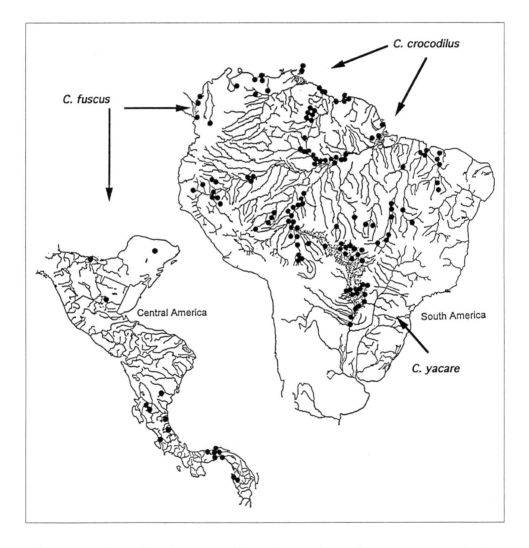

Figure 1. Sample localities for C. crocodilus *studies employing direct sequencing and PCR assays of specific sites.*

were easily carried out in the field since there was not a need for refrigeration or special handling. Total genomic DNA was obtained from the dried blood by a method employing a chelating resin (Chelex 100, BioRad) optimized for forensics samples (Walsh et al., 1991). A standard phenol/chloroform DNA isolation procedure (Caccone et al., 1987) was used for the samples stored in RT buffer. All samples yielded microgram quantities of total genomic DNA.

A small sample of DNA sequences was used to identify nucleotide sites that vary within the *C. crocodilus* complex. Approximately 1000 nucleotides of mitochondrial (mt) DNA [fragments of 12S and 16S mt ribosomal DNA and mt cytochrome b – Kocher et al. (1989; Irwin et al., 1991; Gatesy and Amato, 1992)] were sequenced in both directions from 23 individuals of *C. crocodilus*. These samples include the three "subspecies", *C. c. crocodilus*, *C. c. yacare* and *C. c. fuscus*, from across the broad range of the species complex (Fig. 1).

An additional 134 samples were surveyed for four polymorphic sites by use of PCR assays. Reactions were carried out using approximately 250 ng of template DNA, a magnesium concentration of 1.5 mM, primer concentration of 0.1 µm, and 0.5 units of Taq polymerase in 50-µl reaction volumes. PCR was performed in a Perkin-Elmer Cetus DNA thermal cycler at 94 °C for 1 min, 52 °C for 1.5 min and 72 °C for 2 min for 40 cycles. Two 17-base primers were constructed for each surveyed site with the most 3' base specific to the two alternate bases identified by direct sequencing. These primers were designed from previously sequenced *C. crocodilus*. Base-specific primers were paired with universal vertebrate primers (Kocher et

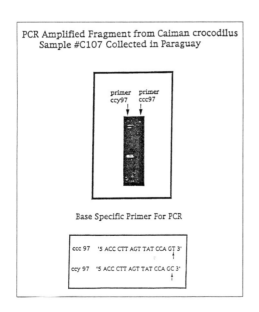

Figure 2. 12S ribosomal mtDNA fragment of C. c. yacare *amplified with base-specific primers (ccy 97 and ccc 97).*

al., 1989), amplifying fragments of approximately 150 bases. Each sample was amplified with both primers, providing a positive and a negative control (Fig. 2).

Samples of C. crocodilus were assigned to a subspecies according to morphological criteria and geographic position (Brazaitis et al., 1990, 1992). We then surveyed for character states, inferred from direct sequencing and PCRASS information, that were unique to each "subspecies" (Davis and Nixon, 1992). If there were no fixed differences between subspecies, the division of C. crocodilus into several phylogenetic species/conservation units would not be warranted.

Results

There were 22 variable nucleotide sites apparent from the alignment of the 23 C. crocodilus sequences. Only two of these sites are homoplastic within this sample (Fig. 3). The four sites that were more extensively sampled by PCRASS showed con-

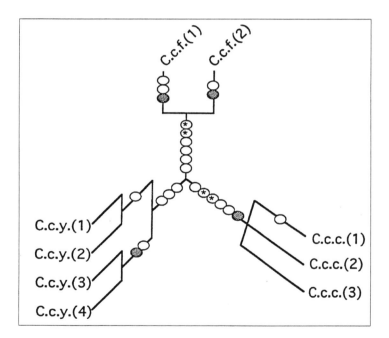

Figure 3. The most parsimonious network (24 steps; consistency index 0.889) for the C. crocodilus mtDNA sequences (Swofford, 1990). Homoplastic changes are shown in gray. Sites that changed once are represented by white circles. The stars mark sites checked by PCRASS in 134 individuals. C.c.f. = C. fuscus, C.c.c. = C. crocodilus, C.c.y. = C. yacare.

sistency across the ranges of each subspecies. Sites 1 and 2 from the 12S mtDNA sequence display a unique state in *C. c. crocodilus*, while sites 3 and 4 from the 16S mtDNA sequence are unique in *C. c. fuscus*. These sites remain monomorphic within subspecies despite the increase in sample size from 23 to 134.

Discussion

An independent study was subsequently conducted examining *C. crocodilus* subspecies using morphological characters. In this study, 52 morphological traits were analyzed using a sample size of 565 (Brazaitis et al., 1997a, b). Numerous multivariate, statistical approaches were used. The reults from these studies were entirely concordant with our PCR assay results (Brazaitis et al., 1997a, b) supporting the utility of the PCRASS/PSC approach.

While we strongly feel that the PSC approach is the best framework for conservation unit analysis, it is worthwhile to examine inherent weaknesses. In PSC, characters are what identify phylogenetic species. However, the choice of characters may sometimes be subjective. Davis and Nixon (1992) suggest a methodology for assessing which attributes are phylogenetic characters and which are traits. This methodology (population aggregation analysis) involves successive searches for fixed differences between aggregations of local populations. Characters are attributes that are not polymorphic and are unique within populations. Traits are attributes that may be polymorphic or are not unique to a population. Davis and Nixon (1992) point out that cladistic methodology allows for creating a hierarchy out of the terminal units by scoring attributes, but that only characters, not traits, can be used for determining a hierarchy that has phylogenetic information. Furthermore, they observe that small sample sizes can affect our ability to discriminate between a character and a trait. This is especially problematic for many endangered species that exist in small, fragmented, remnant populations (Amato et al., 1995). In the most rigorous application of the PSC approach, assessing the significance of diagnostic characters would be subjective. For this reason, we are especially concerned about discriminating patterns of attributes in a species that now only exists in highly fragmented, small populations in comparison to its historic distribution. In these cases, all of the individuals in a population may be closely related and share attributes that reflect those familial relationships. These attributes would not be informative about the populations' evolutionary distinctiveness or history. For example, the black rhinoceros population was estimated to be over 60,000 in 1970 (already greatly reduced from 1900), and now is less than 2500. To survey scattered populations that may number as few as 10 animals for diagnostic characters may yield patterns that do not reflect evolutionary events (Amato et al., 1993). Character data may have to be discussed in the context of additional data from museum specimens representing areas where the animals have been locally extirpated.

In addition to character assessment, identifying what constitutes a population is critical, and often subjective, for PAA. Davis and Nixon (1992) acknowledge the subjective nature of operationally identifying what constitutes a population. However, some assumptions are necessary to employ any framework, and PSC requires fewer such assumptions than many other proposed strategies such as using overall similarity or genetic distances which have been shown to be largely uninformative for identifying taxonomic rank or units of conservation (Vogler and DeSalle, 1994; Avise and Acquadro, 1982). We have used the study of the *C. crocodilus* complex to demonstrate a case that is amenable to the outlined framework. By having a large number of samples from a group still present in much of its original range, many of the acknowledged problems could be avoided. A wide distribution of diagnostic molecular characters that are congruent with morphological and biogeographical data support the division of the *C. crocodilus* complex into three phylogenetic species/conservation units.

The use of PCRASS allows us to increase our survey of attributes and test hypotheses about phylogenetic species. Only continued research using this approach, followed by decision making and action by managers, will ultimately demonstrate its usefulness. Systematics provides us with an important framework to aid in identifying conservation units while population genetics provides models for managing these units. It is the managers in zoological parks, governments and international conservation organizations that must use these results to implement the important management actions.

Acknowledgements

We especially thank G. Robelo, and C. Yamashita for research that provided the sample materials. We also thank R. DeSalle, P. Goldstein, A. Vogler, J. Powell and D. Wharton for useful discussion and comments on the manuscript. The work was supported by a Conservation Grant from the Institute for Museum Services, a grant from the American Zoo and Aquarium Conservation Endowment Fund, the Wildlife Conservation Society and an AMNH postdoctoral fellowship to J. Gatesy.

References

Amato, G. D. (1991) Species hybridization and protection of endangered animals. *Science* 253 (5017): 250.

Amato, G. D. (1994) *A systematic approach for identifying evolutionarily significant units for conservation: the dilemma of subspecies*, PhD dissertation, Yale University, New Haven, CT.

Amato, G. D., Ashley, M. V. and Gatesy, J. (1993) Molecular evolution in living species of rhinoceros: implications for conservation. *In*: Ryder, O. A. (ed.) *Proceedings of the International Conference on Rhinoceros Conservation and Biology*, Zoological Society of San Diego, San Diego, CA.

Amato, G., Wharton, D., Zainuddin, Z. Z. and Powell, J. R. (1995) Assessment of conservation units for the Sumatran rhinoceros (*Dicerorhinus sumatrensis*). *Zoo Biol.* 14: 395–402.

Avise, J. C. and Aquadro, C. F. (1982) A comparative summary of genetic distances in the vertebrates. *Evolut. Biol.* 15: 151–185.

Avise, J. C. and Ball, R. M. (1990) Principles of genealogical concordance in species concepts and biological taxonomy. *In*: Antonovics, J. and Futuma, D. (eds) *Oxford Surveys in Evolutionary Biology*, Oxford University Press, London, pp.45–67.

Avise, J. C. (1996) Toward a regional conservation genetics perspective: Phylogeography of faunas in the Southeastern United States. *In*: J. C. Avise and J. L. Hamrick (eds) *Conservation Genetics: Case Histories from Nature*, Chapman and Hall, New York, pp. 431–470.

Barrowclough, G. F. and Flesness, N. R. (1996) Species, subspecies, and races:The problem of units of management in conservation. *In*: Kleinman, G. G., Allen, M. and Harris, H. (eds) *Wild Mammals in Captivity*, University of Chicago Press, pp. 247–254.

Brazaitis, P. (1973) The identification of living crocodilians. *Zoologica* 58: 59–101.

Brazaitis, P., Rebelo, G. and Yamashita, C. (1990) A summary report of the CITES central South American caiman study. Phase I: Brazil. *In*: *Crocodiles, Proceedings of the 9th Working Meeting of the Crocodile Specialist Group*, vol. 1, IUCN, Gland, Switzerland, pp. 100–115.

Brazaitis, P., Robelo, G. and Yamashita, C. (1992) Report of the WWF/Traffic USA Survey of Brazilian Amazonia Crocodilians: Survey period July 1988 to January 1992, (unpublished).

Brazaitis, P., Madden, R., Amato, G. and Watanabe, M. (1997a) Morphological characteristics, statistics, and DNA evidence used to identify closely relaed crocodilian species for wildlife law enforcement. *In*: *The Proceedings of the American Academy of Forensic Sciences*, D28, American Academy of Forensic Sciences, New York, pp. 92–93.

Brazaitis, P., Madden, R., Amato, G., Robelo, G., Yamashita, C. and Watanabe, M. (1997b) The South American and Central American caiman (*Caiman*) complex. Systematics of the *Caiman*: Results from morphological, statistical, molecular genetics, and species delimitation studies. Special report to the U. S. Fish and Wildlife Service, (unpublished).

Caccone, A., Amato, G. D. and Powell, J. R. (1987) Intraspecific DNA divergence in *Drosophila*: a case study in parthenogenetic *D. mercatorum*. *Molec. Biol. Evol.* 4: 343–350.

Carvalho, A. L. (1955) *Os Jacares do Brasil*. Arquivos do Museu Nacional (Rio de Janeiro). 42 (1): 125–139.

Cracraft, J. A. (1983) Species concept and speciation analysis. *Curr. Ornithol.* 1: 159–187.

Cracraft, J. A. (1989) Speciation and ontology: the empirical consequences of alternative species concepts for understanding patterns and processes of differentiation. *In*: Otte., D. and Endler, J. A. (eds) *Speciation and ist Consequences*, Sinauer Associates, Sunderland, MA, pp. 28–59.

Cracraft, J. (1991) Systematics, species concepts and conservation biology. Abstract. Society for Conservation Biology, 5th Annual meeting, Society of Conservation Biologists, Madison, Wisconsin.

Cracraft, J. (1997) Species concepts in systematics and conservation biology – an ornithological viewpoint. *In*: Claridge, M. F., Dawah, H. A. and Wilson, M. R. (eds) *Species: The Units of Biodiversity*, Chapman and Hall, New York.

Daudin, F. (1802) *Histoire naturelle, generale et particuliere des reptiles*, vol. 2, Paris, p. 399.

Davis, J. and Nixon, K. C. (1992) Populations, genetic variation and the delimitation of phylogenetic species. *Syst. Biol.* 41: 421–435.

DeSalle, R. and Birstein, V. J. (1996) PCR identification of black caviar. *Nature* 381: 197–198.

Farris, J. S. (1989) The retention index and the rescaled consistency index. *Cladistics* 5: 417–419.

Felsenstein, J. (1990) *PHYLIP: Phylogeny Inference Package*, version 3.3, University of Washington, Seattle, WA.

Frair, W. and Behler, J. (1983) Book Review: *Liste der Rezenten Amphibien und Reptilien. Testudines, Crocodylia, Rhynchocephalia* by Wermuth, H. and Mertens, R., *Herp. Rev.* 14(1): 23–25.

Fuchs, K. (1974) *Die Krokodilhäute: Ein wichtiger Merkmalträger bei der Identifizierung von Krokodil-Arten*, Eduard Roether Verlag, Darmstadt.

Garza, J. C. and Woodruff, D. S. (1992) A phylogenetic study of the gibbons (Hylobates) using DNA obtained noninvasively from hair. *Mol. Phylogenet. Evol.* 1: 202–210.

Gatesy, J. and Amato, G. D. (1992) Sequence similarity of 12S ribosomal segment of mitochondrial DNAs of gharial and false gharial. *Copeia* 1: 241–243.

Hennig, W. (1966) *Phyogenetic Systematics*, University of Illinois Press. Urbana, IL.

Hillis, D. M., Moritz, C. and Mable, B. K. (eds) (1996) *Molecular Systematics*, Sinauer Associates, Sunderland, MA.

Irwin, D. M., Kocher, T. D. and Wilson, A. C. (1991) Evolution of the cytochrome b gene of mammals. *J. Mol. Evol.* 32: 128–144.

King, F. W. and Roca, V. (1987) The caimans of Bolivia. A preliminary report on a CITES and Centro de Dessarrollo Forestal sponsored survey of species distribution and status. Report to CITES Secretariat, Lausanne, Switzerland.

King, F. W. and R. L. Burk (1989) *Crocodilian, Tuatara, and Turtle Species of the World: A Taxonomic and Geopraphic Reference*, Association of Systematics Collections, Washington DC, pp. 1–15.

Kocher, T. C., Thomas, W. K., Meyer, A., Edwards, S. V., Paabo, S., Villablanca, F. X. and Wilson, A. C. (1989) Dynamics of mitochondrial evolution in animals: amplification and sequencing with conserved primers. *Proc. Natl. Acad. Sci. USA* 86: 377–382.

Lessa, E. P. and Applebaum, G. (1993) Screening techniques for detecting allelic variation in DNA sequences. *Molec. Ecol.* 2: 119–129.

Medem, F. (1960) Notes on the Paraguay caiman, *Caiman yacare* Daudin. *Mitt. Zool. Mus. Berlin* 36(1): 129–142.

Medem, F. (1983) *Los Crocodylia de Sur America*, vol. 2, Universidad Nacional de Colombia and Colciencias, Bogota, Clombia.

Medem, F. and Marx, H. (1955) An artificial key to the New-World species of crocodilians. *Copeia* 1: 1–2.

Moritz, C. (1994) Defining "evolutionary significant units" for conservation. *Trends. Ecol. Evolut.* 9: 373–375.

Myers, R. M., Maniatis, T. and Lerman, L. S. (1986) Detection and localization of single base changes by denaturing gradient gel electrophoresis. *Methods Enzymol.* 155: 501–527.

Myers, R. M., Sheffield, V. C. and Cox, D. R. (1989a) Mutation detection, GC-clamps, and denaturing gradient gel electrophoresis. *In*: Erlich, H. A. (ed.) *PCR technology: Principles and Applications for DNA Amplification*, Stockton Press, New York, pp. 71–88.

Myers, R. M., Sheffield, V. C. and Cox, D. R. (1989b) Polymerase chain reaction and denaturing gradient gel electrophoresis. *In*: Erlich, H. A., Gibbs, R. and Kazazian, H. H. (eds) *Polymerase Chain Reaction*, Cold Spring Harbor Laboratory Press, Cold Spring Harbor, New York, pp. 177–181.

Nelson, G. J. and Platnick, N. I. (1981) *Systematics and Biogeography: Cladistics and Vicariance*, Columbia University Press, New York.

Newton, C. R., Graham, A., Heptinstall, I. E., Powell, S. J., Summers, C. and Kalsheker, N. (1989) Analysis of any point mutation in DNA. The amplification refractory mutation system (ARMS). *Nucl. Acid. Res.* 17: 2503–2515.

Nichols, W. C., Liepnicks, J. J., McKusick, V. A. and Benson, M. D. (1989) Direct sequencing of the gene for Maryland/German familial amyloidotic polyneuropathy type II and genotyping by allele-specific amplification. *Genomics* 5: 535–540.

Nixon, K. C. and Wheeler, Q. D. (1990) An amplification of the phylogenetic species concept. *Cladistics* 6: 212–223.

O'Brien, S. J. and E. Mayr (1991) Bureaucratic mischief: recognizing endangered species and subspecies. *Science* 251: 1187–1188.

Okayama, H., Curiel, D. T., Brantly, M. L., Holmes, M. D. and Crystal, R. G. (1989) Rapid nonradioactive detection of mutations in the human genome by allele-specific amplification. *J. Lab. Clin. Med.* 114: 105–113.

Orita, M., Iwahana, H., Kanazawa, H., Hayashi, K. and Sekiya, T. (1989) Detection of polymorphisms of human DNA by gel electrophoresis and single-strand conformation polymorphisms. *Proc. Natl. Acad. Sci. USA* 86: 2766–2770.

Polzien, R. O., Strobeck, C., Sheraton, J. and Beech, R. (1995) Bovine mtDNA (mitochondrial DNA) discovered in North American bison populations. *Conserv. Biol.* 9(6): 1638–1643.

Ruano, G. and Kidd, K. K. (1991a) Coupled amplification and sequencing of genomic DNA. *Proc. Natl. Acad. Sci. USA* 88: 2815–2819.

Ruano, G. and Kidd, K. K. (1991b) Genotyping and haplotyping of polymorphisms directly from genomic DNA via coupled amplification and sequencing (CAS). *Nucl. Acid Res.* 19: 6877–6882.

Sommer, S. S., Groszbach, A. R. and Bottema, C. D. K. (1992) PCR amplification of specific alleles (PASA) is a general method for rapidly detecting known single-base changes. *BioTechniques* 12: 82–87.

Swofford, D. L. (1993) *PAUP: Phylogenetic Analysis Using Parsimony*, version 3.0, Illinois Natural History Survey, Champaign, IL.

Vogler, A. P. and DeSalle, R. (1993) Phylogeographic patterns in coastal North American Tiger Beetles (*Cicindela dorsalis* Say) inferred from mitochondrial DNA sequences. *Evolution* 47: 1192–1202.

Vogler, A. P. and DeSalle, R. (1994) Diagnosing units of conservation management. *Conserv. Biol.* 8(2): 354–363.

Vrana, P. and Wheeler, W. (1992) Individual organisms as terminal entities: laying the species problem to rest. *Cladistics* 8: 67–72.

Walsh, P. S., Metzger, D. A. and Higuchi, R. (1991) Chelex 100 as a medium for simple extraction of DNA for PCR-based typing from forensic material. *BioTechniques* 10(4):

506–513.

Wartell, R. M., Hosseini, S. H. and Moran, C. P. (1990) Detecting base pair substitutions in DNA fragments by temperature gradient gel electrophoresis. *Nucl. Acid Res.* 18: 2699–2705.

Wermuth, H. and Mertens, R. (1977) Testudines, Crocodylia and Rhyncocephalia. *Das Tierreich* 100: 1–174.

Wu, W. Y., Ugozzoli, L., Pal, B. K. and Wallace, R. B. (1989) Allele-specific enzymatic amplification of β-globin genomic DNA for diagnosis of sickle cell anemia. *Proc. Natl. Acad. Sci. USA* 86: 2757–2760.

Extinction and the evolutionary process in endangered species: What to conserve?

Alfried P. Vogler

Department of Entomology, The Natural History Museum, London, SW7 5BD, and
Department of Biology, Imperial College at Silwood Park, Ascot, Berkshire, SL5 7PY, UK

Summary

The study of intraspecific variation is important for the management of endangered populations. Current approaches for establishing the value of particular populations for conservation management have been contentious, largely because they draw on different, possibly irreconcilable, scientific concepts. The approach taken here is based on the pattern of variation in molecular or morphological traits. Basic entities for conservation are (groups of) populations whose members exhibit diagnosable characters, consistent with the phylogenetic species concept. Cladistic theory predicts that the extinction of populations can result in formation of allopatric, diagnosable entities from formerly contiguous patterns of variation, a significant event in the evolution of new species. Hence, the effect of extinction on the origination of new diagnosable groups should be taken into account when assessing what to conserve. A procedure for evaluating the potential for cladogenesis within existing species is proposed, as a basis for conservation decisions. The procedure determines if populations represent "linkers" which, because of the alleles present, connect otherwise diagnosable entities; loss of these populations will leave new phylogenetic lineages of high conservation value. The procedure is illustrated using a case study of clinally distributed variation in the endangered tiger beetle, *Cicindela dorsalis*. Two geographic areas along the Atlantic coast of North America are identified where extinction of linker population could produce new diagnosable entities. In one of these cases this process apparently has already taken place after the extinction of numerous populations, leaving a diagnosable relict population of high conservation priority. The procedure will be useful to anticipate the effects of future population loss on the distribution and magnitude of the remaining variation, and could provide testable hypotheses of biodiversity loss under conditions of regional and global environmental change.

Introduction

Populations are the basic entities of conservation management. Given a choice, however, which populations should we aim to conserve, and can systematics contribute the conceptual framework for making these decisions? A recurring issue in

R. DeSalle, B. Schierwater (eds) Molecular Approaches to Ecology and Evolution
©1998, Birkhäuser Verlag Basel

these discussions is the problem of how to best conserve the intraspecific variation present across the geographic range of an endangered species or lineage. Two criteria are usually quoted as the determinants for giving priority to particular populations: (i) the uniqueness of a population or lineage and its level of divergence from other taxa, and (ii) the potential for maintaining evolutionary process for the "protection of future maximum biodiversity" (Erwin, 1991).

The first criterion has received extensive coverage: the basic assertion is that populations or groups of populations that are distinct in their DNA and allozyme markers or their morphological traits should have high conservation status (Avise, 1989b; Bowen et al., 1991; Dizon et al., 1992; Ryder, 1986; Waples, 1991; Woodruff, 1989). On higher taxonomic levels, a variety of cladogram-based measures have been proposed to select taxa for conservation based on their phylogenetic uniqueness (Nixon and Wheeler, 1992b; Vane-Wright et al., 1991) or based on the amount of variation (or "feature diversity") contained in a lineage (Crozier, 1992; Crozier and Kusmierski, 1994; Faith, 1992). However, the discussion about which parameter to maximize and the best way of achieving the specified goal is still contentious (Crozier, 1997). The second criterion for selecting populations for conservation, with the goal of maintaining the evolutionary process to maximize diversity in the future, is currently less well specified. It includes the proposition that diversity has to be maintained so as to preserve the evolutionary potential for survival in a constantly changing environment or in the presence of disease agents (O'Brien, 1994). This is achieved following principles of population genetics, for example by maintaining high levels of heterozygosity or allelic diversity to avoid inbreeding (reviewed in Frankham, 1995; O'Brien, 1994) and other genetic causes that precipitate extinction (Lande, 1988; Lande and Barrowclough, 1987; Soulé, 1987). Preserving the evolutionary process is also the goal of conservation on a higher taxonomic level; of particular conservation value are those lineages that are in a process of fast radiation ("evolutionary fronts"; Erwin, 1991). Under these hypotheses, evolution of a lineage proceeds from a particular center of radiation from where taxa expand and radiate outwards. Phylogenetically "older" lineages tend to radiate less and are eventually replaced by more derived fast-radiating lineages (Erwin, 1985; Wilson, 1961).

Although most of these approaches operate in a narrow range of problems concerned with population-level processes and their effect on the differentiation of lineages and species, they differ in their philosophical underpinnings. The multitude of approaches to specify conservation goals and their various implementations for determining what should have priority has resulted in a confusing variety of definitions, recommendations and measures, many of which are largely *ad hoc*. In the following sections, after a brief discussion of the conceptual differences of current approaches, I will look at the effects of extinction on the level and distribution of diversity in populations and endangered species. I will propose that the pattern of marker distribution can be used to anticipate the effects of future extinctions and demonstrate how this knowledge can be used as a rationale for conservation measures. A case study using data for the endangered tiger beetle, *Cicindela dorsalis*, will be presented to illustrate the effects of past extinctions and anticipated population decline in the future, with an eye towards modeling the future fate of communities and ecosystems under changing environmental conditions on a regional scale.

Local differentiation and conservation units

Studies of variation within traditionally recognized taxa have been carried out for a large number of endangered species. The initial step of these studies is usually to describe the patterns of differentiation across the geographic range of a species. In many widespread species these studies revealed clear subdivision of geographical isolates which is interpreted as the result of paleoclimatic or geological events that affected populations in the past (Avise, 1994). The study of geographic separation of differentiated groups using phylogenetic reconstruction of gene trees is commonly referred to as "phylogeography" (Avise, 1994). This approach has been very influential in its application to issues of intraspecific conservation for several endangered taxa, and the principal ideas are guiding the current use of molecular markers in conservation biology (Avise, 1994; Cronin, 1993; Dizon et al., 1992; Moritz, 1994a, b).

The presumption of most phylogeographic studies is that a gradual evolutionary process is acting which results in the slow accumulation of differences between allopatric populations to account for the eventual differentiation of populations (Avise et al., 1987). Beginning with a largely homogeneous gene pool, the separation of populations results in a slow accumulation of neutral variation, which is subject to lineage sorting and the eventual separation of lineages recognizable as monophyletic allele trees (Moritz, 1994a, Fig. 3; redrawn in Moritz, 1994b; Avise, 1994). When this level of differentiation has been reached, and in particular when the differentiation process has resulted in concordant gene genealogies of mitochondrial and nuclear markers (Avise and Ball, 1990), this indicates the existence of two "species" or "subspecies". Under the presumptions of the biological species concept, the endpoint in this process of gradual divergence results in the reproductive isolation of populations. The presumed process is illustrated schematically in Figure 1. Five hypothetical haplotypes distinguished by DNA polymorphisms are encountered in four continuously distributed populations A through D. Gene flow is eventually severed between populations A plus B and C plus D, followed by changes in gene frequencies and subsequent stochastic loss of haplotypes in one but not the other subgroups (e.g. haplotype 3 is lost in population C of Fig. 1). Eventually, base changes occur in both subgroups independently, which may result in the establishment of two monophyletic (with respect to the single gene marker under investigation) clades originating from the contiguously distributed single assemblage.

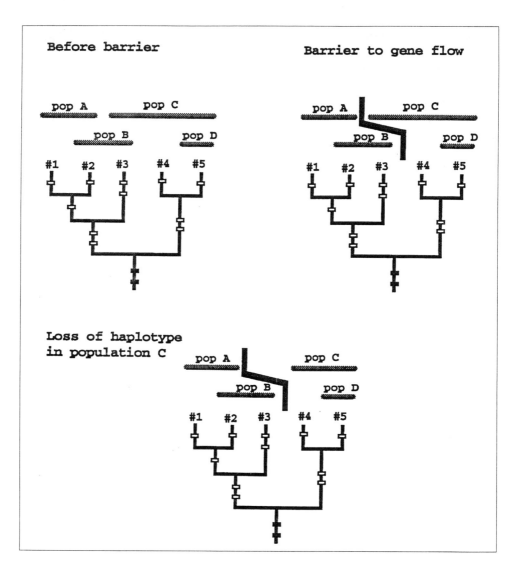

Figure 1. The standard model for the development of phylogenetic structure (after Moritz, 1994b and Avise, 1994). Haplotypes 1 through 5 are found in populations A through D, and traits that define haplotypes are shown by hashmarks. Continuously distributed populations are separated into two after a physical barrier prevents gene flow. Stochastic variation of gene frequencies results in the eventual loss of haplotypes in one but not in the other assemblage. Note that in this example haplotype 3 is lost in population C; stochastic processes could also have caused the loss of haplotype 3 in population B but not population C, in which case the assemblage on the right side of the barrier would be paraphyletic with respect to the other assemblage.

Conservation biology has struggled with the question at which stage of the presumed diversification process entities should be conserved as separate groups. The term *evolutionarily significant unit* (ESU) has been coined to refer to populations within traditionally accepted taxa that are recognizable as "differentiated" (Ryder, 1986). The term has not been applied consistently, however, because authors differ in what type and what amount of differentiation they consider "significant" (Moritz, 1994a, b). Various protocols have been proposed to delimit such units. Some of these maintain that the amount of sequence divergence be used to establish categories with different levels of conservation priority (Dizon et al., 1992), others propose that "significantly" different allele frequencies are considered as conservation targets (Waples, 1991; see also Pennock and Dimmick, 1996 for the implementation of this proposal under the Endangered Species Act of the U.S.), whereas yet other protocols assert that mtDNA haplotypes (which because of their smaller effective population size should go to fixation faster than the diploid nuclear genome) should be of different information content than Mendelian nuclear markers (Moritz, 1994a). These various protocols have led to the recognition of "stocks" (Dizon et al., 1992; Waples, 1991) and "management units" (MU) (Moritz, 1994a) which are characterized by "significant" differences in allele frequencies as targets of conservation management. These entities are distinguished from ESUs, which only refer to groups of populations that are monophyletic with respect to mitochondrial DNA (mtDNA) gene trees.

In summary, phylogeography and the definitions of conservation units derived from this concept build scenarios about historical processes that explain the current pattern of marker distribution by reconstructing gene phylogenies and using a combination of frequency and geographic distribution of haplotypes. The approach draws rather casually on both population genetics (the study of gene frequencies) and phylogenetics (the study of hierarchic structure in nature) to describe and interpret variation. This is problematic, as both population genetic and phylogenetic approaches operate from different and irreconcilable ontological paradigms (Brower et al., 1996).

ESUs as phylogenetic species

A different approach to define ESUs based on the phylogenetic species concept (Cracraft, 1983; Nelson and Platnick, 1981) avoids these conceptual inconsistencies (Amato, 1991; Barrowclough and Flesness, 1996; Dowling et al., 1992; Frost and Hillis, 1990; Legge et al., 1996; Vogler and DeSalle, 1994). ESUs under this definition are diagnosable entities, that is they comprise populations or groups of populations in which all individuals share a (set of) unique character(s) not present in any individual of other such groups. The discovery of such diagnostic characters is taken as evidence that, first, these individuals represent a coherent group sharing a common gene pool and, second, no exchange has occurred between this gene pool and others. Phylogenetic species, therefore, are simply the smallest detectable samples of self-perpetuating organisms (Nelson and Platnick, 1981).

In practice, these samples constitute populations of individuals which exhibit unique attributes not found in other populations (Cracraft, 1983). Operationally, population aggregation analysis (Davis and Nixon, 1992) can be applied for the detection of such conservation units (Vogler and DeSalle, 1994). This method provides an iterative procedure for grouping individuals drawn from different populations into one or more phylogenetic species. Individuals identical for the attributes analyzed are considered to belong to the same phylogenetic species, and all those populations are aggregated into a single entity that also comprises individuals which are identical for that attribute. Those populations whose members exhibit different attributes are diagnosably distinct, and hence represent separate phylogenetic species. In the terminology of Nixon and Wheeler (1990), those attributes that confer distinctions between phylogenetic species are "characters" and those attributes that distinguish individuals within phylogenetic species are "traits". We and others have advocated the application of the phylogenetic species concept to delimit conservation units (Amato, 1991; Barrowclough and Flesness, 1996; Cracraft, 1997; Dowling et al., 1992; Vogler and DeSalle, 1994) as the most objective and practical method which also has a sound philosophical underpinning (Vogler and DeSalle, 1994).

Extinction and the formation of phylogenetic species

I will now investigate the effects of extinction on the pattern of marker distribution and the recognition of diagnosable entities. Extinction of a population will always result in the reduction of total diversity if the extinct population comprises individuals with unique alleles. Loss (and gain) of alleles from populations is one of the main principles and is a prerequisite for the Darwinian understanding of evolution by natural selection, and thus the extinction of an allele *per se* is no reason for concern to conservation. However, the loss of particular populations and alleles does not always affect the level and distribution of diversity in the same way, with significant implications for the purpose of maximizing the preservation of biological diversity.

As Nixon and Wheeler (1992a) demonstrated, the extinction of lineages can result directly in the formation of species when traits are lost from certain sets of individuals. To illustrate the various effects of population extinction on the level and distribution of diversity, we again consider the hypothetical example of four populations and five different haplotypes (see Fig. 2). All of these haplotypes are members of a monophyletic group defined by two synapomorphies (black hashmarks on the cladogram). All other nucleotides are distributed so that populations either contain more than one haplotype or the haplotypes are widespread. Polymorphisms that

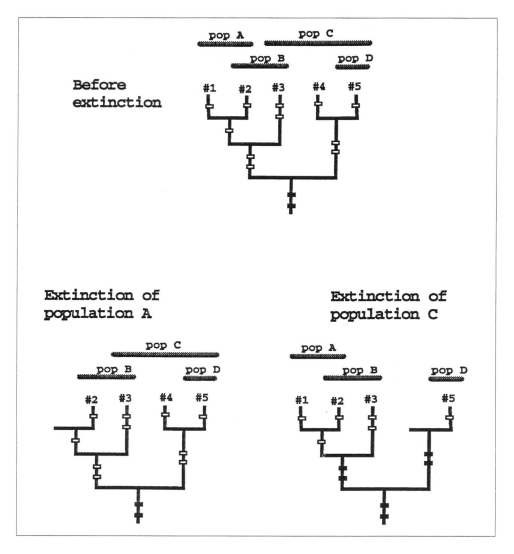

Figure 2. The extinction model for the development of phylogenetic structure. The different effects of extinction of populations A and C on the remaining variation. Certain traits (sensu Nixon and Wheeler, 1990; open hashmarks) would be recognized as true characters (black hashmarks) after the extinction of population C but not population A. As a consequence, more than one diagnosable conservation group can be recognized.

distinguish these haplotypes from each other (white hashmarks in Fig. 2) are not limited to single populations, and thus none of the changes are diagnostic. Beginning with this scenario, hypothetical extinction of populations can have different effects.

Elimination of either population A or C, both of which contain unique haplotypes, results in the complete loss of a particular haplotype. In addition, the disappearance of population C, but not of population A, also has a profound effect on the distribution of variation in the remaining populations. Because the composition of haplotypes in population C overlaps with those in other populations, the loss of this population results in the disconnecting of previously contiguous variation. Consequently, the remaining populations and haplotypes represent two diagnosable groups, consisting of populations A + B and population D. Those synapomorphies that are grouping some of the haplotypes therefore become diagnostic and have to be considered "characters" (*sensu* Nixon and Wheeler, 1990).

The extinction model and the recognition of "linker populations"

If this model of the extinction process is accepted, how can it be applied to practical conservation management? Specifically, how can this model be used to guide the selection of populations that best capture the intraspecific structure of variation to maintain "evolutionary process"? The important event during population extinction is not the loss of a particular haplotype or the loss of overall diversity, but the effect the loss of a population has on the remaining diversity. The extinction model presented above provides a possibility to project what will happen as a result of future loss of populations in an endangered species. If the current patterns of marker distribution are known, this effect can be tested by removing particular populations from the analysis. In cases where the distribution of markers is geographically structured, different outcomes of this removal of populations can be expected. If populations have unique alleles, these are inevitably lost, thus reducing the level of overall diversity. What is more important for the evolutionary future of the species, however, are those population extinctions that result in an altered structure of the total variation, in particular those which result in the formation of additional diagnosable entities. Populations that, if extinct (such as populations B or C in Fig. 2), have such effects are therefore of particular relevance for the assessment of conservation units. These populations, however, themselves may not be the most important targets for conservation, because they do not represent unique variation or a unique component of the (under the extinction model) anticipated process.

I propose the term *linker population* for those populations which connect otherwise diagnostic entities. These population contain a combination of traits that are allopatrically distributed in the remainder of the geographic range. Removing the linker populations, for example as the result of local extinction, will reveal a pattern of diagnosable subdivision. It is this pattern of diagnosable variation that needs to be preserved to ensure the continued existence of evolutionarily significant entities. In practice, the analysis of marker variation has to include a test for the presence of populations which link the otherwise diagnosably distributed variation. This can be done in a modification of population aggregation analysis by sequentially removing particular populations from the analysis for diagnostic markers. In this test, population aggregation analysis is conducted repeatedly, in each round eliminating a single population from the analysis. If the removal of a population results in the recognition of an additional diagnosable unit, this population is designated a linker population. The two diagnosable entities that result from this procedure deserve priority for conservation.

In many real examples where the geographic distribution of clinal variation is less steep, such as the case of the endangered *C. dorsalis* presented below, the removal of a single population may not result in the recognition of additional diagnosable entities. However, if several adjacent populations are eliminated from the population aggregation analysis at the same time, this may result in clearly recognizable groups that should be treated accordingly as independent entities. Though this treatment may appear somewhat subjective (how many populations have to be eliminated before an effect is observed?), it is clear that the aggregation of adjacent populations for this analysis allows the detection of larger-scale patterns in the data. This is a desirable feature of the procedure, because the final goal of the analysis is to identify regional conservation needs which can most easily be detected when regional entities (such as a number of adjacent populations) are pooled.

What to conserve: evolutionary process or pattern of marker distribution?

Inferences about the processes at the "species boundary" are central to evolutionary biology (Avise, 1989a; Harrison, 1990; Harrison, 1991). They are largely dependent on our interpretations of the patterns of marker distribution within and between populations across the geographic range of a species. What generated the patterns is generally thought to have taken long periods of evolutionary time; the phylogeographic scenarios which are commonly inferred from intraspecific phylogenies and allele frequencies are generally considered to have been produced since the Pliocene and Pleistocene, thus dating back several thousand to a few million years. Inferences from the observed patterns (gene phylogenies and allele frequencies) are therefore strongly affected by our background knowledge about the underlying evolutionary processes. Given how hard it is to "know" the processes that have shaped the pattern of variation in the past, it may be equally, if not more, problematic to make projections about how they will affect a lineage in the future (Goldstein et al., 1998). The objective of conservation programs which aim to preserve the future of the evolutionary process or determine "evolutionary potential" and the measure that best "predicts the ability of populations to evolve" (Frankham, 1995) may be quite unrealistic.

The proposed procedure to determine conservation groups, while not free from the problems associated with any process model, may permit some predictions of what lies ahead for an endangered species. The phylogenetic species concept was conceived and has been interpreted in the light of an assumed process, speciation, here understood as the splitting (cladogenesis) of a lineage marked by transformed characters in at least one of the newly formed sublineages. This cladogenetic event may be caused by the extinction of traits, thus changing irreversibly the course of evolutionary history (Nixon and Wheeler, 1992a). The elimination of populations, as shown in Figure 2, is the basic process of allopatric speciation (Nixon and Wheeler, 1992a). Knowledge of the patterns of marker distribution can be used to investigate how extinction will affect variation and its distribution. Accepting this model as the basic process resulting in the formation of phylogenetic lineages, a hypothetical post-extinction pattern can be predicted if the pre-extinction patterns are known from the study of extant populations. In anticipation of this extinction process, it is possible to determine those populations that are potentially most relevant for the evolutionary future of the lineage.

The model does not require any assumptions about what caused the existing patterns of variation, be it selection, differentiation resulting from geographic distance, stochastic processes such as drift, differences in mutation rate or population size, ancestral polymorphisms or others. This is different from models of evolutionary differentiation which assume a continuum from quantitative variation to qualitative variation (Avise, 1994; Dizon et al., 1992; Moritz, 1994a) that begins with the differentiation of allele frequencies and ends with the formation of new species (recognizable by diagnostic characters, monophyletic mtDNA gene phylogenies, reproductive isolation or otherwise). Obviously, this represents only one of many process explanations consistent with the data, only valid under a narrow set of assumptions of a stochastic model of neutral variation which is almost certainly violated on many accounts in natural populations. Similarly, conservation measures aiming to maximize the total amount of variation are not necessarily grounded on the historical process that generated this variation, as the level of variation in a population is affected by factors such as population size, differences in the mutation rate or several other neutral and non-neutral parameters, and thus the observed variation can have different historical causes. Gene frequency data are inherently non-historical and not amenable to reconstructing evolutionary history (Crother, 1990) or, for that matter, for predicting the evolutionary future of endangered lineages. For inferences about what shaped the evolution of endangered populations we are limited to using only the historical information contained in hierarchical relationships and congruent patterns of vicariance [i.e. historical biogeography; Nelson and Platnick (1981)] to make inferences about the geological, paleoclimatic and other factors in earth history which shaped the geographic structure of lineages.

It is this type of information that also should be used predominantly for the choice of differentiated populations in conservation management. The extinction model is a convenient tool as a procedure to extend this historical approach for a likely path in the future. Its value, however, is not to provide guidance for the management of individual populations; for example, the procedure does not address issues like inbreeding depression, metapopulation structure, demographic effects and others, all of which are dependent on gene frequencies and temporary effects that may well be relevant for the short-term management of particular endangered populations. The focus on the preservation of historical patterns advocated here is geared towards the preservation of local differentiation. The maintenance of local variation, as a proxy for local adaptations and local evolutionary processes, seems a more important conservation objective than to aim for maximal overall variation. This is an important consideration, because it is the objective of *in situ* conservation to maintain a given population in a given locality. The selection of such populations should target centers of local "endemism" which potentially are the nucleus for re-colonization of the area from which they had been extirpated.

Conservation units in *C. dorsalis*

The analysis of geographic variation using the described procedures will be illustrated in a case study using the endangered tiger beetle, *C. dorsalis*. The *C. dorsalis* species complex consists of four subspecies and a closely related species (*C. curvata*) which have been distinguished based on morphometric and coloration data (Boyd and Rust, 1982). *C. dorsalis* is distributed along the Atlantic and Gulf of Mexico coast from Massachusetts to Veracruz, Mexico. In large parts of its geographic range this formerly very common and widespread species is currently reduced to a few remnant populations. The affected areas are mostly in the northeastern United States, where the nominate *C. d. dorsalis* is listed under the Federal Endangered Species Act.

Variation in mtDNA has been studied extensively using a total of 434 specimens in a combined sequencing and PCR-RFLP approach (DeSalle and Vogler, 1994; Vogler and DeSalle, 1993a, b). The topology (Fig. 3) of an mtDNA parsimony tree which has been discussed in detail elsewhere is characterized by a complete subdivision of Atlantic Ocean and Gulf of Mexico haplotypes. This subdivision is in accordance with a common biogeographic pattern observed in a number of unrelated species with similar geographic distribution along the shores of the Atlantic and the Gulf of Mexico (Avise, 1992, 1996). The large number of "characters" differentiating these two groups clearly defines them also as two different conservation groups. Within the Atlantic lineage, 15 different haplotypes could be separated based on 25 variable nucleotide positions. The phylogenetic relationships of most of these haplotype sequences are poorly resolved except for a single node that defines haplotypes A9 through A12. The boundary between the two Atlantic coast subspecies, *C. d. dorsalis* and *C. d. media*, was not discernible in the pattern of mtDNA distribution. This suggests that the relationships between the populations (as indicated by the mtDNA sequences) are not well described as a hierarchy. Also, these data clearly fail to define both subspecies as diagnosable groups with respect to mtDNA.

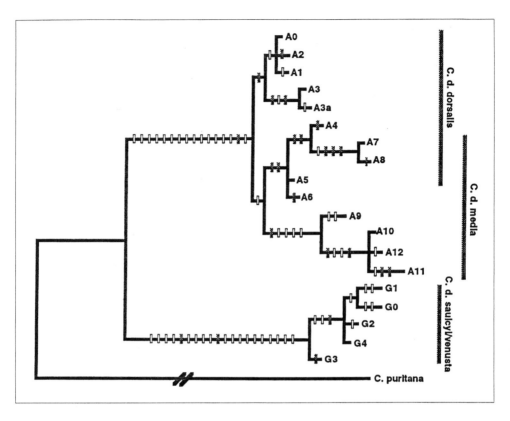

Figure 3. Gene phylogeny for C. dorsalis. *Homoplastic changes are shown by stippled hashmarks. Vertical bars indicate the subspecies in which these haplotypes occur.*

Despite the overall lack of hierarchical structure, there is clear structure in the geographic distribution of markers, as many nucleotide polymorphisms are confined to particular geographic regions. For example, members of the A9–A12 mtDNA haplotype clade occur largely in the southern part of the range, with a slowly increasing frequency in the mid-Atlantic States in Georgia and North Carolina (Fig. 4). The population aggregation analysis (Tab. 1) suggested that distribution of markers was continuous and that essentially no diagnosable subgroupings exist within a wide geographic area covering coastal North America from 28 to 41 degree latitude. However, a single population in Martha's Vineyard (MV), at the extreme northern end of the geographic distribution, is an exception to this because it exhibits a single diagnostic character change and thus represents a recognizable entity (Vogler and DeSalle, 1993a, b).

We then investigated the effects of extinction of particular populations on distribution of diversity in the *C. dorsalis* complex (Tab. 2). The most severe effect on the

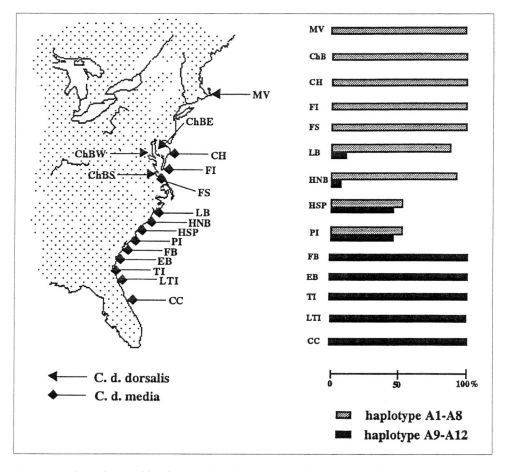

Figure 4. The "cline" of haplotypes A0–A8 and A9–A12 along the Atlantic coast. Populations of C. d. dorsalis *and* C. d. media *exhibit the two main groups of haplotypes in the frequencies shown.*

overall level of diversity resulting from the extinction of a single population is observed if the MV population disappears: not only are three unique haplotypes lost which are found nowhere else, but the disappearance of this population will also result in the loss of a diagnosable entity (Tab. 2). Since we are interested in the effects of extinction on a wider regional scale, we also investigated how the loss of several adjacent populations will effect the level of overall diversity. If three adjacent populations (a focal population and its two nearest neighbors) are assumed to be extinct, the loss of diversity was fairly minor except at the northern end of the distribution, where the loss of the MV population and the Chesapeake Bay populations combined would result in the loss of up to 6 (of 15 total) haplotypes. We also in-

Table 1. Number of mtDNA haplotypes and number of specimens (n) tested in Atlantic populations of C. dorsalis

Popu-lation	A0	A1	A2	A3	A3a	A4	A5	A6	A7	A8	A9	A10	A11	A12	n
							Haplotypes								
MV	20	1	1												22
ChE				43	5		9		7	10					74
ChW				60		29	20								89
ChS							5		11	5					21
CH							21								21
FI							28	1							29
FS							2		1						3
LB									15	6		2			23
HNB									2	13		1			16
HSP							1		12	1	6	3	3	1	26
PI									10	4		12			26
FB												3			3
EB												4			4
TI												15	3		18
LTI												18	4		22

Population codes as in Figure 4.

vestigated if further diagnosable entities can result from the extinction of these populations (Tab. 2). The elimination of any single population did not result in the recognition of further diagnostic populations, nor did the elimination of their neighboring populations on either side. However, if a total of five populations are lost (the focal population plus the two nearest neighbors on either side), we observe two cases, populations HSP and PI, in which the continuous variation in the group is subdivided into two diagnosable entities. Under these conditions these two populations are linker populations. Their extinction will be of particular significance to the remainder of the assemblage.

The isolated position of the MV population is intriguing. This population located at the very northern end of the geographic range represents a relict occurrence of C. dorsalis after numerous populations in the northeastern United States have gone to extinction as a result of habitat destruction since the beginning of the century when the species was common and widespread on the barrier beaches of the coastline from Massachusetts southward. It is conceivable that the mtDNA haplotype distribution in the extinct populations was also clinal and provided a link between the Chesapeake Bay and MV populations. In this case the linker populations would be all extinct, leaving a newly formed diagnosable entity of high conservation sta-

Table 2. *Effect of population extinction and test for linker populations in C. dorsalis*

Population	No. individ.	No. alleles	No. alleles lost if extinct	No. lost if neighbors extinct	No. lost if two neighbors extinct	Diagno- sable?	New diagnos. unit if 2 neighbors extinct
MV	22	3	3	4	6	Y	N
ChBWest	89	3	1	6	6	N	N
ChBEast	74	5	1	3	6	N	N
ChBSouth	21	3	0	1	4	N	N
CH	21	1	0	0	2	N	N
FI	29	2	1	1	1	N	N
FS	3	2	0	0	1	N	N
LB	23	3	0	0	2	N	N
HNB	16	3	0	1	1	N	Y
HSP	26	6	1	1	1	N	Y
PI	26	3	0	1	1	N	Y
FB	3	1	0	0	1	N	N
EB	4	1	0	0	0	N	N
TI	18	2	0	0	0	N	N
LTI	22	2	0	0	0	N	N

Note that the extinction of populations HNB or HSP (including two neighboring populations on each side) results in the recognition of two diagnosable entities (plus the population MV), although the total loss of haplotype diversity resulting from theses extinctions is small. See Figure 4 for the geographic location of populations.

tus. This hypothesis can be tested by analyzing haplotypes of museum specimens that have been collected in large number before the populations went extinct. For example, we have already analyzed a single specimen from Block Island, an offshore island some 40 miles southwest of the MV population, collected in 1972 shortly before this population disappeared. The analysis of a small fraction of the mtDNA genome of this specimen (unpublished data) clearly indicated the presence of the unique G to A32 nucleotide change in the COIII gene diagnostic for the haplotype A0–A2 (Vogler and DeSalle, 1993b), indicating that these haplotypes, now endemic to MV, were more widespread along the coastline and were possibly overlapping with the extant haplotypes further south. As the result of the extinction of linker populations, cladogenesis has apparently already occurred in MV. While these studies have yet to be carried out in more detail, the extinction model provides clearly testable hypotheses about the historical distribution of variation and provides a rationale for a sampling strategy for the use of museum specimens in such analysis.

The analysis of the extinction pattern is partly dependent on the sampling. For example, in the Chesapeake Bay where several large populations of *C. d. dorsalis* still exist, we combined several adjacent populations for the purposes of the analysis presented here (see Vogler, 1994, for complete data set). If the analysis for linker populations was carried out on these individual populations separately, even the removal of five adjacent populations did not result in new diagnosable populations; however, at the very northern part of the range of *C. dorsalis* in the Chesapeake Bay, two populations become recognizable as diagnosable entities after all other populations in the Chesapeake Bay are extinct. While these fulfill the criteria for recognition as conservation units only under very specific conditions, this information can be used as a hint for conservation priorities within the Chesapeake Bay where clinal variation was also found in the killifish *Fundulus heteroclytus* (Powers and Schulte, 1996; Powers et al., 1993), indicating congruent patterns of North–South clinal variation (rather than, e.g., differentiation of the western and eastern shores of the Bay).

The distribution of populations in the Chesapeake Bay also points to the problem of scale in these analyses: The sampling needs to be sufficiently comprehensive in terms of the number of populations, the number of individuals per population and the number of variable attributes scored in each individual to detect potentially diagnosable entities. But more important for the sampling regime is the scale of the area in which diagnosable variation occurs. The approach of scoring hypothetical extinctions of populations can be used to determine the size of these regions. This information defines the size of a region that should be considered a single conservation area on which to manage local diversity. The regional approach to determining diagnosable variation can also be applied across different taxa with similar geographic distribution and similar distribution of genetic variation to analyze congruent patterns of variation. The southeastern United States is an excellent region for testing for these congruent patterns, because a large number of terrestrial, marine and freshwater species have been studied for the distribution of genetic markers (Avise, 1996). Determining the location of linker populations and the size of the area they occupy provides information on the geographic scale of conservation measures.

Conclusion

Patterns are the prerequisite for inferring evolutionary process. Cladistic theory provides a framework for a process model of cladogenesis which requires that alleles and populations go extinct locally when a lineage is splitting. This model is extended here to derive conservation priorities from patterns of marker distribution across the geographic range of a species. The entities that remain after the hypothetical extinction of linker populations will maintain a large part of the unique diversity that is contained in geographically structured assemblages. In addition, they will also serve the goal of potentially maintaining cladogenesis as the evolutionary process that resulted in the ever-increasing species numbers in Earth history. Using this procedure for determining conservation priorities, it is possible to select much of the existing intraspecific variation by conserving a minimum set of populations (conservation units). It is obvious that the lineages so preserved face a precarious existence, without the buffering effects of a close network of populations (metapopulations), recombination between related lineages and limited geographic distribution. Such lineages therefore can be predicted to be short-lived (Nixon and Wheeler, 1992a), and the preservation of the presumed evolutionary process using such a limited set of populations is inevitably faced with a heightened risk for the assemblage to disappear completely.

Loss of biodiversity frequently happens on a regional scale, and selecting potentially diagnostic variation as proposed here is a way of detecting the effects of disturbance across large geographic regions. Because of the widespread loss of natural areas, continental habitats are being reduced to scattered island-like remnants. It is likely that most of the planet will be affected by such fragmentation and local population extinctions, making the described scenario a common occurrence. This will "inevitably disrupt the evolutionary process as we know it" (Erwin, 1991). The analysis of extinction and its effect on the overall diversity of lineages will permit predictions about the future of biological diversity. The study of broad patterns of communities and ecosystems using congruent distribution of genetic markers – with improved DNA technology now a realistic possibility – will provide a trajectory of predicted processes in the face of global environmental change.

Acknowledgments

I am indebted to Rob DeSalle, who contributed in many ways to the ideas presented in this paper and who first suggested the term "linker population". I thank Paul Goldstein for valuable discussions and inspiration. My work on *Cicindela* tiger beetles was supported by the National Science Foundation (DEB 9225074) and the Natural Environmental Research Council of the United Kingdom (GR3/10632).

References

Amato, G. D. (1991) Species hybridization and protection of endangered animals. *Science* 253: 250.

Avise, J. C. (1989a) Gene trees and organismal histories: A phylogenetic approach to population biology. *Evolution* 43: 1192–1208.

Avise, J. C. (1989b) A role for molecular genetics in the recognition and conservation of endangered species. *Trends Ecol. Evol.* 4: 1–10.

Avise, J. C. (1992) Molecular population structure and the biogeographic history of a regional fauna: a case history with lessons for conservation biology. *Oikos* 63: 62–76.

Avise, J. C. (1994) *Molecular Markers, Natural History and Evolution*, Chapman and Hall, New York.

Avise, J. C. (1996) Toward a regional conservation genetics perspective: Phylogeography of faunas in the Southeastern United States. *In:* J. C. Avise and J. L. Hamrick (eds) *Conservation Genetics: Case Histories from Nature*, Chapman and Hall, New York, pp. 431–470.

Avise, J. C., Arnold, J., Ball, R. M., Bermingham, E., Lamb, T., Neigel, J. E., Reeb, C. A. and Saunders, N. C. (1987) Intraspecific phylogeography: The mitochondrial bridge between population genetics and systematics. *Annu. Rev. Ecol. Syst.* 18: 489–522.

Avise, J. C. and Ball, R. M. (1990) Principles of genealogical concordance in species concepts and biological taxonomy. *Oxf. Surv. Evol. Biol.* 7: 45–68.

Barrowclough, G. F. and Flesness, N. R. (1996) Species, subspecies and races: the problem of units of management in conversation. *In:* Kleiman, D. G. (ed.) Wild Mammals in Captivity: Principles and Techniques, University of Chicago Press, Chicago, IL, pp. 247–254.

Bowen, B. W., Meylan, A. B. and Avise, J. C. (1991) Evolutionary distinctiveness of the endangered Kemp's ridley sea turtle. *Nature* 352: 709–711.

Boyd, H. P. and Rust, R. W. (1982) Intraspecific and geographical variations in *Cicindela dorsalis* (Coleoptera: Cicindelidae). *Coleopts. Bull.* 36: 221–239.

Brower, A. V. Z, DeSalle, R. and Vogler, A. P. (1996) Gene trees, species trees and systematics. *Annu. Rev. Ecol. Syst.* 27: 423–450.

Cracraft, J. (1983) Species concept and speciation analysis. *Curr. Ornithol.* 1: 159–187.

Cracraft, J. (1997) Species concepts in systematics and conservation biology – an ornithological viewpoint. *In:* Claridge, M. F., Dawah, H. A. and Wilson, M. R. (1997) *Species: The Units of Biodiversity*, Chapman and Hall, New York, pp. 325–340.

Cronin, M. A. (1993) Mitochondrial DNA in wildlife taxonomy and conservation biology: cautionary notes. *Wildlife Soc. Bull.* 21: 339–248.

Crother, B. I. (1990) Is "some better than none" or do allele frequencies contain phylogenetically useful information? *Cladistics* 6: 277–281.

Crozier, R. H. (1997) Preserving the information content of species: genetic diversity, phylogeny, and conservation worth. *Annu. Rev. Ecol. Syst.* 28: 243–268.

Crozier, R. H. (1992) Genetic diversity and the agony of choice. *Biol. Conserv.* 61: 11–15.

Crozier, R. H. and Kusmierski, R. M. (1994) Genetic distances and the setting of conservation priorities. *In:* Loeschcke, V., Tomiuk, J. and Jain, S. K. (ed.) *Conservation Genetics*, Birkhäuser, Basel, Switzerland, pp. 227–237.

Davis, J. I. and Nixon, K. C. (1992) Populations, genetic variation and the delimitation of phylogenetic species. *Syst. Biol.* 41(4): 421–435.

DeSalle, R. and Vogler, A. P. (1994) Phylogenetic analysis on the edge: the application of cladistic techniques at the population level. *In*: Golding, B. (ed.) *Non-neural evolution. Theories and molecular data*, Chapman and Hall, New York, pp. 154–172.

Dizon, A. E., Lockyer, C., Perrin, W. F., Demaster, D. P. and Sisson, J. (1992) Rethinking the stock concept. *Conserv. Biol.* 6: 24–36.

Dowling, T. E., DeMarais, B. D., Mickley, W. L. and Douglas, M. E. (1992) Use of genetic characters in conservation biology. *Conserv. Biol.* 7–8.

Erwin, T. L. (1985) The taxon pulse: a general pattern of lineage radiation and extinction among carabid beetles. *In*: Ball, G. E. (ed.) *Taxonomy, Phylogeny and Zoogeography of Beetles and Ants*, W. Junk, Dordrecht, pp. 437–472.

Erwin, T. L. (1991) An evolutionary basis for conservation strategies. *Science* 253: 750–752.

Faith, D. P. (1992) Conservation evaluation and phylogenetic diversity. *Biol. Conserv.* 61: 1–10.

Frankham, R. (1995) Conservation genetics. *Annu. Rev. Genet.* 29: 305–327.

Frost, D. R. and Hillis, D. M. (1990) Species in concept and practice: *Herpetological* applications. *Herpetologica* 46: 87–104.

Goldstein, P. Z.(1997) Functioning ecosystems and biodiversity buzzwords. *Conserv. Biol.* ; *in press.*

Harrison, R. G. (1990) Hybrid zones: windows on evolutionary process. *Oxf. Surv. Evol. Biol.* 7: 69–128.

Harrison, R. G. (1991) Molecular changes at speciation. *Annu. Rev. Ecol. Syst.* 22: 281–308.

Lande, R. (1988) Genetics and demography in biological conservation. *Science* 241: 1455–1460.

Lande, R. and Barrowclough, G. F. (1987) Effective population size, genetic variation, and their use in population management. *In*: Soulé, M. E. (ed.) Viable Populations for Conservation, University Press, Cambridge, pp. 87–123.

Legge, J.T, Roush, R., DeSalle, R., Vogler, A. P. and May, B. (1996) Genetic criteria for establishing evolutionarily significant units in Cryan's buckmoth. *Conserv. Biol.* 10: 85–90.

Moritz, C. (1994a) Applications of mitochondrial DNA analysis in conservation: a critical review. *Molec. Ecol.* 3: 401–411.

Moritz, C. (1994b) Defining "Evolutionary Significant Units" for conservation. *Trends Ecol. Evol.* 9: 373–375.

Nelson, G. J. and Platnick, N. I. (1981) *Systematics and Biogeography: Cladistics and Vicariance*, Columbia University Press, New York.

Nixon, K. C. and Wheeler, Q. D. (1990) An amplification of the phylogenetic species concept. *Cladistics* 6: 212–223.

Nixon, K. C. and Wheeler, Q. D. (1992a) Extiction and the origin of species. *In*: Novacek, M. J. and Wheeler, Q. D. (eds) *Extinction and Phylogeny*, Columbia University Press, New York, pp. 117–143.

Nixon, K. C. and Wheeler, Q. D. (1992b) Measures of phylogenetic diversity. *In*: Novacek, M. J. and Wheeler, Q. D. (eds) *Extinction and Phylogeny*, Columbia University Press, New York, pp. 216–234.

O'Brien, S. J. (1994) Genetic and phylogenetic analyses of endangered species. *Annu. Rev. Genet.* 28: 467–489.

Powers, D. A. and Schulte, P. M. (1996) A molecular approach to the selectionist/ neutralist controversy. *In*: Ferraris, J. D. and Palumbi, S. R. (eds) *Molecular Zoology: Advances, Strategies and Protocols*, Wiley-Liss, New York, pp. 327–352.

Powers, D. A., Smith, M., Gonzalez-Villasenor, I., DiMichelle, L, Craford, D. L., Bernardi, G. and Lauerman, T. A. (1993) Multidisciplinary approach to the selectionist/neutralist controversy using the model teleost *Fundulus heteroclitus*. *Oxf. Surv. Evol. Biol.* 9: 43–109.

Ryder, O. A. (1986) Species conservation and systematics: The dilemma of subspecies. *Trends Ecol. Evol.* 1: 9–10.

Soulé, M. E. (1987) *Viable Populations for Conservation*, Cambridge University Press, Cambridge.

Vane-Wright, R. I., Humphries, C. J. and Williams, P. H. (1991) What to protect? – Systematics and the agony of choice. *Biol. Conserv.* 55: 235–254.

Vogler, A. P. (1994) Extinction and the formation of phylogenetic lineages: diagnosing units of conservation management in the tiger beetle *Cicindela dorsalis*. *In*: Schierwater, B., Streit, B., Wagner, G, P. and DeSalle, R. (eds) *Molecular Ecology and Evolution: Approaches and Applications*, Birkhäuser, Basel, pp. 261–273.

Vogler, A. P. and DeSalle, R. (1993a) Mitochondrial DNA evolution and the application of the phylogenetic species concept in the *Cicindela dorsalis* complex (Coleoptera: Cicindelidae). *In*: Desender, K. (ed.) *Carabid Beetles: Ecology and Evolution*, Kluwer, Dordrecht, The Netherlands, pp.79–85.

Vogler, A. P. and DeSalle, R. (1993b) Phylogeographic patterns in coastal North American Tiger Beetles, *Cicindela dorsalis* inferred from mitochondrial DNA sequences. *Evolution* 47: 1192–1202.

Vogler, A. P. and DeSalle, R. (1994) Diagnosing units of conservation management. *Conserv. Biol.* 8: 354–363.

Waples, R. S (1991) Pacific salmon, *Oncorhynchus* spp. and the definition of a "species" under the Endangered Species Act. *Marine Fisheries Rev.* 53: 11–22.

Wilson, E. O. (1961) The nature of the taxon cycle in the Melanesian ant fauna. *Amer. Naturalist.* 95: 169–193.

Woodruff, D. S. (1989) The problem of conserving genes and species. *In*: Western, D. and Pear, M. C. (ed.) *Conservation for the Twenty-First Century*, Oxford University Press, New York, pp. 76–78.

Quantitative trait loci: a new approach to old evolutionary problems

Eric Routman[1] and James M. Cheverud[2]

[1]Department of Biology, San Francisco State University, 1600 Holloway Ave., San Francisco, CA 94132, USA
[2]Department of Anatomy & Neurobiology, Washington University School of Medicine, St. Louis, MO 63110, USA

Summary

Because of innovations in quantitative and molecular genetics, it is becoming possible to study the individual loci (QTL) affecting quantitative traits. We review the results of our empirical and theoretical research on QTL effects and interactions. We believe this area of investigation, already in widespread use in the agricultural and medical sciences, will become increasingly important in evolutionary biology as a way of addressing some very old but stubborn questions.

Introduction

Modern genetics began with Gregor Mendel's discovery of the particulate nature of inheritance. However, at the time, many people working in animal and plant breeding and evolutionary biology doubted that Mendel's laws were universal. The vast majority of the traits with which they were familiar varied continuously, rather than in discrete Mendelian ratios. It was not until Fisher (1918) reconciled particulate inheritance with continuous phenotypic variance that genes were recognized as the basis of all inheritance. Because multiple loci are difficult to detect with traditional Mendelian methods, evolutionary biologists would still have been unable to take advantage of this knowledge if Fisher had not also developed methods that used the covariance among relatives to dissect the genetic basis of quantitative traits. These methods, known as biometrical or quantitative genetics, allowed prediction of the change in mean phenotype and response to selection on phenotypic values. There followed an enormously fruitful period of theoretical and empirical elaboration of Fisher's ideas. It has come to be recognized that for accurate understanding of evolutionary change, knowledge of the quantitative genetics of the phenotype of interest is equal in importance to understanding the phenotype/environment relation-

R. DeSalle, B. Schierwater (eds) Molecular Approaches to Ecology and Evolution
©1998, Birkhäuser Verlag Basel

ship. Analyses based on the covariance of phenotype among relatives have led to a mature and highly predictive theory of evolution.

Yet many controversies remain in evolutionary biology. Perhaps suprisingly, quite a few of these controversies revolve around genetic architecture, the number of genes involved in evolutionarily important phenotypes and the nature of their interactions (Templeton, 1980, 1982). Fisher originally derived (and every student of quantitative genetics initially learns) biometrical genetics from the perspective of individual locus effects on phenotype. But the elegance of quantitative genetic methods lies in their ability to ignore the complexity of many individual loci in favor of the simplicity of phenotypic covariances among relatives. This requires simplifying assumptions about gene action that some models of evolutionary change explicitly violate. For example, take the notion that population bottlenecks or founding events can cause drastic change in phenotypes, perhaps leading to speciation (Mayr, 1963; Carson, 1968; Templeton, 1980). This long and occasionally bitter debate hinges almost entirely on whether fitness is determined by a large number of additive loci or a few epistatic loci (e.g. Barton and Charlesworth, 1984 vs. Carson and Templeton, 1984). Other examples of evolutionary controversies centered on genetic architecture are the maintenance of additive genetic variance in selected phenotypes [mutation-selection balance vs. antagonistic pleiotropy (Rose, 1982)], the causes of increased additive genetic variance in some experimentally bottlenecked populations (Bryant and Meffert, 1993; Lynch, 1988), the relative evolutionary importance of genes of minor vs. major phenotypic effect (Coyne and Lande, 1985; Carson and Templeton, 1984; Barton and Turelli, 1989) and the importance of the shifting balance process to evolution (Wright, 1931). Some of these controversies are over 60 years old – remarkable stubborness given the current pace of scientific advances.

Of course, the reason that many of these arguments have persisted unresolved for so long is that it was not possible until recently to isolate the individual loci affecting complex traits (quantitative trait loci, or QTL). However, modern developments in molecular genetics and analytical methods have greatly improved our ability to localize and quantify these genes (Lander and Botstein, 1989; Tanksley, 1993). Development of high-throughput laboratory techniques for detection of large numbers of highly variable loci has made it relatively easy to saturate the genome with genetic markers that can be correlated with phenotypic variation. When combined with new analytical algorithms for estimating the effects and location of QTL, the molecular techniques are beginning to shed new light on some very old arguments. Our chapter in the previous edition of this book (Routman and Cheverud, 1994a) reviews the molecular techniques and analytical methods relevant to QTL analysis (although the field is progressing so rapidly that the coverage of both topics is already out of date). This chapter will review our own empirical and theoretical studies identifying QTL and gene interaction effects. We will present this work as examples of the kinds of old but still unresolved questions that QTL analysis can help answer.

Antagonistic pleiotropy and genetic modularization

A negative genetic correlation between early and late growth in mice (Riska et al., 1984) led to the conclusion that alleles that cause faster than average early growth are in linkage disequilibrium with alleles that cause slower than average late growth. One way this can occur is when individual loci influence both early and late growth, and alleles within loci exhibit opposite pleiotropic effects on the two growth periods. An alternative cause of the negative correlation could be linkage disequilibrium between loci that affect early growth and other loci that affect late growth. Selection for a specific intermediate range of adult body size could cause "fast" alleles at loci affecting early growth to be associated (within individuals) with "slow" alleles at other loci that influence late growth. The relative contribution of these two mechanisms will determine the ability of the organism to respond to directional selection for body size, and can only be resolved when the individual loci are known.

We conducted an experiment to discover the QTL influencing body weight in two strains of inbred mice (Cheverud et al., 1996). The strains, LG (Large) and SM (Small) originate from selection experiments (Chai, 1956a,b, 1961, 1968) on body size, and so are expected to have allelic differences at large numbers of loci affecting growth. Details of the experimental design are given elsewhere (Routman and Cheverud, 1994b; Cheverud et al., 1996), so we will merely summarize here. LG females were mated to SM males, and the resulting F_1 were intercrossed to produce 534 F_2 offspring segregating for all loci that differ between the two parental strains. The F_2 hybrids were weighed weekly for 10 weeks, sacrificed and autopsied. Each individual was scored for 75 microsatellite loci dispersed througout the genome, except for the sex chromosomes (Fig. 1). Interval mapping, using the program MAP-MAKER (Lincoln et al., 1992a,b), was used to identify QTL location and effect on growth. Early growth was defined as growth from 1 to 3 weeks of age; late growth was defined as growth from 6 to 10 weeks of age.

We detected 11 QTL affecting early growth and 12 QTL affecting late growth, explaining 39 and 38%, respectively, of the variance among the F_2 hybrids (table 5 of Cheverud et al., 1996). Early growth QTL map to different chromosomal locations than do late growth QTL (Fig. 1). [Although four chromosomal intervals have the potential to hold QTL affecting both early and late growth, the loci do not map to the same location. Further, one expects independent early and late growth loci to map to the same interval five times by chance alone (Cheverud et al., 1996)]. The distinct nature of the loci affecting early and late growth decisively falsifies the hypothesis of antagonistic pleiotropic effects for growth in these strains.

However, we also found a very low phenotypic correlation between early and late growth, contrary to the results of Riska et al. (1984) for a random-bred strain of mouse. The difference in the two experiments could be due to the evolutionary history of the strains involved, specifically, directional artificial selection in the lines leading to LG and SM and presumed stabilizing natural selection in the random-

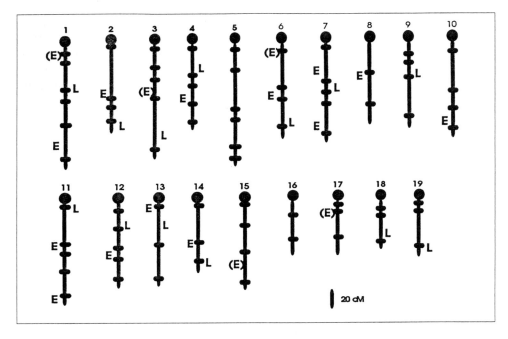

Figure 1. Chromosome map of the lab mouse, showing location of marker loci (hatch marks) and estimated QTL affecting early (E) and late (L) growth. Letters in parentheses are estimated QTL that did not quite reach statistical significance.

bred strain. Directional selection on body size could select for modifier loci whose action is to eliminate the undesirable pleiotropic effects of alleles at growth QTL, whereas stabilizing natural selection in the random-bred strains would maintain antagonistic pleiotropy. This model of the evolution of pleiotropy was formally developed by Riedl (1978) and Wagner (1989, 1996). QTL analysis for growth in crosses of inbred strains that are not the result of selection on body size will provide an important test of this model.

Another prediction of the aforementioned models of the evolution of pleiotropy is that pleiotropic effects of single genes will be restricted to traits that are functionally and/or developmentally related (genetic modularization). Traits that are not related will be under the control of separate sets of genes. We were able to test this prediction with our mouse experiment (Cheverud et al., 1997). In addition to body weight, we also measured aspects of mandibular morphology. The mandible consists of two parts, the corpus and the ramus, each formed from a separate mesenchymal condensation during ontogeny and maintained by association with separate sets of surrounding tissues. We measured 15 mandibular landmarks from the right mandible with a digital video data collection system. We used the landmark data to construct a set of 21 linear distances chosen to delineate the corpus and the ramus. We estimated the location and effect of 37 QTL affecting the corpus and ramus trait groups. Of these, 11 QTL only affect a single distance, and therefore have no opportunity to show pleiotropic effects either within or between trait groups. Of the 26 QTL affecting more than one distance, only 6 (23%) influence traits across the entire mandible. The rest have effects restricted to either the ramus or the corpus. This restricted pleiotropy is precisely what is predicted by the models of genetic modularization.

Epistasis

Like the models of the evolution of pleiotropy, many of the controversial hypotheses in evolutionary biology assume epistasis – interaction among genes at different loci, such that the effects of alleles at one locus depend on the genotypes at other loci. When the loci affecting the trait are unknown, epistasis can only be detected indirectly, by its effects on the covariance among relatives. However, because of the influence of allele frequencies in the calculation of epistatic variance, there is not a one-to-one correlation between epistatic variance at the population level and interaction effects among genes. Epistasis at the gene level contributes mainly to additive genetic variance, and epistatic variance is usually small, even if plenty of epistasis exists among loci. Cheverud and Routman (1995) developed an analytical technique that allows gene-level two-locus epistasis to be detected directly in an F_2 generation, without using epistatic variance, when QTL are known.

When QTL (or linked markers) have been identified, it is possible to estimate the genotypic effects at each locus unweighted by allele frequencies. The nonepistatic genotypic value for any two-locus genotype is given by

$$ne_{ijkl} = G_{ij..} + G_{..kl} - G_{....} \tag{1}$$

where G refers to the genotypic value, subscripts i and j refer to one locus and subscripts k and l refer to the other locus, and the standard dot notation indicates averages over the genotypes at the dotted locus. The epistatic genotypic value (e_{ijkl}) is simply the difference between the actual mean phenotype of a two locus genotype (G_{ijkl}) and the nonepistatic genotypic value.

Interaction deviations are given by

$$I_{ijkl} = e_{ijkl} - e_{ij..} - e_{..kl} + e_{....} \tag{2}$$

where $e_{....}$ is the frequency-weighted average epistasis in the population. The variance of the interaction deviations is the interaction or epistatic variance. Full derivation of these values and associated statistical tests are given in Cheverud and

Routman (1995) and Routman and Cheverud (1997). Calculation of these values for QTL is a direct and powerful way to detect the presence of gene interactions.

We used our microsatellite marker data and estimated QTL locations to examine two-locus epistasis for 10-week body weight in our F_2 mouse population (Routman and Cheverud, 1997). Body weight is considered the archtypal quantitative trait, affected by many loci with mainly additive effects. And indeed, using interval mapping we discovered 19 QTL, on 13 chromosomes, affecting 10-week body weight. However, we also detected considerable epistasis among these QTL. All QTL were involved in statistically significant epistatic interactions, and several exhibited pairwise epistasis with multiple loci. Epistatic values were generally higher than 0.75 g, indicating a 1.5-g deviation from additivity (nonepistatic values were in the area of 30 g). Most of these interactions would not be detected by testing for significant epistatic variance.

When we examined epistasis among marker loci, we of course detected the epistasis caused by interaction among the estimated QTL. In addition, however, we found highly significant additive-by-additive epistasis among six marker loci not linked to any estimated QTL. This suggests that body weight QTL do in fact exist near these markers, but the effects in an interval mapping analysis are masked by the epistatic interactions. We have shown that the suppression of additive effects by additive-by-additive epistasis ("epistatic nullification") can occur at intermediate allele frequencies (Cheverud and Routman, 1995, 1996). If this effect is common, it means that QTL analyses may underestimate the number of loci affecting phenotypes unless gene interactions can be accounted for in the analysis.

Much of the argument about the effects of population bottlenecks hinges on how bottlenecks will affect additive genetic variance (V_a). If locus interaction is strictly additive, severe reductions in population size reduce V_a. As a result, natural selection will not be able to change phenotypes, and bottlenecks are unlikely to be involved in the speciation process (Barton and Charlesworth, 1984) or cause peak shifts in an adaptive landscape (Barton and Rouhani, 1987). If epistasis and dominance exist, then it is possible that altering allele frequencies via population size reduction will increase additive genetic variance (Templeton, 1980; Cheverud and Routman, 1996). We used the two-locus epistasis model described above (Cheverud and Routman, 1995) to simulate the effects of bottlenecks on V_a (Cheverud and Routman, 1996). We calculated V_a of populations undergoing several bottleneck sizes (2, 8, 16, 32 and 64 individuals) drawn from a parental population with intermediate allele frequencies at two loci. Frequency distributions of populations displaying all possible combinations of allele frequencies at the two loci were calculated using the Markov chain model (Crow and Kimura, 1970) and V_a was averaged over all populations within each of many generations with no population size increase. This estimates the average effect of two-locus epistasis on V_a for bottlenecks of varying severity. We simulated the effects of arbitrarily chosen values of pure forms of additive-by-additive, additive-by-dominance, and dominance-by-domi-

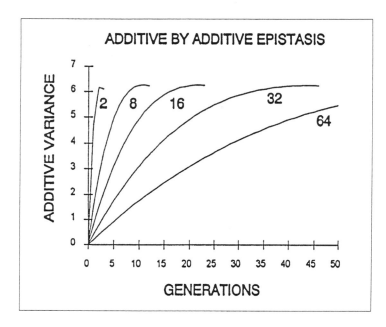

Figure 2. The effects of pure additive by additive epistasis on additive genetic variance in populations experiencing a bottleneck. Bottleneck size is indicated next to each curve. Curves are truncated one generation after peak additive genetic variance is reached. Additive variances have been multiplied by 100.

nance epistasis, as well as for empirical data from two pairs of markers from our mouse body weight experiment.

Our results with both theoretical and empirical examples show that two-locus epistasis can indeed result in increases in additive genetic variance, not only for a few particular sets of allele frequencies but on average across populations experiencing the same bottleneck size. A surprising discovery was that in many cases V_a continued to increase for many generations after the initiation of the bottleneck (e.g. Fig. 2). Many of the theories regarding bottleneck speciation, such as Templeton's genetic transilience or Carson's founder-flush models, have postulated that a period of rapid population growth is necessary immediately after the bottleneck, in order to prevent the decay of V_a. Our results suggest that the increase of V_a may continue for many generations of low population size, and that an immediate population flush may actually inhibit the action of selection. This was particularly true for additive-by-dominance epistasis. For pure forms of this type of epistasis, population size reduction intially cause a decline in V_a, but if the small population size is maintained, V_a increases substantially in subsequent generations. If populations were to experience a flush shortly after the bottleneck occurred, V_a would be "frozen" at its low level, and selection would be ineffective in causing speciation.

Conclusions

The ability to estimate the position and effects of QTL greatly increases our ability to address evolutionary controversies that center on gene action. Obviously, testing models of the evolution of gene action, such as those for pleiotropy described above, requires that QTL be known. Founder effect speciation theories and Wright's shifting balance theory of evolution all depend on the existence of epistasis for fitness that can cause shifts in additive genetic variance when allele frequencies change due to drift. Our studies of epistasis among body weight QTL have shown that such epistasis exists even for "additive" traits like size. QTL studies of fitness components may reveal even more epistasis, since fitness is more likely to involve gene interaction (Wright, 1969; Wade, 1992). Recently, QTL studies of the morphological differences among species of plants have shown that genes of major effect can be responsible for reproductive isolation and drastic morphological evolution (Bradshaw et al., 1995; Doebley and Stec, 1991; Doebley, 1996). While these early studies have not resolved the disagreement over whether or not genes of major effect are frequently important in evolution, they have at least forced some skeptics to recognize that major genes must be considered (Coyne and Lande, 1985 vs. Orr and Coyne, 1992). Many more QTL analyses of relevant phenotypes must be conducted before many of these disagreements can be fully resolved. However, the technology now exists to estimate QTL locations and effects, so the resolution of these old questions is finally within our grasp.

Acknowledgements

This work was supported by NSF grant BSR 9106565 to JMC.

References

Barton, N. and Charlesworth, B. (1984) Genetic revolutions, founder effects and speciation. *Annu. Rev. Ecol. Syst.* 15: 133–164.

Barton, N. and Rouhani, S. (1987) The frequency of peak shifts between alternative equilibria. *J. Theor. Biol.* 125: 397–418.

Barton, N. and Turelli, M. (1989) Evolutionary quantitative genetics: How little do we know? *Annu. Rev. Genet.* 23: 337–370.

Bradshaw, H. D. Jr., Wilbert, S. M., Otto K. G. and Schemske, D. W. (1995) Genetic mapping of floral traits associated with reproductive isolation in monkeyflowers (Mimulus) *Nature* 376: 762–765.

Bryant, E. and Meffert, L. (1993) The effect of serial founder-flush cycles on quantitative genetic variation in the housefly. *Heredity* 70: 122–129.

Carson, H. L. (1968) The population flush and its genetic consequences. *In:* Lewontin, R. C.

(ed.) *Population Biology and Evolution*, Syracuse University Press, New York, pp. 123–137.

Carson, H. L. and Templeton, A. R. (1984) Genetic revolutions in relation to speciation phenomena: the founding of new populations. *Annu. Rev. Ecol. Syst.* 15: 97–131.

Chai, C. (1956a) Analysis of quantitative inheritance of body size in mice. I. Hybridization and maternal influence. *Genetics* 41: 157–164.

Chai, C. (1956b) Analysis of quantitative inheritance of body size in mice. II. Gene action and segregation. *Genetics* 41: 167–178.

Chai, C. (1961) Analysis of quantitative inheritance of body size in mice. IV. An attempt to isolate polygenes. *Genet. Res.* 2: 25–32.

Chai, C. (1968) Analysis of quantitative inheritance of body size in mice. V. Effects of small numbers of polygenes on similar genetic backgrounds. *Genet. Res.* 11: 239–246.

Cheverud, J. M. and Routman, E. (1993) Quantitative trait loci: individual gene effects on quantitative characters. *J. Evol. Biol.* 6: 463–480.

Cheverud, J. M. and Routman, E. J. (1995) Epistasis and its contributions to genetic variance components. *Genetics* 139: 1455–1461.

Cheverud, J. M. and Routman, E. J. (1996) Epistasis as a source of increased additive genetic variance at population bottlenecks. *Evolution* 50: 1042–1051.

Cheverud, J. M., Routman, E. J., Duarte, F. A. M., Cothran, van Swinderin, B., Cothran, K. and Perel, C. (1996) Quantitative trait loci for murine growth. *Genetics* 142: 1305–1319.

Cheverud, J. M., Routman, E. J. and Irschick, D. K. (1997) Pleiotropic effects of individual gene loci on mandibular morphology. *Evolution* 51: 2006–2016.

Crow, E. J. and Kimura, M. (1970) *An Introduction to Population Genetics Theory*, Burgess Publishing Company, Minneapolis, MN.

Coyne, J. A. and Lande, R. (1985) The genetic basis of species differences in plants. *Amer. Naturalist* 126: 141–145.

Doebley J. (1996) Genetic dissection of the morphological evolution of maize. *Aliso* 14: 297–304.

Doebley, J. and Stec, A. (1991) Genetic analysis of the morphological differences between maize and teosinte. *Genetics* 129: 285–295.

Fisher, R. A. (1918) The correlations between relatives on the supposition of Mendelian inheritance. *Trans. R. Soc. Edinb.* 52: 399–433.

Lander, E. and Botstein, D. (1989) Mapping Mendelian factors underlying quantitative traits using RFLP linkage maps. *Genetics* 121: 185–199.

Lander, E. S., Green, P., Abrahamson, J., Barlow, A., Daly, M., Lincoln, S. and Newburg, L. (1987) MAPMAKER: An interactive computer package for constructing primary genetic linkage maps of experimental and natural populations. *Genomics* 1: 174–181.

Lincoln, S., Daly, M. and Lander, E. (1992a) *Constructing Genetic Maps with MAPMAKER/EXP 3.0*. Whitehead Institute Technical Report, 3rd ed.

Lincoln, S., Daly, M. and Lander, E. (1992b) *Mapping Genes Controlling Quantitative Traits with MAPMAKER/QTL 1.1*. Whitehead Institute Technical Report, 3rd ed.

Lynch, M. (1988) Design and analysis of experiments on random drift and inbreeding depression. *Genetics* 120: 791–807.

Mayr, E. (1963) *Animal Species and Evolution*, Belknap Press, Cambridge, MA.

Orr H. A. and Coyne J A. (1992) The genetics of adaptation: a reassessment. *Amer. Naturalist* 140: 725–742.

Riedl, R. (1978) *Order in Living Organisms*, Wiley, New York.

Riska, B., Atchley, W. and Rutledge, J. (1984) A genetic analysis of targeted growth in mice. *Genetics* 107: 79–101.

Rose, M. (1982) Antagonistic pleiotropy, dominance and genetic variation. *Heredity* 48: 63–78.

Routman, E. and J. Cheverud (1994a) Individual genes underlying quantitative traits: Molecular and analytical methods. *In*: Schierwater, B., Streit, B., Wagner, G, P. and DeSalle, R. (eds) *Molecular Ecology and Evolution: Approaches and Applications*, Birkhäuser, Basel, pp. 593–606.

Routman, E. and Cheverud, J. (1994b) A rapid method of scoring simple sequence repeat polymorphisms with agarose gel electrophoresis. *Mamm. Genome* 5: 187–188.

Routman, E. R. and Cheverud, J, (1997) Genetic effects on a quantitative trait: two-locus epistatic effects measured at microsatellite markers and at estimated QTL. *Evolution* 51: 1654–1662.

Templeton, A. R. (1980) The theory of speciation via the founder principle. *Genetics* 94: 1011–1038.

Templeton, A. R. (1982) Genetic architecture of speciation. *In*: Barigozzi, C. (ed.) *Mechanisms of Speciation*, Alan R. Liss, New York, pp. 411–433.

Wade, M. J. (1992) Sewall Wright: gene interacion and the shifting balance theory. Oxford Surveys of *Evolut. Biol.* 8: 33–62.

Wagner, G. (1989) Multivariate mutation-selection balance with constrained pleiotropic effects. *Genetics* 122: 223–234.

Wagner, G. (1996) Homology, natural kinds and the evolution of modularity. *Amer. Zool.* 36: 36–43.

Wright, S. (1931) Evolution in mendelian populations. *Genetics* 16: 97–159.

Wright, S. (1969) *Evolution and Genetics of Populations*, vol. 2. University of Chicago Press, Chicago.

Seeking the genetic basis of phenotypic differences among bacterial species

Howard Ochman[1] and Eduardo A. Groisman[2]

[1]Department of Biology, University of Rochester, Rochester, NY 14627, USA
[2]Howard Hughes Medical Institute, Department of Molecular Microbiology, Washington University School of Medicine, St. Louis, MO 63110, USA

Summary

The enteric bacteria *Escherichia coli* and *Salmonella enterica* are closely related, but differ with respect to a variety of phenotypic characters, including the ability to utilize certain compounds as a sole carbon source and their capacity to cause disease. While the observed differences could be due to allelic variation in the same gene complement, the majority of the species-specific characters are encoded by genes that are unique to one of the two species. We analyze the strategies presently used to identify DNA regions that are specific to a bacterial species, discuss the role of these sequences in the evolution and physiology of a microbe, and describe the methods employed to investigate the origin and function of species-specific genes.

Introduction

Bacteria have traditionally been differentiated from one another according to their biochemical, metabolic and growth properties. For example, members of the family Enterobacteriaceae can be distinguished from other bacteria by their ability to ferment glucose and their inability to reduce nitrite (Brenner, 1992). And within the family Enterobacteriaceae, species differ in their capacity to utilize particular sugars as sole carbon sources, to resist antibiotics and phages, and to cause disease in various animal hosts. Because these species-specific traits are significant to the evolution, epidemiology and diversification of bacteria, recent studies have focused on two related issues: (i) What are the genetic bases of the traits that distinguish bacterial species? and (ii) How do these unique phenotypic characteristics arise in bacterial genomes?

Genetic variation in microorganisms can be mediated by several mechanisms, but the changes resulting from these processes are of two general classes – those that modify the genetic information preexisting within the genome, and those that result

R. DeSalle, B. Schierwater (eds) Molecular Approaches to Ecology and Evolution
©1998, Birkhäuser Verlag Basel

in the acquisition of genetic information from another organism. The first class encompasses changes that occur through point mutations (including nucleotide substitutions and frameshift mutations), homologous exchanges that modify extant sequences, and chromosomal rearrangements associated with the deletion and/or reorganization of segments of the genome. The second class includes conjugation-, transduction- and transformation-mediated events that introduce foreign sequences into the genome by lateral gene transfer (Ochman and Bergthorsson, 1995). While allelic variation in homologous sequences can account for some of the phenotypic diversity observed among bacterial strains or species, the majority of characteristics unique to a particular lineage is associated with horizontally acquired sequences.

In this chapter, we discuss the approaches taken to determine the genetic basis of the characteristics that distinguish the enteric bacteria *Escherichia coli* and *Salmonella enterica*. First, we describe the salient phenotypic properties that characterize these two bacterial species. Then, we outline the molecular and genetic techniques used to identify segments of the genome that are specific to particular bacterial lineages, and we describe how to assess the function of these species-specific DNA segments. And in the final section of this chapter, we discuss how to establish the origin of foreign sequences in bacterial genomes.

Differentiating *E. coli* and *S. enterica*

The enteric species *E. coli* and *S. enterica*, as well as strains within these species, differ with respect to their metabolic characteristics, host ranges and virulence properties. The extensive phenotypic differences between these species suggests that their genomes have incorporated substantial changes since diverging from a common ancestor more than 100 million years ago (Ochman and Wilson, 1987). *E. coli* has traditionally been distinguished from *S. enterica* by its ability to ferment lactose and produce indole, and its failure to utilize citrate as a carbon source. *E. coli* is isolated principally from mammals, which is not surprising given its ability to ferment the milk sugar lactose. In contrast, the salmonellae have a much broader host range and have been recovered from both warm- and cold-blooded vertebrates. Perhaps the most relevant characteristic of *Salmonella*, at least with respect to mammalian hosts, is its potential to cause disease (Jones and Falkow, 1996), whereas *E. coli* is typically a benign constituent of the intestinal flora.

The traits that distinguish *E. coli* from *S. enterica* could potentially be conferred by genes that are present in both organisms but expressed differentially (due either to mutations that generate a null allele in one of the species or to the presence of distinct regulatory circuits in these two species) (Groisman and Ochman, 1994). However, in most cases studied to date, unique traits are associated with genes that are present in only one of the two species. For example, the ability to use lactose as the sole carbon source is encoded by the *lac* operon, which resides in a segment of the *E. coli* chromosome not present in the *Salmonella* genome; and similarly, the genes conferring the ability to transport citrate are part of a region of the chromosome that is unique to *Salmonella* (Ochman and Groisman, 1994). This contrasts the situation observed in *Shigella flexneri*, a pathogenic member of the *E. coli* species complex that cannot metabolize lactose due to point mutations in the *lac* operon.

The chromosomes of *E. coli* and *S. enterica* are roughly of the same size and base composition (Normore and Brown, 1970), and when the genetic maps of these two chromosomes are aligned, the arrangement of genes (i.e. the order, orientation and spacing of mapped loci) is highly conserved (Riley and Krawiec, 1987; Riley and Sanderson, 1990). Yet, the chromosomes of these enteric species differ in the presence of a large inversion encompassing 10% of the chromosome and in some 30 regions that are unique to only one of the species. These species-specific regions range from 15 to >50 kb in length and have been designated "loops" as a relict of early heteroduplex analyses whereby mismatched regions were recognized as loops by electron microscopy. Considering all of the incongruities between the genetic maps of *E. coli* K-12 and *S. enterica* serovar Typhimurium LT2, these species differ by at least 600 kb, which represents over 10% of the genome (Riley and Krawiec, 1987). And, as noted above, in certain cases these unique regions harbor genes that confer the metabolic properties that are specific to each of these two species (Riley and Sanderson, 1990; Groisman et al., 1992, 1993).

The chromosome loops originally identified by comparing genetic maps need not contain species-specific sequences because certain chromosomal rearrangements (i.e. inversions, transpositions and translocations) can produce inconsistencies in the observed linkage relationships but do not generate sequences particular to one species. Sequences that appear to be specific to one of two species can arise from either the acquisition of foreign DNA or the deletion of a region from one of the species. And these two possibilities can be resolved phylogenetically, by analyzing the distribution of a region among species of known genetic and genealogical relationships.

The identification and recovery of species-specific sequences

The initial identification of DNA segments specific to either *E. coli* or *S. enterica* came from aligning the genetic maps of laboratory strains of each of these two species. These comparisons revealed that there were several large regions confined to each species; however, neither the number nor the function of most genes residing within these portions of the chromosome were revealed. Some of these species-specific regions are responsible for the unique metabolic characteristics of a species, whereas others are restricted to *Salmonella* and have been of particular interest because they confer *Salmonella* its virulence properties and serve as a diagnostic tool to detect the presence of *Salmonella* in food and tissues.

In an early attempt to obtain DNA fragments specific to *Salmonella*, Fitts (1985) screened a library of clones containing *Salmonella* DNA against of panel of chromosomal DNAs from other enteric species. To avoid the potential for background hybridization, the library of *Salmonella* DNA was constructed in a yeast-*E. coli* shuttle vector, and the hybridization experiments were performed with DNA retrieved from the yeast host. Fitts reported the recovery of five clones that did not hybridize to many enteric species but were present in the majority of *Salmonella* serovars (Fitts, 1985). (The genus *Salmonella* includes two species – *S. bongori* and *S. enterica* – and the vast majority of strains are typed to *S. enterica*, which contains >2000 serovars that differ in host specificity and the disease condition provoked in different host animals.) Several features of the *Salmonella*-specific clones recovered by Fitts – base composition, map location, phylogenetic distribution, nucleotide sequence and function – suggest they were acquired independently during the evolution of *Salmonella* (Groisman et al., 1993).

The systematic screening of a genomic libraries as performed by Fitts (1985) is tedious and subject to numerous technical problems. In fact, subsequent analyses have shown that many of the original clones identified by Fitts as "*Salmonella*-specific" have homologs in other enteric species (Groisman et al., 1993). Thus, several other approaches have been developed to recover DNA sequences that differ between two genomes.

Genomic subtraction

Brown and Curtiss (1996) used subtractive hybridization to identify DNA sequences present in an avian pathogenic strain of *E. coli* but absent from a nonpathogenic laboratory strain of *E. coli* K-12. Their procedure, which was based on a technique developed by Straus and Ausubel (1990), entailed the mixing of chromosomal DNA from *E. coli* K-12 with phage lambda DNA at a ratio of 100:1, shearing this DNA to lengths of 1 to 3 kb and biotinylating the resulting fragments. The resulting DNA mixture was hybridized to *Sau*3A-digested DNA from the avian pathogenic strain, and hybrid molecules were removed using streptavidin-coated magnetic beads. [The addition of lambda DNA promotes the removal of the lambdoid sequences present in the avian pathogenic strain, which would otherwise be recovered as unique strain-specific DNA since the *E. coli* K-12 strain used is devoid of these sequences. However, the removal of lambdoid sequences might also result in the elimination of the relevant virulence determinants since bacteriophages can mediate the incorporation of and are associated with virulence genes (Cheetham and Katz, 1995; Waldor and Mekalanos, 1996)]. Adaptors were ligated to the DNA fragments that remained unbound after five rounds of subtraction, and these fragments were amplified by the polymerase chain reaction (PCR). Southern hybridization experiments using labeled PCR products confirmed that the sequences recovered following these procedures were present in the avian-pathogenic genome but absent from the *E. coli* K-12 chromosome (Brown and Curtiss III, 1996).

To determine the map location of the sequences specific to the avian pathogenic strain, the labeled PCR products were also used to probe a cosmid library prepared from this strain. The hybridizing cosmids recovered from this screen were of two types: clones containing solely sequences specific to the avian pathogenic strain, and clones carrying both sequences unique to the avian pathogenic strain along with up-stream or downstream regions that are also found in *E. coli* K-12. These "hybrid" clones were then used as probes against an ordered library of *E. coli* K-12; and, as a result, Brown and Curtiss III (1996) established that sequences specific to the avian pathogen mapped to at least 12 positions in the chromosome.

Lan and Reeves (1996) employed an analogous genomic subtraction procedure to estimate the amount of DNA sequences contained in the laboratory strain LT2 of *S. enterica* serovar Typhimurium that is not present in the genomes of four diver-gent strains of *S. enterica*. These experiments established that the Typhimurium LT2 genome accommodates some 84 to 106 kb of DNA absent from another Typhimurium strain (SARA 21), but as much as 778 to 1286 kb of unique DNA when compared with the more distantly related *S. bongori*. By analyzing the genet-ic content and base composition of several clones specific to the Typhimurium LT2 genome, they concluded that nearly half of the DNA specific to this strain was gained through lateral transfer (Lan and Reeves, 1996).

Representational difference analysis

An alternate genomic subtraction procedure has been used to identify and recover the differences between two genomes. This method, called "representational differ-ence analysis" (or RDA), has the advantage of sequentially enriching the sequences unique to an organism and is particularly useful in cases where the genomes have a high degree of complexity (Lisitsyn et al., 1993; Lisitsyn, 1995). For RDA, DNAs from the two sources – the "tester" and the "driver" – are digested with a restric-tion enzyme, and the resulting restriction fragments are ligated to short oligonu-cleotide adaptors which are used to selectively amplify a portion of the fragments. Because most of the DNA fragments are too long to be amplified, there is reduced complexity in the driver and tester DNA samples, which, in turn, increases the ef-fectiveness of subtractive hybridizations. Through repeated rounds of reassociation and selective amplification of DNA fragments (after redigesting DNA, and using dif-ferent adaptors in alternate rounds), the samples are enriched for unique tester DNA fragments.

Due to the high degree of sensitivity of this technique, RDA has been applied to uncover differences between mammalian genomes, whereas less efficient methods have been used for comparisons of small bacterial genomes. However, Swain et al.

(unpublished results) employed RDA to recover unique DNA fragments among strains of *Helicobacter pylori* associated with differing disease pathologies. *H. pylori* colonizes the stomach of mammals and has been implicated in ulcers or gastric cancer in about 10% of infected humans (Cover and Blaser, 1995). Several coding and noncoding regions of the *Helicobacter* genome were obtained through RDA, and in two cases, the clones contained the *cag* gene, which is necessary for the development of clinical symptoms in humans (Covacci et al., 1997).

Examining regions of the chromosome targeted by foreign sequences

An alternate approach to identify species-specific sequences is to examine those regions of the chromosome commonly used as integration sites for foreign sequences (Campbell, 1992). For example, certain phages are known to insert at transfer RNA (tRNA) genes (Inouye et al., 1991; Cheetham and Katz, 1995), and these loci often contain foreign DNA sequences: a 40-kb pathogenicity island is present at the $tRNA^{val}$ locus of *Salmonella* (Hensel et al., 1997); and in *E. coli*, the PAI-2, PAI-4 and PAI-5 pathogenicity islands have integrated at the $tRNA^{leuX}$, $tRNA^{pheV}$ and $tRNA^{pheR}$ genes, respectively (Blum et al., 1994; Swenson et al., 1996) (Fig. 1). Pathogenicity islands are regions of the chromosome harboring virulence genes in pathogenic organisms but absent from related nonpathogenic bacterial species (Groisman and Ochman, 1996; Lee, 1996; Hacker et al., 1997). The frequent insertion of foreign DNA sequences at tRNA genes is presumably due to the high degree of sequence conservation of tRNA genes across species and the fact that these genes occur in multiple copies within a genome (Campbell, 1992; Cheetham and Katz, 1995).

There is one tRNA gene that has been repeatedly targeted by horizontally acquired sequences: the *selC* locus is the integration site for two different pathogenicity islands and for a bacteriophage in *E. coli* (Groisman and Ochman, 1996) (Fig. 2). Because the $tRNA^{selC}$ locus often harbors pathogenicity islands in *E. coli*, Blanc-Potard and Groisman (1997) examined the homologous locus in *Salmonella* for the presence of foreign sequences. Their experimental strategy was to initially carry out PCR reactions with primers corresponding to the *selC* gene and to a sequence downstream of *selC* that is conserved between the *E. coli* K-12 and *S. enterica* serovar Typhimurium chromosomes. In *E. coli* K-12, these primers yielded a 2-kb fragment, whereas no product was obtained for *S. enterica*, suggesting that the distance between primers was too long for PCR amplification under the tested conditions and that a large segment of DNA had integrated at this site. Plasmid clones containing the *Salmonella selC* locus were then isolated and used to identify a 17-kb region specific to the *Salmonella* chromosome. Subsequent mutational analysis of these clones revealed that this region encodes essential virulence attributes; and therefore, this 17-kb region was designated SPI-3 for *Salmonella* pathogenicity island 3 (Blanc-Potard and Groisman, 1997).

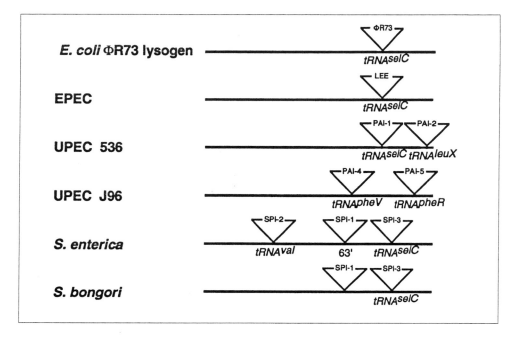

Figure 1. Location of selected pathogenicity islands and phages in E. coli *and* Salmonella. *Bacterial chromosomes and pathogenicity islands are depicted as horizontal lines and triangles, respectively, and are not drawn to scale. tRNA genes are often the site of insertion of horizontally acquired sequences. The presence of repeated sequences at the site of insertion is indicated by short grey lines.*

The identification of horizontally acquired DNA at the *selC* locus of *S. enterica* serovar Typhimurium suggests that a similar mechanism is operating to integrate sequences at the *selC* locus in several enteric species. The transfer of these sequences is likely to be phage mediated because *selC* is the attachment site for the retronphage ΦR73 (Inouye et al., 1991) and a ΦR73-related integrase is encoded within PAI-1, a pathogenicity island which has also inserted at the *selC* locus of certain uropathogenic strains of *E. coli* (Hacker et al., 1997). It is interesting to note that the genetic contents of pathogenicity islands detected at *selC* are not identical and that the insertion events have occurred at different distances from the *selC* gene. For example, both the 70-kb PAI-1 and the 35-kb LEE pathogenicity islands of *E. coli* are present 16 bp to the 3' side of the *selC* gene, whereas SPI-3 of *Salmonella* inserted 11 bp downstream of *selC* and ΦR73 is found 9 bp to the 3' side of *selC* in *E. coli* strains lysogenic for this retronphage (Blanc-Potard and Groisman, 1997). And although short direct repeats flank PAI-1 and ΦR73, no such repeats are found at the insertion sites of SPI-3 and LEE.

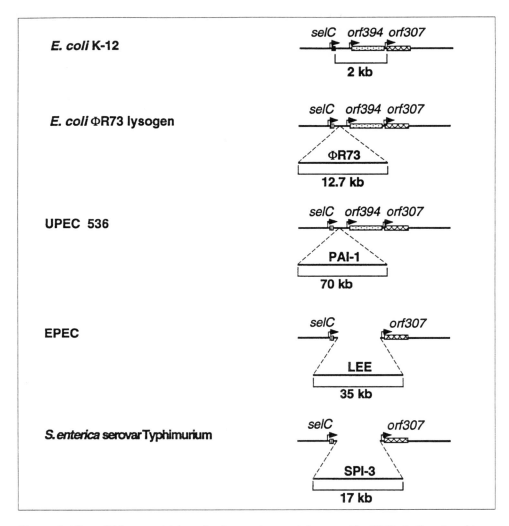

Figure 2. The selC locus, which codes for a selenocysteine-specific tRNA, is the site of insertion of different foreign sequences in E. coli and S. enterica. In laboratory strains of E. coli K-12, the selC gene is followed by orf394 and located 2 kb upstream of orf307. Lysogeny with the retronphage ΦR73 results in the incorporation of the 12.7-kb phage genome 9 bp 3' to the selC gene. The uropathogenic E. coli strain 536 harbors a 70-kb pathogenicity island, designated PAI-1, 16 bp 3' to the selC gene. In contrast to the presence of orf394 in strain 536 and ΦR73 lysogens, acquisition of the 35-kb LEE island in enteropathogenic E. coli and of the 17-kb SPI-3 island in S. enterica resulted in the deletion of orf394.

In sum, several horizontally acquired sequences have been incorporated at tRNA loci in phage-mediated events. Although these loci are most often associated with virulence traits in *E. coli* and *Salmonella*, there is little doubt that tRNA loci probably harbor several other types of foreign sequences. In addition to tRNA-targeted phage-mediated events, DNA can be incorporated into bacterial chromosomes by other vehicles, such as plasmids, transposons and insertion sequences (Protsenko et al., 1991; Campbell, 1992; Fetherston et al., 1992). Therefore, by screening regions adjacent to these translocatable elements, it should be possible to recover additional genes gained through horizontal gene transfer and to establish the relative contribution of each of these elements to the evolution of bacterial genomes.

Cloning DNA segments conferring novel phenotypic properties

The approaches described above allow for the recovery of species-specific sequences independently of the properties encoded by these sequences. However, when one is interested in isolating the genes responsible for a particular trait, genetic screens and/or selections can be used to identify the gene of interest based on its ability to confer a scorable phenotype upon a microorganism that does not express that trait. This strategy entails introducing DNA from the organism expressing the trait into a bacterial host (typically, a laboratory strain of *E. coli* K-12 used for recombinant DNA experiments) and selecting for those clones that express the phenotype of interest.

To identify DNA sequences conferring *Yersinia pseudotuberculosis* the ability to invade eukaryotic cells, Isberg and Falkow (1985) introduced a cosmid library of *Y. pseudotuberculosis* DNA into *E. coli* K-12 and selected for clones that could invade host cells. (*Y. pseudotuberculosis* is an invasive pathogen of mammals closely related to *Y. pestis*, the causative agent of bubonic plague.) They demonstrated that a single gene – *inv* – is sufficient to allow entry of *E. coli* K-12 into epithelial cells and that this gene is absent from the *E. coli* K-12 genome (Isberg and Falkow, 1985). Applying the same approach, Miller and Falkow (1988) identified another *Yersinia*-specific gene – *ail* – that can also mediate entry of *E. coli* K-12 into mammalian cells.

The recovery of clones conferring a novel phenotype upon *E. coli* K-12 need not indicate that the isolated sequences are absent from the *E. coli* genome: the isolation of a particular gene may simply reflect the higher level of expression of a plasmid-borne sequence as opposed to the same regions when present on the *E. coli* chromosome. For example, the *slyA* gene of *S. enterica* serovar Typhimurium was once believed to encode a hemolysin (i.e. a protein with the ability to lyse red blood cells) because it could render *E. coli* K-12 hemolytic (Libby et al., 1994). However, it has recently been established that the *slyA* gene is present in both *Salmonella* and *E. coli* K-12 and that it encodes a regulatory protein controlling the expression of a hemolytic factor encoded in the *E. coli* K-12 chromosome (Oscarsson et al., 1996). In contrast to *Salmonella*, *E. coli* K-12 is nonhemolytic when grown under laboratory conditions, but the presence of *slyA* on a multicopy vector allows expression from a cryptic hemolysin gene.

Strategies that are based on the ability of a DNA segment to confer a dominant phenotype upon a recipient organism require the necessary species-specific DNA sequences, the appropriate levels of gene expression and the proper subcellular localization of the encoded products. Hence, this approach may not result in the recovery of clones if the phenotype of interest is encoded by unlinked genes or if the characteristic levels of gene expression in the source microorganism are different from those of the recipient species.

Functional genomics: evaluating the biological role of species-specific sequences

The identification of species-specific sequences provides an opportunity to address several physiological and evolutionary questions about the genetic differences between bacterial genomes. In particular, we can begin to evaluate: (i) how these sequences were incorporated into a recipient's genome, and (ii) the role that these sequences play in the lifestyle of a microorganism.

The function of foreign genes can be determined by a variety of techniques, but perhaps the simplest approach is to determine the nucleotide sequence of a species-specific DNA fragment and then to search the sequence databases for the presence of related genes or encoded proteins. While similarity to a known sequence does not unequivocally establish the function of a gene, it can provide information about the potential biochemical activity of the encoded proteins and suggest a role for these sequences in microbial adaptation to particular environments. However, assessing the function of species-specific genes demands testing the relevant phenotype in mutant strains in which the gene of interest has been inactivated. A variety of molecular and genetic strategies have been developed to generate mutations in specific DNA sequences (Link et al., 1997), and the choice of methods depends, in part, on the procedure used to recover the species-specific region and the bacterial species of interest.

Typically, a copy of the gene of interest present in a plasmid or cosmid vector is inactivated by inserting a DNA fragment harboring an antibiotic resistance determinant within its coding region. The mutated gene is then transferred back to the chromosome by introducing the plasmid into a strain that cannot sustain replication of the plasmid, selecting for resistance to the antibiotic encoded by the marker used to inactivate the gene of interest, and screening for sensitivity to the antibiotic encoded by the vector.

We have applied this approach to establish that the *Salmonella*-specific gene clusters mapping to the 31- and 63-minute regions of the *S. enterica* serovar Typhimurium chromosome are essential for survival within macrophages and invasion of mammalian cells, respectively (Groisman and Ochman, 1993; Ochman et al., 1996). Plasmids harboring pMB1 replicons and carrying mutated DNA fragments were introduced into a *polA* mutant strain, and the resulting chromosomal mutations were transferred by phage P22-mediated transduction into the wild-type strain to generate strains that were isogenic with the wild-type parent except for the mutated genes. This approach has allowed both the inactivation of specific genes and the deletion of large DNA segments encompassing several open reading frames. Furthermore, it does not require the use of special vectors, because the majority of the plasmids and cosmids used for recombinant DNA experiments harbor pMB1-based replicons, which require a functional *polA* gene product for replication. The chromosomal inactivation of cloned genes can also be carried out by using plasmids that are conditional for replication due to their harboring replicons that are either temperature-sensitive or that depend on a specific polymerase for replication and maintenance (Link et al., 1997).

An alternative approach for constructing strains deleted for species-specific DNA segments takes advantage of the conservation in order and sequence among strains of a given bacterial species (Riley and Krawiec, 1987; Riley and Sanderson, 1990; Ochman and Groisman, 1994). Brown and Curtiss III (1996) constructed mutant derivatives of an avian pathogenic strain of *E. coli* that were lacking any one of four pathogen-specific regions. They used phage P1-mediated transduction to introduce regions of the *E. coli* K-12 chromosome equivalent to those flanking the pathogen-specific sequences in the avian pathogenic strain and which were genetically marked with an antibiotic resistance determinant. As a result, only two of the four mutant strains (in which unique DNAs were replaced by *E. coli* K-12 sequences) showed a reduction in virulence in chickens. The potential problem with this approach is that mutant strains harbor a least two mutations – the deletion of the species-specific sequences, and the linked antibiotic resistance determinant. And these modifications to the genome may be compounded by any allelic differences in regions conserved between the pathogenic and nonpathogenic strains, because large DNA segments from the *E. coli* K-12 strain are introduced into the pathogenic strain during the removal of the species-specific sequences.

Evaluating the origin of foreign sequences in bacterial genomes

Sequences that are incorporated into a bacterial genome through horizontal gene transfer are likely to have retained features of the donor genome; thus, if genes are acquired from an organism having a different genomic base composition, the introgressed sequences will bear unusual characteristics that distinguish them from the rest of the genes in the recipient genome. For example, the 35-kb region conferring the characteristic virulence properties of enteropathogenic strains of *E. coli* has a G+C content that is much lower than the overall G+C content of the *E. coli* chromosome (39 vs. 50%). (The G+C content of bacterial genomes ranges from 25 to 75% and the genes within a particular species' genome are fairly homogeneous in base composition). In addition, genes from a given species have characteristic codon usage patterns that reflect, in part, the overall base composition of the organism (Muto and Osawa, 1987; Wada et al., 1992). Therefore, the identification of regions of the chromosome with atypical nucleotide contents and/or sequence characteristics is often cited as evidence of horizontal transfer, even when the source or mode of transmission of these sequences is not known (Lawrence and Ochman, 1997).

The incorporation of foreign sequences may create an obstacle for the recipient strain because expression of the acquired genes must be coordinated with that of the rest of the genome (Groisman and Ochman, 1997; Groisman, 1998). And not surprisingly, many of the horizontally acquired genes that have been implicated in the virulence properties of *Salmonella* are transcriptionally controlled by PhoP/PhoQ (Groisman and Heffron, 1995), a regulatory system widely distributed in the family Enterobacteriaceae (Groisman et al., 1989), or by the RpoS alternative sigma factor of RNA polymerase (Fang et al., 1992).

Conclusions

Despite the overall similarity in the size, structure and organization of *E. coli* and *S. enterica* chromosomes, it has been estimated that over 10% of the genomes of species arose through horizontal transfer. Although many of the sequences gained through gene transfer have little effect on the organism, some of these sequences, such as those conferring novel virulence attributes or metabolic properties, have allowed *E. coli* and *S. enterica* to explore ecological niches that were previously unavailable. Although point mutational evolution has traditionally been viewed as the major force driving the divergence and adaptation of species, a substantial amount of change in bacteria occurs through horizontal transfer. And in contrast to point mutations, which change the existing genetic information, the type of information introduced by horizontal transfer involves the traits that differentiate bacterial species and can, in effect, change the character of a bacterial species.

Acknowledgments

We thank Sella Garlich for the preparation of the figures in this chapter. Work in our laboratories is supported by grants GM56120 and GM55535 (HO), and GM54900 (EAG) from the National Institutes of Health (NIH), and grant 9602159 (EAG) from the United States Department of Agriculture. HO is a recipient of a Research Career Development Award (AI01353) from the NIH. EAG is the recipient of a Research Career Development Award (AI01310) from the NIH and an associate investigator of the Howard Hughes Medical Institute.

References

Blanc-Potard, A.-B. and Groisman, E. A. (1997) The *Salmonella selC* locus contains a pathogenicity island mediating intramacrophage survival. *EMBO J.* 16: 5376–5385.

Blum, G., Ott, M., Lischewski, A., Ritter, A., Imrich, H., Tschäpe, H. and Hacker, J. (1994) Excision of large DNA regions termed pathogenicity islands from tRNA-specific loci in the chromosome of an *Escherichia coli* wild-type pathogen. *Infect. Immun.* 62: 606–614.

Brenner, D. J. (1992) Introduction to *Enterobacteriaceae*. *In*: Balows, H., Trüper, H. G., Dworkin, M., Harder, W. and Schliefer, K.-H. (eds) *The Prokaryotes*, vol. 3, Springer, New York, pp. 2673–2696.

Brown, P. K. and Curtiss III, R. (1996) Unique chromosomal regions associated with virulence of an avian pathogenic *Escherichia coli* strain. *Proc. Natl. Acad. Sci. USA* 93: 11149–11154.

Campbell, A. M. (1992) Chromosomal insertion sites for phages and plasmids. *J. Bacteriol.* 174: 7495–7499.

Cheetham, B. F. and Katz, M. E. (1995) A role for bacteriophages in the evolution and transfer of bacterial virulence determinants. *Mol. Microbiol.* 18: 201–208.

Covacci, A., Falkow, S., Berg, D. E. and Rappuoli, R. (1997) Did the inheritance of a pathogenicity island modify the virulence of *Helicobacter pylori*? *Trends Microbiol.* 5: 205–208.

Cover, T. L. and Blaser, M. J. (1995) *Helicobacter pylori*: a bacterial cause of gastritis, peptic ulcer, and gastric cancer. *ASM News* 61: 21–26.

Fang, F. C., Libby, S. J., Buchmeier, N. A., Loewen, P. C., Switala, J., Harwood, J. and Guiney, D. G. (1992) The alternative σ factor KatF (RpoS) regulates *Salmonella* virulence. *Proc. Natl. Acad. Sci. USA* 89: 11978–11982.

Fetherston, J. D., Schuetze, P. and Perry, R. D. (1992) Loss of the pigmentation phenotype in *Yersinia pestis* is due to the spontaneous deletion of 102 kb of chromosomal DNA which is flanked by a repetitive element. *Mol. Microbiol.* 6: 2693–2704.

Fitts, R. (1985) Development of a DNA-DNA hybridization test for the presence of *Salmonella* in foods. *Food Technol.* 39: 95–102.

Groisman, E. A. (1998) The ins and outs of virulence gene expression: Mg^{2+} as a regulatory signal. *BioEssays* 20: 96–101.

Groisman, E. A. and Heffron, F. (1995) Regulation of *Salmonella* virulence by two-component regulatory systems. *In*: Hoch, J. A. and Silhavi, T. J. (eds) *Two-Component Signal Transduction*, ASM Press, Washington, D.C., pp. 319–332.

Groisman, E. A. and Ochman, H. (1993) Cognate gene clusters govern invasion of host epithelial cells by *Salmonella typhimurium* and *Shigella flexneri*. *EMBO J.* 12: 3779–3787.

Groisman, E. A. and Ochman, H. (1994) How to become a pathogen. *Trends Microbiol.* 2: 289–294.

Groisman, E. A. and Ochman, H. (1996) Pathogenicity islands: bacterial evolution in quantum leaps. *Cell* 87: 791–794.

Groisman, E. A. and Ochman, H. (1997) How *Salmonella* became a pathogen. *Trends Microbiol.* 5: 343–349.

Groisman, E. A., Chiao, E., Lipps, C. J. and Heffron, F. (1989) *Salmonella typhimurium phoP* virulence gene is a transcriptional regulator. *Proc. Natl. Acad. Sci. USA* 86: 7077–7081.

Groisman, E. A., Saier, M. H. Jr. and Ochman, H. (1992) Horizontal transfer of a phosphatase gene as evidence for mosaic structure of the *Salmonella* genome. *EMBO J.* 11: 1309–1316.

Groisman, E. A., Sturmoski, M. A., Solomon, F. R., Lin, R. and Ochman, H. (1993) Molecular, functional and evolutionary analysis of sequences specific to *Salmonella*. *Proc. Natl. Acad. Sci. USA* 90: 1033–1037.

Hacker, J., Blum-Oehler, G., Mühldorfer, I. and Tschäpe, H. (1997) Pathogenicity islands of virulent bacteria: structure, function and impact on microbial evolution. *Mol. Microbiol.* 23: 1089–1098.

Hensel, M., Shea, J. E., Bäumler, A. J., Gleeson, C., Blattner, F. and Holden, D. W. (1997) Analysis of the boundaries of *Salmonella* pathogenicity island 2 and the corresponding chromosomal region of *Escherichia coli* K-12. *J. Bacteriol.* 179: 1105–1111.

Inouye, S., Sunshine, M. G., Six, E. W. and Inouye, M. (1991) Retronphage φR73: an *E. coli* phage that contains a retroelement and integrates into a tRNA gene. *Science* 252: 969–971.

Isberg, R. R. and Falkow, S. (1985) A single genetic locus encoded by *Yersinia pseudotuberculosis* permits invasion of cultured animal cells by *Escherichia coli* K-12. *Nature* 317: 262–264.

Jones, B. D. and Falkow, S. (1996) Salmonellosis: host immune responses and bacterial virulence determinants. *Annu. Rev. Immunol.* 14: 533–561.

Lan, R. and Reeves, P. R. (1996) Gene transfer is a major factor in bacterial evolution. *Molec. Biol. Evol.* 13: 47–55.

Lawrence, J. G. and Ochman, H. (1997) Amelioration of bacterial genomes: rates of change and exchange. *J. Mol. Evol.* 44: 383–397.

Lee, C. A. (1996) Pathogenicity islands and the evolution of bacterial pathogens. *Infect. Agents Dis.* 5: 1–7.

Libby, S. J., Goebel, W., Ludwig, A., Buchmeier, N., Bowe, F., Fang, F. C., Guiney, D. G., Songer, J. G. and Heffron, F. (1994) A cytolysin encoded by *Salmonella* is required for survival within macrophages. *Proc. Natl. Acad. Sci. USA* 91: 489–493.

Link, A. J., Phillips, D. and Church, G. M. (1997) Methods for generating precise deletions and insertions in the genome of wild-type *Escherichia coli*: application to open reading frame characterization. *J. Bacteriol.* 179: 6228–6237.

Lisitsyn, N. A. (1995) Representational difference analysis: finding the differences between two complex genomes. *Trends Genet.* 11: 303–307.

Lisitsyn, N., Lisitsyn, N. and Wigler, M. (1993) Cloning the differences between two complex genomes. *Science* 259: 946–951.

Miller, V. L. and Falkow, S. (1988) Evidence for two genetic loci in *Yersinia enterocolitica* that can promote invasion of epithelial cells. *Infect. Immun.* 56: 1242–1248.

Muto, A. and Osawa, S. (1987) The guanosine and cytosine content of genomic DNA and bacterial evolution. *Proc. Natl. Acad. Sci. USA* 84: 166–169.

Normore, W. M. and Brown, J. R. (1970) G+C composition in bacteria. *In*: Sober, H. A. (eds) *Handbook of Biochemistry: Selected Data for Molecular Biology*, CRC Press, Cleveland, pp. 24–74.

Ochman, H. and Bergthorsson, U. (1995) Genome evolution in enteric bacteria. *Curr. Opinion Genet. Develop.* 5: 734–738.

Ochman, H. and Groisman, E. A. (1994) The origin and evolution of species differences in *Escherichia coli* and *Salmonella typhimurium*. *In*: Schierwater, B., Streit, B., Wagner, G, P. and DeSalle, R. (eds) *Molecular Ecology and Evolution: Approaches and Applications,*

Birkhäuser, Basel, pp. 479–493.

Ochman, H. and Wilson, A. (1987) Evolutionary History of Enteric Bacteria. *In*: Neidhardt, F. C., Ingraham, J. L., Low, K. B., Magasanik, B., Schaechter, M. and Umbarger, H. E. (eds) *Escherichia coli and Salmonella typhimurium: Cellular and Molecular Biology*, vol. 2, ASM Press, Washington, DC, pp. 1649–1654.

Ochman, H., Soncini, F. C., Solomon, F. and Groisman, E. A. (1996) Identification of a pathogenicity island required for *Salmonella* survival in host cells. *Proc. Natl. Acad. Sci. USA* 93: 7800–7804.

Oscarsson, J., Mizunoe, Y., Uhlin, B. E. and Haydon, D. J. (1996) Induction of haemolytic activity in *Escherichia coli* by the *slyA* gene product. *Mol. Microbiol.* 20: 191–199.

Protsenko, O. A., Filippov, A. A. and Kutyrev, V. V. (1991) Integration of the plasmid encoding the synthesis of capsular antigen and murine toxin into *Yersinia pestis* chromosome. *Microb. Pathogen.* 11: 123–128.

Riley, M. and Krawiec, S. (1987) Genome organization. *In*: Neidhardt, F. C., Ingraham, J. L., Low, K. B., Magasanik, B., Schaechter, M. and Umbarger, H. E. (eds) *Escherichia coli and Salmonella typhimurium: Cellular and Molecular Biology*, vol. 2, ASM Press, Washington, DC, pp. 967–981.

Riley, M. and Sanderson, K. E. (1990) Comparative genetics of *Escherichia coli* and *Salmonella typhimurium*. *In*: Drlica, K. and Riley, M. (eds) *The Bacterial Chromosome*, American Society for Microbiology, Washington, D.C., pp. 85–95.

Straus, D. and Ausubel, F. M. (1990) Genomic subtraction for cloning DNA corresponding to deletion mutations. *Proc. Natl. Acad. Sci. USA* 87: 1889–1893.

Swenson, D. L., Bukanov, N. O., Berg, D. E. and Welch, R. A. (1996) Two pathogenicity islands in uropathogenic *Escherichia coli* J96: cosmid cloning and sample sequencing. *Infect. Immun.* 64: 3736–3743.

Wada, K., Wada, Y., Ishibashi, F., Gojobori, T. and Ikemura, T. (1992) Codon usage tabulated from the Genbank genetic sequence data. *Nucl. Acid. Res.* 18: 2367–2413.

Waldor, M. K. and Mekalanos, J. J. (1996) Lysogenic conversion by a filamentous phage encoding cholera toxin. *Science* 272: 1910–1914.

Part 3. Higher taxa and systematics

Rob DeSalle and Bernd Schierwater

Systematics has a rich tradition of empirical and theoretical work, and molecular data have become commonplace in most systematic studies. Systematic approaches can easily be divided into character-based and distance-based methods. These different methods retain different philosophical approaches (see below), and even within character-based methods different philosophical approaches are debated (such as between maximum likelihood and maximum parsimony). The distance perspective is clearly outlined in the writings of Sokal and Sneath (1963) with more recent elaboration using neighbor joining by Nei and colleagues (Nei, 1988; Kumar et al., 1993). Maximum likelihood methods are most clearly outlined in Felsenstein (1988), Edwards (1992) and Swofford et al. (1996). Maximum parsimony methods are discussed in a wide variety of volumes such as Hennig (1966), Eldredge and Cracraft (1980), Nelson and Platnick (1983), Forey et al. (1992), Panchen (1992) and Wiley et al. (1991).

Although the philosophical arguments and issues surrounding the appropriateness of the various approaches used in systematics is beyond the scope of this book, philosophical issues are not irrelevant to the way the chapters in this section should be read. The logical and philosophical underpinnings of all the various approaches used in phylogenetic analysis should be examined and made familiar to the researchers who use them. Introduction to the philosophical issues in systematic theory centered on the appropriateness of various techniques can be found in Farris (1983), Edwards (1992), Sokal (1987), Sokal and Sneath (1963), Hennig (1966), Sober (1983, 1988), Felsenstein (1988), Panchen (1992), Nelson and Platnick (1983), Kluge (1997) and Sanderson and Hufford (1996) to name a few. The chapters in this section take a character-based look at systematics, and hence this approach should be explicitly examined and scrutinized while reading these chapters.

In our previous edited volume we discussed the major problems that face the systematist using molecular characters. DeSalle et al. (1994) discussed three areas of modern systematics that are problematic in data analysis (alignment, data combination as it pertains to robustness of phylogenetic inference and computational difficulty of analyzing large numbers of taxa) and in applying molecular information in systematics. Here, we include two more areas (biogeography and the impact of development on evolutionary and developmental studies) that we feel are important to discuss in the context of molecular methods in systematics.

Alignment is a tricky business. Currently there are many ways that systematists treat raw data to obtain character columns for phylogenetic analysis. These methods range from entirely "ocular" approaches to complete reliance on computer-generated results. The most popular methods involve multiple alignment approaches to establish character columns. Hein (1998) and Wheeler (1994) summarize these alignment methods and the theory behind multiple sequence alignment and discuss its utility in phylogenetic analysis. In this volume Wheeler elaborates upon a phylogenetic analysis procedure that does not involve multiple sequence alignment. This method, called "direct optimization", uses dynamic programming procedures implicit in multiple sequence alignment. The procedure does not result in a multiple alignment matrix, but rather a phylogenetic hypothesis based on the direct optimization is produced.

The second problem discussed by DeSalle et al. (1994) involves the search for optimal trees from data sets with extremely large numbers of taxa. Felsenstein (1978) discussed the problem of NP completeness and the number of phylogenetic trees to be evaluated in a parsimony search for N taxa. Exact solution of data sets with more than 20 or so taxa is in general intractable due to the astronomically large number of trees that need to be evaluated in a parsimony search. Even heuristic searches for large data sets use incredibly lengthy computer searches, and such searches can result in solutions that are not optimal (Maddison, 1991; Templeton, 1993). This problem has become more and more acute with the addition of reams of sequence data to various data bases. At last count (January 1998) a search of GenBank resulted in the recovery of large numbers of sequences for various frequently used gene regions in molecular systematics studies. For instance, there are nearly 3000 entries for the plant ribulose-1,5-bisphosphate carboxylase large subunit gene (rbcl), over 3400 entries for bacterial 16S ribosomal RNA (rRNA) sequences, over 1500 entries for 18S rRNA sequences (includes plants and animals), over 1300 entries for 28S rRNA gene sequences (includes plants and animals) and over 350 entries each for animal mitochondrial cytochrome b (cytb) and cytochrome oxidase II (COII) gene sequences. It seems only likely that the problem of analyzing large numbers of taxa in data sets will increase as time goes by. Goldstein and Specht address this problem in their detailed review of "large molecular data sets". Particularly important in this review is an examination of the logical and philosophical problems involved in analyzing large data sets. These authors come to the conclusion that if one wants to distill this rather complex problem to simple choices, then rooting, taxon sampling and appropriate gene choice become paramount issues in treating large molecular data sets.

Patterson (1988) attempted to take stock of the "torrent" of molecular data that had inundated systematics for nearly two decades. The title of his edited volume *Molecules and Morphology in Evolution: Conflict or Compromise* embodies the polarization that existed between morphological systematists and molecular systematists at that time. Although there is still a great deal of healthy skepticism from both camps today, the large majority of current systematists see the merit of both kinds of data. Since the publication of our first edited volume in 1994, several papers have appeared that examine these merits, in general, and issues of combining the information from both kinds of data in systematic analysis in particular. In the current volume Larson discusses the subject of comparing molecular and morphological data. He discusses the methods of taxonomic congruence (consensus methods) and character congruence (total evidence) and describes several statistical tests that allow the assessment of congruence within both frameworks between morphology and molecular data.

Two subjects that were covered in our 1994 volume that are reprised in the current volume are biogeographic analyses and developmental considerations in systematics. Biogeographic methods have been a part of character-based analyses since the beginning of the use of these methods (Hennig, 1966; Nelson and Platnick, 1983). For an interesting example of the application of these classical biogeographic approaches, the reader should examine Wagner and Funk (1995), *Evolution on a Hotspot: Biogeography of the Hawaiian Archipelago*. More recently, the utility of genealogical analysis of DNA sequence data within a "phylogeographic" framework (Avise et al., 1987; Avise, 1994) has been recognized. Modern molecular biology aids in the ease with which population genetic data and phylogenetic data can be collected to address biogeographic questions. Cunningham and Collins suggest that focusing on vicariance events in a biogeographic framework may result in the oversight of other interesting biological complexities. A novel approach of "inverting the priorities" of vicariance biogeography is taken by these authors to examine biogeographic patterns in two well-studied geographic areas – the southeastern United States and the trans-Arctic exchange.

The recent proliferation of interest in developmental biology and how this discipline contributes to our understanding of evolution has intensified since the publication of our first volume (see Gilbert et al., 1996, and Raff, 1996, for cogent reviews). We included several chapters on this subject in our previous volume (Tautz, 1994; Muller, 1994; Jacobs, 1994; Wagner, 1994) that address developmental issues that are relevant to evolution and systematics. The interest in developmental data seems to have rekindled many of the classic arguments about homology and homology assessment. Excellent discussions of this debate can be found in Panchen (1992), Hall (1994), Nelson (1994), Wagner (1989), Roth (1988) and Abhouief (1997), among others. Jacobs et al., advance the arguments about homology in the context of several recent examples from developmental biology (eyes, hearts and dorsoventral inversion). These authors recognize the typological nature of developmental biology and discuss the appropriateness of this approach in systematics and evolution.

In addition to the impact of molecular data, the field of systematics has been revolutionized by the development of computer packages that implement alignment and phylogenetic analysis. Many of these programs are listed and are described in Seberg and Petersen (1998; NYSYS, PHYLIP, MEGA, ClustalW, MALIGN, PAUP, Hennig86, PeeWee, NoNa, JACK, MacClade, CLADOS and DADA). Random Cladistics (Siddall, 1997) and Autodecay (Eriksson, 1996) should also be added to this list.

Several important suggestions about the future of the field of molecular systematics have been made by Zimmer (1994). In addtion to the points raised by Zimmer, the present volume presents chapters in five areas that will most likely be targets of intense future research. In particular, computational methods with respect to alignment and large data sets will have to be developed in order to handle the large amount of empirical data being collected by this generation of systematists and the next. The application of molecular data to addressing questions about evolution such as biogeographic and adaptational questions will become more and more intense, and methods of analysis of these kinds of data sets will need more theoretical development (see Maddison and Maddison, 1992; Brooks and McLennan, 1992; Harvey and Paigel, 1991; Wenzel and Carpenter, 1994; Givnish and Sytsma, 1997). Finally, further examination of developmental issues in evolution should resolve the utility of gene localization studies (*in situ* hybridization and antibody methods) in embryos that have been so influential the past few years. Certainly, broader taxonomic sampling using gene product localization techniques will aid greatly in proving the utility of this approach. Other methods of examining developmental processes in the context of evolution and systematics such as deciphering regulatory networks (Arnone and Davidson, 1997) will be important approaches in the synthesis of development and evolution.

References

Abhouief, E. (1997) Developmental genetics and homology: a hierarchical approach. *Trends Ecol. Evol.* 12: 405–408.

Arnone, M. and Davidson, E. (1997) The hierarchy of development: organization and function of genomic regulatory systems. *Development* 124: 1851–1864.

Avise, J. C., Arnold, J., Ball, R. M., Bermingham, E., Lamb, T., Neigel, J. E., Reeb, C. A. and Saunders, N. C. (1987) Intraspecific phylogeography: The mitochondrial bridge between population genetics and systematics. *Annu. Rev. Ecol. Syst.* 18: 489–522.

Avise, J. C. (1994) *Molecular Markers, Natural History and Evolution*, Chapman and Hall, New York.

Brooks, D. R. and McLennan, D. A. (1991) *Phylogeny, Ecology, and Behavior: A Research Program in Comparative Biology*, University of Chicago Press, Chicago.

DeSalle, R., Wray, C. and Absher, R. (1994) Computational problems in molecular systematics. *In*: Schierwater, B., Streit, B., Wagner, G, P. and DeSalle, R. (eds) *Molecular Ecology and Evolution: Approaches and Applications*, Birkhäuser, Basel, pp. 353–370.

Edwards, A. W. F. (1992) *Likelihood*, Johns Hopkins University Press, Baltimore.

Eldredge, N. and Cracraft, J. (1980) *Phylogenetic Patterns and the Evolutionary Process*, Columbia University Press, New York.

Eriksson, T. (1996) *Autodecay*, v2.9.2, Stockholm, Sweden.

Farris, J. S. (1983) The logical basis of phylogenetic analysis. *In*: Platnick N. I. and Funk, V. A. (eds) *Advances in Cladistics* 2, Columbia University Press, New York, pp. 7–36.

Felsenstein, J. (1978) The number of evolutionary trees. *Syst. Zool.* 27: 27–33.

Felsenstein, J. (1988) Phylogenies from molecular sequences. Inferences and reliability. *Annu. Rev. Genet.* 22: 521–565.

Forey, P. L., Humphries, C. J., Kitching, I. J., Scotland, R. W., Siebert, D. J. and Williams, D. M. (1992) *Cladistics: A Practical Course in Systematics*, Clarendon Press, Oxford.

Gilbert, S. F., Opitz, J. M. and Raff, R. (1996) Resynthesizing evolutionary and developmental biology. *Dev. Biol.* 173: 357–372.

Givnish, T. and Sytsma, K. (eds) (1997) *Molecular Evolution and Adaptive Radiation*, Cambridge University Press, Cambridge.

Hall, B. K. (1994) *Homology: The Hierarchical Basis of Comparative Biology*, Academic Press, San Diego.

Harvey, P. H. and Pagel, M. D. (1991) *The Comparative Method in Evoluionary Biology*, Oxford University Press, Oxford.

Hein, J. (1998) Multiple alignment. *In*: Karp, A., Isaac, P. G. and Ingram, D. S. (eds) *Molecular Tools for Screening Biodiversity*, Chapman and Hall, London, pp. 334–340.

Hennig, W. (1966) *Phyogenetic Systematics*, University of Illinois Press. Urbana, IL.

Jacobs, D. J. (1994) Developmental genes and the origin and evolution of Metazoa. *In*: Schierwater, B., Streit, B., Wagner, G, P. and DeSalle, R. (eds) *Molecular Ecology and Evolution: Approaches and Applications*, Birkhäuser, Basel, pp. 537–550.

Kluge, A. (1997) Testability and the refutation and corroboration of cladistic hypotheses. *Cladistics* 13: 81–96.

Kumar, S., Tamura, K. and Nei, M. (1993) *MEGA (Molecular Evolutionary Genetic Analysis)*, Pennsylvania State University, University Park, PA.

Maddison, D. (1991) The discovery and importance of multiple islands of most parsimonious trees. *Syst. Zool.* 40: 315–328.

Maddision, W. P. and Maddison, D. R. (1992) *MacClade: Analysis of Phylogeny and Character Evolution*, Sinauer Associates, Sunderland MA.

Muller, W. A. (1994) To what extent does genetic information determine structural characteristics and document homology. *In*: Schierwater, B., Streit, B., Wagner, G, P. and DeSalle, R. (eds) *Molecular Ecology and Evolution: Approaches and Applications*, Birkhäuser, Basel, pp. 551–558.

Nei, M. (1988) *Molecular Evolutionary Genetics*, Columbia University Press, New York.

Nelson, G. (1994) Homology and systematics. *In*: Hall, B. K. (ed.) *Homology: The Hierarchical Basis of Comparative Biology*, Academic Press, San Diego, pp. 102–138.

Nelson, G. J. and Platnick, N. I. (1981) *Systematics and Biogeography: Cladistics and Vicariance*, Columbia University Press, New York.

Panchen, A. (1992) *Classification, Evolution and the Nature of Biology*, Cambridge University Press, Cambridge.

Patterson, C. (1988) *Molecules and Morphology in Evolution: Conflict or Compromise*, Cambridge University Press, Cambridge.

Raff, R. (1996) *The Shape of Life: Genes Development and the Evolution of Animal Form*, University of Chicago Press, Chicago.

Roth, E. L. (1988) The biological basis of homology. *In*: Humphries, C. J. (ed.) *Ontogeny and Systematics*, Columbia University Press, New York, pp. 1–26.

Sanderson, M. and Hufford, L. (1996) *Homoplasy and the Evolutionary Process*, Academic Press, San Diego.

Seberg, O. and Petersen, G. (1998) Selected software packages for personal computers. *In*: Karp, A., Isaac, P. G. and Ingram, D. S. (eds) *Molecular Tools for Screening Biodiversity*, Chapman and Hall, London, pp. 341–343.

Siddall, M. (1997) *Random Cladistics: Program and Documentation*, University of Michigan, Ann Arbor.

Sober, E. (1983) Parsimony is systematics: philosophical issues. *Annu. Rev. Ecol. Syst.* 14: 335–358.

Sober, E. (1988) *Reconstructing the Past: Parsimony, Evolution and Inference*, MIT Press, Cambridge, MA.

Sokal, R. (1987) Phenetic taxonomy: theory and methods. *Annu. Rev. Ecol. Syst.* 17: 423–442.

Sokal, R. and Sneath, P. H. A. (1963) *Principles of Numerical Taxonomy*, W.H. Freeman Press, San Francisco.

Swofford, D. L., Olsen, G. J., Waddel, P. J. and Hillis, D. M. (1996) Phylogenetic inference. *In*: Hillis, D. M., Moritz, C. and Mable, B. K. (eds) *Molecular Systematics*, Sinauer Associates, Sunderland, MA, pp. 407–514.

Tautz, D. (1994) Evolutionary analysis of genes involved in early embryonic pattern formation. *In*: Schierwater, B., Streit, B., Wagner, G, P. and DeSalle, R. (eds) *Molecular Ecology and Evolution: Approaches and Applications*, Birkhäuser, Basel, pp. 579–592.

Templeton, A. R. (1993) The "Eve" hypothesis: a genetic critique and reanalysis. *Amer. Anthropol.* 95: 51–72.

Wagner, G. P. (1989) The origin of morphological characters and the biological basis of homology. *Evolution* 43: 1157–1171.

Wagner, G. (1994) Evolution and multi-functionality of the chitin system. *In*: Schierwater, B., Streit, B., Wagner, G, P. and DeSalle, R. (eds) *Molecular Ecology and Evolution: Approaches and Applications*, Birkhäuser, Basel, pp. 559–578.

Wagner, W. L. and Funk, V. (1995) *Hawaiian Biogeography: Evolution on a Hot Spot Archipelago*, Smithsonian Institution Press, Washington D.C.

Wenzel, J. and Carpenter, J. (1994) Comparing methods: adaptive traits and tests of adaptation. *In*: Eggleton, P. and Vane-Wright, R. I. (eds) *Phylogenetics and Evolution*, Academic Press, London, pp. 79–101.

Wheeler, W. (1994) Sources of ambiguity in nucleic acid sequence alignment. *In*: Schierwater, B., Streit, B., Wagner, G, P. and DeSalle, R. (eds) *Molecular Ecology and Evolution: Approaches and Applications*, Birkhäuser, Basel, pp. 323–354.

Wiley, E. O., Siegel-Causey, D., Brooks, D. R. and Funk, V. A. (1991) *The Compleat Cladist: A Primer of Phylogenetic Procedures*, The University of Kansas Museum of Natural History, Lawrence, KS.

Zimmer, E. A. (1994) Perspetives on future applications of experimental biology to evolution. *In*: Schierwater, B., Streit, B., Wagner, G, P. and DeSalle, R. (eds) *Molecular Ecology and Evolution: Approaches and Applications*, Birkhäuser, Basel, pp. 607–616.

Alignment characters, dynamic programming and heuristic solutions

Ward Wheeler

Department of Invertebrates, American Museum of Natural History, Central Park West @ 79th, Street, New York, NY 10024, USA

Summary

The method of direct optimization of nucleic acid sequences proposed by Wheeler (1996) is elaborated and explained in light of dynamic programming procedures. An exact solution to the problem of phylogenetic reconstruction of unequal-length sequences is described, and its impracticality demonstrated. A branch-and-bound procedure is elucidated to accelerate this process. Additionally, a series of heuristic solutions are defined for this general problem, allowing for both significant decrease in computational effort and integration with existing algorithmic economies. Finally, potential implications of this method are discussed in light of putative long-branch attraction problems.

Introduction

Phylogenetic reconstruction of molecular sequence data requires two types of transformational events: (i) nucleotide (or amino acid) substitution and (ii) nucleotide (or amino acid) insertion and deletion. Standard procedures of character optimization and diagnosis (Farris, 1970; Fitch, 1971; Sankoff and Rousseau, 1975; Sankoff and Cedergren, 1983) accommodate character state transformation easily. Insertion-deletion events, however, are not so simply explained. Normally, a sequence alignment procedure is performed to establish the putative homologies which are required for standard character analysis (Feng and Doolittle, 1987, 1990; Hein, 1989, 1990; Higgins and Sharp, 1988, 1989; Wheeler and Gladstein, 1992, 1994), and gaps are inserted and treated as a fifth state. A heuristic method has been proposed (Wheeler, 1996) to diagnose cladogram topologies directly without the intervening multiple-alignment step. This discussion seeks to place the direct optimization method within the context of dynamic programming and to present both exact and heuristic solutions to the problem.

R. DeSalle, B. Schierwater (eds) Molecular Approaches to Ecology and Evolution
©1998, Birkhäuser Verlag Basel

The problem

Unlike many sources of information, molecular sequence data present not only variation in character state, but also in character number. That is, the number of characters presented by terminals may vary because sequences frequently differ in length. A cartoon of this situation is illustrated in Figure 1. Normally, the four terminal sequences (A, AA, AG and A), would undergo multiple alignment (Fig. 2), and some sort of dynamic programming (Sankoff and Rousseau, 1975; Sankoff and Cedergren, 1983) or short-cut procedure (Fitch, 1971) would be performed to diagnose the length or cost (in evolutionary steps) of any dendrogram (Fig. 3). Part of this process involves the insertion of placeholders – gaps ("-") to make the individual characters comparable. The sequence gaps are not observations, but the residue of insertion-deletion events required by the variation in sequence length. This introduces a certain epistemological inconsistency (Wheeler, 1996), treating what are in essence transformational events as equal to observations (such as A, G etc.).

The optimization procedure of Wheeler (1996) seeks to avoid this inconsistency and simplify the process by generalizing optimization to include insertion-deletion events. Direct optimization yields more intelligible and frequently more parsimonious results (Fig. 4).

Dynamic programming

An exact solution to the problem of sequence length variation can be achieved by recasting the diagnosis of cladograms from one of sequentially optimizing a series of simple characters (including gaps) to one of relating a single immensely (but not infinitely) complex character. In essence, all imaginable sequences (of all lengths) are possible states of this single character. The objective, then, is to create the most parsimonious character transformation series.

Within this framework, dynamic programming can be applied to determine the exact solution. As with the steps involved in optimizing Sankoff-type characters, the first issue is to define all the possible character states at each node. For the standard approach based on multiple alignment this would be simple – five states: A, C, G, T and gap. Here, however, the character correspondences and ancestral sequence length are unknown. The length of the hypothetical ancestral sequences is bounded by length zero at one extreme – no bases (sequences arise *de novo* repeatedly), and by the sum of the lengths of all the input sequences, since no parsimony-based operation could yielded anything longer. Since each of the four bases (with nucleic acids – gaps do not exist in real sequences after all) is possible at each position, the total number of possible states is: #states = $\Sigma 4^k$ where k is summed from 0 to the sum length of all input sequences. For three sequences of length four, there would be 22,369,621 possible states. In reality this number would be more tightly bound-

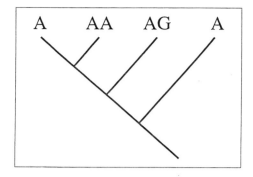

Figure 1. Four simple terminal sequences (A, AA, AG and A) related by cladogram.

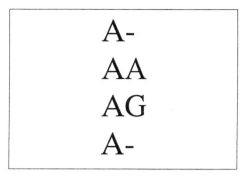

Figure 2. The minimum-cost multiple alignment for the four sequences of Figure 1.

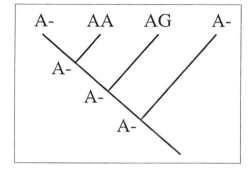

Figure 3. Standard method diagnosis of the cladogram and taxa of Figure 1 using the multiple alignment of Figure 2.

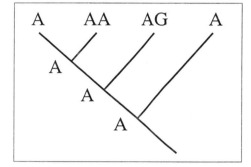

Figure 4. Diagnosis of taxa and cladogram of Figure 1 by the direct method proposed here.

ed. Since any insertion which would simultaneously occur in two descendent lineages would be transferred to the line leading to their ancestor, the minimum length of an ancestral sequence would be the lower of the two descendants and the maximum length would be their sum. Hence: #states = $\Sigma 4^k$ again, however, where k is summed from min (descendent 1, descendent 2) to the sum length of the two descendants (descendent 1 + descendent 2) sequences. For three sequences of length three, four and five there would be from 21,824 to 349,504. These are still large numbers, but orders of magnitude smaller than the exhaustive case.

Once all the possible cases are enumerated (and a suitable cost matrix relating all the possible states to each other has been specified), standard dynamic programming in a "down pass" will yield the most parsimonious cladogram length and ancestral state assignments. Unfortunately, this will take a long time. Each internal node will require approximately twice the number of states squared operations, from each of the possible combinations of the two descendent states and each of the ancestral states (this will be smaller if the descendants are terminal taxa). Clearly, this is impracticable for all but the smallest cases.

Branch and bound

Although the exhaustive optimization described above is absurdly involved, the number of considered ancestral states and operations can be further limited. Since the cost of each path through each possible state is a monotonically increasing function (lengths cannot go down), an upper bound on cost can exclude the vast majority of possible states (Hendy and Penny, 1982). Longer nucleic acid strings require (in general) more indels, hence the longer character states (i.e. sequences) are likely to be excludable very early in the process.

The procedure outlined earlier (Wheeler, 1996) can act as such a bound since it yields upper-bound estimates of tree length. In the example of Figure 1, an upper bound of two insertion-deletion events and one base transformation (assuming insertion-deletion events are assigned greater cost than base substitution) would be postulated (Fig. 5). This would exclude ancestral state reconstructions of length greater than two. Any higher number would exceed the bounded cost. The number of possible character states would be limited to six (A, G, AA, GG, AG and GA). Although this would result in a further dramatic reduction in computational complexity, for real-world cases the reduction would likely be from an extremely absurd situation to one which is merely absurd. Most likely, we will be limited to heuristics.

Heuristics

Wheeler (1996) presented a procedure to improve the initial estimates of cladogram lengths. The optimization procedure is a straightforward generalization of nonadditive or unordered optimization (Farris, 1970; Fitch, 1971). The down pass optimization is depicted in Figure 6. In this case, there are five sequences of unequal lengths. Without prior knowledge of base correspondences (alignment), it is impossible to construct a hypothetical ancestor or determine how costly that operation is (in terms of transformations). Hence, correspondences (putative homologies) must be constructed as we go down the tree for the comparisons made at each node. In essence, all possible schemes of comparison must be examined for each node and that scheme which minimizes the number of minimum-cost union events (weighted by the cost of a base transformation) and insertions and deletions (weighted by the gap cost) is assigned to the node. In this way, the most efficient (i.e. lowest cost) hypothetical ancestor is constructed.

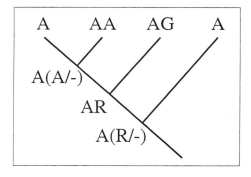

Figure 5. Down-pass optimization of Figure 1 using the method of Wheeler (1996) used as an upper bound for exact analysis.

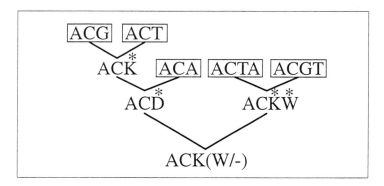

Figure 6. A more complex example of the direct-optimization procedure. Boxes surround terminal taxa, asterisks (*) denote base substitutions, parentheses insertion-deletion events.

As with non-additive analysis, the procedure begins at the top of the tree (or more specifically at an ancestral node of two terminal taxa) with the sequences ACG and ACT. The construction of the hypothetical ancestor can be broken down into two operations. The first can be thought of as an alignment step. The sequences are aligned to minimize the weighted cost of indels and base transformations as determined by union/intersection counts. This is performed with the proviso that if a gap is inserted in one sequence to correspond to a gap in the other, this is done at no cost (the sequences would have a nonempty intersection). Each possible alignment is considered (via dynamic programming) as in the Needleman and Wunsch (1970) procedure. In this example, the best alignment contains one base transformation and no gaps (Fig. 6). In the second operation, the hypothetical ancestor is constructed from this alignment by taking the union/intersection position by position along the sequence yielding ACK (K = T or G in IUPAC parlance). This hypothetical ancestral sequence is then compared with the next terminal ACA yielding another hypothetical ancestral sequence, ACD at the cost of another base transformation.

On the other side, a similar operation comparing ACTA with ACGT yields ACKW. Proceeding to the next node, ACK is compared with ACKW. Here, the alignment step requires no nucleotide transformations but does require an insertion-

247

deletion event, yielding an ancestral sequence with ambiguities in both base assignment and length. The first three bases are reconstructed as before ACK. The reconstruction of the fourth, however, is more complicated. This optimization would allows each of the three possibilities A, T, or GAP. This signifies that the position may contain either an A or a T or just not be there at all. The entire topology has been diagnosed at a cost of one insertion deletion events and four base transformations.

Greediness and shortcuts

In assigning states to hypothetical ancestral sequences, a method has been used of necessity which may introduce error in calculating the tree length. When proceeding down the tree, nucleotide assignments cases may occur in which a nucleotide base (say A) is faced with a corresponding ambiguity in its putative sister taxon as to whether or not a base exists (say G or GAP). How do we determine the condition of the ancestral sequence? If all transformations were equal (including indel events, transitions and transversions), the ancestral condition would be the union of the three states à la Fitch. However, this is rarely the case. More frequently investigators postulate that indels are less plausible that base substitutions, that is the cost of an insertion or deletion is greater than that of a nucleotide substitution. In this case, the ancestral condition would then be assigned "R" for the union of A and G and the GAP possibility excluded, taking the lower-cost transformation as yielding the ancestral condition. It may be more globally parsimonious that the indel ambiguity not be removed at this stage, but the procedure will not foresee this. Hence, the operations described here may overestimate tree length. Analogous reasoning holds for choices where transition-transversion bias is involved. Identical behavior could be observed using the optimization procedure described here with sequences of identical length and comparing the results with a dynamically programmed tree length (via a Sankoff step-matrix procedure).

Since we can not know the future (further down) sequences, optimization of unequal length sequences requires this myopia. As an aside, this method is in essence a weighted nonadditive optimization. Given that the various transformation weights are known, all optimization events, their costs, and results can be calculated before the actual tree search. In this way, weighted step matrix parsimony calculations can be accomplished at considerable savings in cost (in my experience the general weighting comes at a cost of a low – ~5% – fixed premium on execution time). As mentioned before, the procedure is local, globally more parsimonious may not be considered (or even rejected). As a result, any error in length should be an overestimate. For the case where the sequences are equal in length, the results can be adjusted by full dynamic programming of individual candidate topologies.

This direct optimization procedure frequently can be improved to get a better (i.e. lower) upper bound on length through rerooting the down-pass network. A

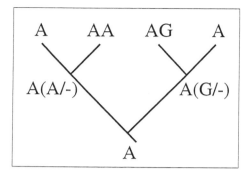

Figure 7. Rerooted version of the example of Figure 5.

"virtual" root is postulated, and each of the $2n - 3$ (for n terminal taxa) possible roots used to determine a down-pass cost and the minimum retained. The estimate of tree length based on the rooting of Figure 5 (two insertion-deletion events and a single base change) is reduced by moving the root (Fig. 7), resulting in a length of two insertion-deletion events. Some of the greediness of the down-pass algorithm can be circumvented this way.

In addition to this rerooting, which requires the examination of $n - 1$ nodes (n terminals), a second or up pass can be added to determine an estimate of the final states for each of the internal nodes (with the proviso that the final sequences not contain any gaps). Although this in itself will not effect the estimate of cladogram length, the determination of "final" states allows the use of other algorithmic speedups (Goloboff, 1996; Gladstein, 1997) which can dramatically improve the efficiency of tree searches.

The determination of final states can be achieved by traversing the tree in the direction opposite from that of the initial pass, away from the root (or virtual root – if rerooting has been performed). The final state set is defined as those states which minimize the total cost summed over the three paths from the node: the current node to each of its descendants and to the final states of its immediate ancestor. As with the initial ("down") pass, the results of all character combinations can be precalculated based on the matrix of transformation costs among each of the four nucleotide states and insertion-deletion costs.

These precalculations assume that only local taxon states matter, and introduce the errors of this heuristic procedure. For the initial pass, a two-dimensional matrix is required containing the resultant state and cost for each of the character combinations of A, C, G, T and "-", and ambiguities– 32×32. The second pass will require a second matrix of three dimensions describing all the possible interactions among the two descendant and one ancestral states.

Conclusions

The procedures outlined here allow the determination of a upper bound on clado-gram length based on direct optimization of nucleotide sequences. This method is a simple extension of parsimony-based cladogram construction to include the origination and disposition of characters. Although this method is more elaborate and time-consuming that standard optimization procedures, the avoidance of multiple sequence alignment should result in both more efficient and parsimonious results. Furthermore, the nonindependence of gaps (i.e. insertion-deletion length), so tribu-lating to phylogenetic analysis based on the necessary assumption of character independence, can be accommodated seamlessly. In fact, any such model in which nucleotide changes interact (e.g. codon effects) can be integrated. Since the transformations are occurring within a single "character", or more specifically among complex character states, the independence of character vectors is not violated. Insertion-deletion costs can be complex, nonlinear functions of length, or codon bias added to analysis without violating this tenet of epistemological unity.

A final note concerns that pit of Dis, the "Felsenstein zone", (Felsenstein, 1978). The basic notion of nonhistorical, stochastically derived character matching based on random similarity. It is postulated that under certain conditions, the most parsimonious result will not be the correct one. Leaving aside the notion of "correctness" for the moment, it is worth noting that the size of this zone is inversely proportional to the number of character states expressed by the phylogenetic data. This is due to the requirement that the "bad" randomly similar character must overwhelm the "good" historically informative ones. S. Farris (personal communication) and others have noted that using characters in n-tuples (supersites) would reduce any long-branch problem by increasing the number of states – thereby decreasing the chance of random similarity. The methodology described here treats entire sequences as characters with huge numbers of possible character states. If the inverse relationship between the size of the Felsenstein zone and number of character states holds, the zone may be small indeed.

Acknowledgments

I would like to thank James Carpenter, Michael Whiting, David Gladstein and Rob DeSalle for many helpful comments and suggestions.

References

Farris, J. S. (1970) A method for computing Wagner trees. *Syst. Zool.* 34: 21–34.

Felsenstein, J. (1978) Cases in which parsimony or compatibility methods will be positively misleading. *Syst. Zool.* 27: 401–410.

Feng, D. and Doolittle, R. F. (1987) Progressive sequence alignment as a prerequisite to correct phylogenetic trees. *J. Mol. Evol.* 25: 351–360.

Feng, D. and Doolittle, R. F. (1990) Progressive alignment and phylogenetic tree construction of protein sequences. *Meth. Enzomol.* 183: 375–387.

Fitch, W. M. (1971) Toward defining a course of evolution: minimum changes for a specific tree topology. *Syst. Zool.* 20: 406–416.

Gladstein, D. G. (1997) Incremental evaluation and the diagnosis of cladograms. *Cladistics* pp. 21–26.

Goloboff, P. A. (1996) NoNa Program and Documentation available from the author.

Hein, J. (1989) A new method that simultaneously aligns and reconstructs ancestral sequences for any number of homologous sequences, when a phylogeny is given. *Molec. Biol. Evol.* 6: 649–668.

Hein, J. (1990) Unified approach to alignment and phylogenies. *Meth. Enzomol.* 183: 626–644.

Hendy, M. D. and Penny, D. (1982) Branch and bound algorithms to determine minimal evolutionary trees. *Math. Biosci.* 59: 277–290.

Higgins, D. G. and Sharp, P. M. (1988) CLUSTAL: a package for performing multiple sequence alignment on a microcomputer. *Gene* 73: 237–244.

Higgins, D. G. and Sharp, P. M. (1989) Fast and sensitive multiple sequence alignments on a microcomputer. *CABIOS* 5: 151–153.

Needleman, S. B. and Wunsch, C. D. (1970) A general method applicable to the search for similarities in the amino acid sequence of two proteins. *J. Mol. Biol.* 48: 443–453.

Sankoff, D. D. and Rousseau, P. (1975) Locating the vertices of a Steiner tree in arbitrary space. *Math. Program.* 9: 240–246.

Sankoff, D. D. and Cedergren, R. J. (1983) Simultaneous comparison of three or more sequences related by a tree. *In*: Sankoff, D. and Kruskal, J. B. (eds) *Time Warps, String Edits, and Macromolecules: The Theory and Practise of Sequence Comparison*, Addison-Wesley, Reading, MA, pp. 253–264.

Wheeler, W. C. (1996) Optimization alignment: the end of multiple sequence alignment in phylogenetics? *Cladistics* 12: 1–9.

Wheeler, W. C. and Gladstein, D. G. (1992) Malign: A Multiple Sequence Alignment Program. New York, NY.

Wheeler, W. C. and Gladstein D. G. (1994) Malign: a multiple nucleic acid sequence alignment program. *J. Hered.* 85: 417.

Pitfalls in phylogenetic analysis of large molecular data sets

Paul Z. Goldstein[1] and Chelsea D. Specht[2]

[1]American Museum of Natural History, Department of Entomology, Central Park West at
79th Street, New York, NY 10024, USA
[2]Department of Biology, New York University, New York, NY 10003, USA

Summary

Fundamental considerations of phylogenetic analysis are reviewed in the context of treating large molecular matrices. While molecular data sets with dozens or hundreds of taxa are increasingly common in phylogenetic inference studies, several computational issues, some unique to such large matrices, others general in phylogenetic inference, nonetheless confront molecular systematists. The most controversial of these, choice among phylogenetic inference methods, bears directly on the analysis of molecular data sets. Maximum likelihood methods have been implemented exclusively for molecular data, but their burdensome computational load becomes acute as the number of taxa being analyzed grows. While there are several reasons to prefer parsimony to maximum likelihood generally, the unfeasibility of using likelihood to treat matrices with many terminals and the desirability of combining morphological and molecular under simultaneous analysis lead to a preference for parsimony more or less by default. Terminal selection and the coding of subset polymorphisms and inapplicable character data are of no less critical concern to molecular systematists than to morphologists. Shortcuts such as collapsing taxa to form "composite" terminals should be viewed with caution. Measures of nodal support, all of which are problematic in one or more ways, may be computationally prohibitive for large matrices. The relatively novel technique of parsimony jacknifing may provide a desirable means of evaluating the robustness of phylogenetic inference, especially as the generation of sequence data becomes increasingly routine.

Introduction

The use of molecular characters in phylogenetic inference has generated an explosion of interest in systematics and evolution while making these fields inviting and accessible to statisticians, molecular biologists and population geneticists. The advent of automated direct DNA sequencing techniques in particular has enabled rapid collection and analysis of phylogenetic data, a task traditionally undertaken

R. DeSalle, B. Schierwater (eds) Molecular Approaches to Ecology and Evolution
©1998, Birkhäuser Verlag Basel

by anatomical or morphological workers perhaps trained more strictly in systematics. Unfortunately, the growing popularity of phylogenetic paradigms among geneticists and molecular biologists has not always been accompanied by an understanding of the fundamental principles of systematics, and to the extent phylogeny reconstruction is pursued in the absence of that understanding, the facile generation of impressive molecular data sets may provide workers with an excuse to avoid important computational and empirical issues. Because of the wide impact of the evolutionary questions being addressed using phylogenetic techniques, it is increasingly imperative that researchers pay special attention to computational issues involved at each step of phylogenetic inference. Indeed, we expect that sequencing technology will grow even more efficient, and that large data sets combined from various source types will become a requisite operating procedure.

Several computational problems become increasingly acute with the size of the matrix. To be sure, one of the early appeals of molecular data was the potential for generating very large data sets involving unprecedented numbers of informative characters that would, it was hoped, resolve any phylogenetic problem (Gould, 1985). Although large data sets involving 50 or more taxa are becoming increasingly common, it is not yet clear whether there exist behavioral properties (other than those related to tractability) that emerge with increased data set size. To the extent such properties do exist, they may not be relevant to the computational aspects of analysis except to illustrate the drawbacks of one or more competing methods.

This essay examines broadly some of the logical and computational debates surrounding the analysis of large molecular data sets and the interpretation of resultant phylogenetic hypotheses. Data set size is of course a function of two factors: the number of terminals and the amount of character data per terminal. Rather few studies have addressed either empirically (Nixon and Davis, 1991; Platnick et al., 1991) or through simulation (Wheeler, 1992a) the effects of taxon vs. character sampling on cladogram construction. And although broad generalizations are possible, the relevance of such statements must be examined on a case-by-case basis unless proved mathematically. Nonetheless in practice at least, there appear to exist some universal limits to the resolving power of character data with n states. We review these fundamentals, and examine two broad issues: (i) the impact and potential pitfalls of taxon selection, including outgroup selection and terminal selection below traditional species boundaries; and (ii) the computational issues associated with the analysis and interpretation of large data sets generally, and molecular data sets in particular. We also emphasize that any interpretation of the relationship between character data and tree support depends ultimately on the framework used to generate the tree from the data, and conclusions based on one paradigm may be contradicted by the analysis of the same data under a different paradigm.

Preliminaries: taxon selection

Ironically, while automated sequencing technology has afforded workers the luxury to generate data for more taxa faster than ever before, some workers may find it tempting to address higher-level phylogenetic questions prematurely, that is before lower-level questions are sufficiently resolved to make higher-level analyses meaningful. One of the risks of molecular systematics has thus been the generation of "phylogenetic" data sets for "groups" that may not actually exist. This is essentially a problem of nomenclature's having to keep pace with phylogenetic hypotheses, but names are nevertheless guidelines, if not benchmarks, for testing monophyly. Paraphyletic genera, tribes and so on are the hobgoblins of higher-level cladistic inference because their presumptive treatment as monophyletic terminals obviates the meaningful interpretation of the relationships among those terminals. Entomological systematics in particular is rife with examples of mixed type series as well as paraphyletic and polyphyletic tribes and genera, some of which were erected without so much as a single uniting feature. The Amphipyrinae (Lepidoptera: Noctuidae) are the largest subfamily within the largest lepidopteran family, yet are not diagnosed by any one feature or combination of features. The amphipyrines have simply been the systematic dumping ground for any group that did not easily fit into one of the other subfamilies. Any putative phylogenetic analyses of "amphipyrine" relationships will not, therefore, contribute in any meaningful way to an understanding of noctuid relationships except to demonstrate the extent of paraphyly.

Plant systematics has begun to yield impressively large and well-sampled matrices (e.g. Chase et al., 1993) that illustrate the prevalence of paraphyly and the utility of large matrices in discovering it. In an analysis of the phylogenetic structure among the monocotyledonous plants, Davis (1995) used generic exemplars to demonstrate that several families and orders are paraphyletic or worse. In this study, only 4 of 10 orders with more than one representative examined were found to be monophyletic in all of the most parsimonious trees; and three of eight families sampled were not monophyletic. As an example, the four representatives of the Araceae formed two monophyletic groups, one of which is nested within a clade containing representatives of the Alismatanae. In other words, the family Araceae *sensu stricto* is paraphyletic with respect to something formally placed in a separate order (!). Furthermore, the Araceae is the only family in the order Arecales and the superorder Arecanae (*sensu* Dahlgren et al., 1985), and its paraphyly thus reflects poorly upon the relationships suggested by ordinal and superordinal names. An analysis of monocot families using nonmonophyletic exemplars would compound higher-level paraphyly and mislead phylogenetic inference.

Because paraphyly is rampant in many described groupings (especially among highly diverse taxa, such as plants and insects), one must approach the problem of exemplar selection carefully in both higher- and lower-level studies. For higher-level work, Nixon and Davis (1991) suggested cleaving terminals to eliminate subset polymorphisms, which can confound phylogenetic analysis even if the polymorphic terminals used are demonstrably monophyletic higher taxa (contra Rice et al., 1997). Such subtaxon recoding has been employed with some success by Struwe et al. (1994).

The applicability of the terminal units to lower-level cladistic analysis is perhaps more controversial. An important corollary to cladistic analysis of any kind is the requirement of underlying nested hierarchy, which becomes a concern in lower-level analyses of many conspecific individuals. As Davis and Nixon (1992), Luckow (1995) and Brower et al. (1996) explain, if the result of a numerical cladistic analysis is to be interpreted as a phylogenetic representation, and since the result of a cladistic analysis is necessarily hierarchic (even if unresolved), then there exists a level below which individual organisms make logically inappropriate terminals. Although cladistic analyses of genealogically (or "tokogenetically") related individuals have become popular, as in the paradigm of "infraspecific phylogeography" christened by Avise et al. (1987), such analyses can be misleading in that they generate nested hierarchic hypotheses of relationships between taxa for which no underlying hierarchy exists.

Analytical issues: competing philosophies

As the importance of phylogenetic hypotheses for classification and macroevolutionary scenario testing has been widely embraced for more than a decade, the focus of debate in the field of phylogenetic inference has shifted to methodological concerns. Perhaps the most contentious issue in current phylogenetics is the competition between tree-building approaches. Philosophical and quantitative arguments aside, tree-building methods are numerous, and many workers appear to view agreement among multiple methods as phylogenetic evidence *per se*, as though abstract rigor were arithmetically additive. We do not share this view, however it might appease members of competing ideologies. The fact that multiple forms of analysis provide the same answer does not in and of itself justify any one of the methods. Moreover, if the preference for the underlying logic of tree-building operations is to be subsumed under the preference for multiple tests regardless of their rationale, then alphabetical and random ordering of taxa would be viewed as legitimate tree-building alternatives, a position we regard as indefensible. Two times two is four, as is two plus two and two squared. One does not conclude, however, that four is a "more" correct answer because it was arrived at by more than one distinct operations designed for quite different purposes.

Most practicing systematists have embraced the paradigm of cladistic parsimony in character analysis, but competing phenetic and "evolutionary taxonomy" schools, often referred to as defunct, have in fact been revived in one or more forms, despite published demonstrations of some rather serious shortcomings. Phenetics survives most commonly in the form of "neighbor joining", a method that not only denies the relevance of apomorphy to systematics but was also shown to be sensitive to taxon input order shuffling, yielding completely contradictory trees (no nodes in common) for the same data set upon switching the order in which taxa are listed (Farris et al., 1996). When more than one solution obtains under neighbor joining, there is no available optimality criterion for choosing among them, and current neighbor-joining software applications simply pick one tree without rationale (i.e. the first one that obtains).

At the fore of the methodological debates, however, is the philosophical divide between the character-based approach on which practitioners of parsimony rely and probabilistic methods advocated by a wider audience generally outside the realm of systematics proper. Probabilistic inference relies on the presumption that phylogenetic character-change data exist as "classes" amenable to statistical inference (Swofford, 1991), rather than as historically unique evolutionary events. This amounts to the *a priori* identification of classes of data presumed suitable for probabilistic inference. These classes may be positional (e.g. first vs. third positions) or chemical (e.g. purine vs. pyrimidine) in nature. The probabilistic school arose specifically in association with electrophoretic and serological data, which appeared to lend themselves to numerical data classification schemes (amenable to probabilistic statistics) more readily than did morphological data. We believe that appearance is deceptive, however. The fact that a DNA sequence character may be one of at most four states leads to the designation of character states (base pairs) as "identical" without being homologous; unrelated characters may share the "same" state. There is thus no more justification for identifying *a priori* classes of data according to base-pair identity than making *a priori* claims about the superiority of, for example, skeletal characters vs. pelage characters.

It is beyond the scope of this paper to compare in detail parsimony and maximum likelihood as phylogenetic inference paradigms. To a large extent, proponents of each school have elected to debate one another on their own terms, and to the degree this has happened many of the arguments have been of dubious value. Detractors of parsimony have criticized it on probabilistic grounds (e.g. Felsenstein, 1978b, 1981), but advocates of parsimony have stressed its correspondence to a rigorous logical interpretation of empirical science, not as a "method" to be preferred for its quantitative sophistication *per se*. Parsimony is a paradigm for interpreting character data along a historical axis, not a means of deducing confidence intervals or P values surrounding historically unique events (Wenzel and Carpenter, 1994). As such, the invocation of parsimony to interpret cladograms as phylogenies does involve assumptions, specifically cladogenesis and descent with modification (the same assumptions inherent in any hierarchic inference of relationships among living things) and the assumption that more character data reflect common ancestry than independent evolutionary events. If that last were not the case, then the comparative basis of all systematic inference would be suspect. Maximum likelihood, on the other hand, relies on models parameterized according to the behavior of data on "known" phylogenies (such as the relationships between a frog, a chicken, a mouse and a human), on "true phylogenies" created by simulating character data on a computer, on parsimonious cladograms involving actual systematic studies or, perhaps least objectionably, through the data under investigation themselves. Adherents of parsimony stress that *a priori* parameterizations of models used to infer common ancestry amount to assumptions about the evolutionary process, and the implication of using such procedures is that one must know how evolution works before one can examine how evolution works (Farris, 1983; Mickevich, 1983). As such, maximum likelihood makes no more sense when viewed from within an empirical parsimony framework than does parsimony when viewed from within a probabilistic framework. The fact that likelihood or other model-based approaches "predict" their own superiority is hardly an argument for their preference (cf. Wheeler, 1992b).

In our view, the controversy distills to fundamentally incompatible views on how science should be approached, and that controversy should therefore be addressed on strictly logical and empirical grounds, preferably relying more heavily on real data than simulated data. We concur with advocates of parsimony that such a framework is the only rational means of comparing the essential suitability of each approach. It should be understood that while a parsimonious interpretation treats characters and characters alone as evidence for common ancestry, maximum likelihood uses assumptions about the evolutionary process to support statements of evolutionary relationships (Carpenter, 1994). As such, it posits unobserved character change. We maintain that workers should have the freedom to use any method they can defend, but that the choice of either method must defend the nature of the quantities treated as "evidence".

Thus, while we find the arguments in favor of parsimony compelling enough to dismiss likelihood altogether as a phylogenetic tool, several treatments of large data sets have involved assertions rooted in the probabilistic literature. For the purposes of this discussion, then, it is not sufficient simply to observe that likelihood will be computationally cumbersome as soon as very many taxa are evaluated. And although most of our discussion centers on quantitative advances in comparing and evaluating phylogenetic data parsimoniously, we will also address probabilistic assertions of the behavior of large data sets to some degree. Computational maneuverability may not in itself be a very "pure" justification of parsimony, but given that there exist misconceptions about the relationship of data size to tractability in both the parsimony and the likelihood literature, we will endeavor to clarify the frameworks from which various claims have been made. It should be understood, however, that because likelihood relies fundamentally on generalized models, empirical evaluations of their performance on real data sets or of their generalized behavior vis-a-vis specific characters are impossible (Carpenter, 1992). Below, we confine our discussion of notions of tree support to the framework of parsimony, which we regard as a fundamental requirement of the concept of apomorphy (Farris, 1983; Farris and Kluge, 1985; Carpenter, 1994).

Matrix size and the number of trees: Is bigger better?

Fundamentals

Computation of best-supported (most parsimonious) topologies depends directly on the number of taxa and the amount of data per taxon under analysis. As such, tree search constitutes an NP-complete problem (Day, 1983; Rice et al., 1997), one in which tree search effort is inextricably tied to the number of taxa under analysis. The addition of taxa forces the operational search to accommodate more topologies; increasing the size of either axis of the matrix (number of taxa or of characters) may encumber approximate tree search by stalling on one or more "islands" of tree space (Maddison, 1991). However, additional data do not necessarily increase computation time, and may even decrease it under certain conditions (Chase et al., 1997), such as when the added data are decisive, and provide a means of bypassing the phenomenon of multiple optimality peaks during tree search. Below, we review some of the quantitative relationships between the number of taxa examined and the number of possible topologies. We then examine the relationship between character state number and tree resolution as one increases the number of taxa for analysis, and the question of the degree to which molecular data may be used for simultaneous treatment of higher- and lower-level analyses.

Following Cavalli-Sforza and Edwards (1967), Dobson (1974) and Phipps (1976a,b), Felsenstein (1978a) summarized measures for calculating the total number of rooted bifurcating (resolved) tree topologies having n unlabelled tips: $(2n - 3)!/2^{n-2}(n - 2)!$

For $n \geq 2$. Thus, the number of possible resolved trees increases by a factor of $(2n - 1)$ for each taxon added; the actual number of additional resolved topologies can be expressed as:

$$(2n - 2)!/2^{n-2}(n - 2)!$$

Thus,

$$f(n)(2n - 1) = f(n) + (2n - 2)!/2^{n-2}(n - 2)!$$

When multifurcations (polytomies) are required – as, of course, they are when data are ambiguous – these numbers increase substantially; Felsenstein (1978a) calculated the number of possible rooted topologies with one or more nodes unresolved with an algebraic matrix.

From the standpoint of resolution alone, the number of terminals bears directly on reconciling alternative topologies. Coddington and Scharff (1994, p. 420) stated, "Large data sets are of special concern because the probability of logically and topologically independent regions of cladistic instability increases with size." We view it as risky to think of nodes or areas on a given tree independent of one another (they are not). To the extent an increase in terminal number is associated with increased ambiguity in the data (a case-specific issue), the number of terminals may indeed influence the ability to interpret consensus trees (Coddington and Scharff, 1996), and "semistrict" consensus trees in particular may be likely to confound the interpretation of large data sets (Nixon and Carpenter, 1996b).

Results from simulation

Many of the assertions in the literature regarding the effects of taxon addition to tree resolution and support have been based on simulations or framed in terms of molecular evolutionary rates. It is difficult to quantify the relationship between character data and tree topology, since the effect of a given character on a given node (the character's distributional behavior) depends ultimately on the weight of evidence provided by the data set as a whole, and since every data set represents a collection of historically unique character information. Nonetheless, some general statements are warranted. Studies have indicated differential effects of various aspects of data sets generally – and the size of molecular data sets in particular – on tree resolution and accuracy.

Wheeler (1992a) used an analysis of variance (ANOVA) of simulated data to examine the effects of the number of taxa, sequence length, missing data, rate of evolution and evolutionary model on the accuracy and resolution of tree topology. Using success of cladogram topology and the resolution of the cladogram as independent variables, Wheeler showed that sequence length carries the greatest influence (85–88%) on variation of resolution, whereas number of taxa contributes the most (47–49%) to cladogram accuracy (under parsimony). The evolutionary model applied in generating the sequence also contributed, albeit a good deal less, to the accuracy of cladogram reconstruction. This effect may vary with the size of the data set, especially the amount of character data per taxon; given clock-like evolution, there is a direct relationship between the number of characters per taxon and the accuracy of the reconstructed topology such that an increase in the amount of sequence data will invariably lead to increased resolution and support. If, however, the sites in question undergo state change with a Poisson distribution, a "plateau" in resolution is reached such that an increase in data no longer increases the reliability of reconstruction.

While several authors have chosen to examine the numerical relationship between taxic representation and character sampling abstractly through simulation

(Huelsenbeck, 1995), some of the most insightful and pertinent quantitative studies have examined potential problems by studying specific data sets in detail (e.g. Coddington and Scharff, 1994, 1996; Farris et al., 1996). In general, however, much of the molecular systematics literature devoted to character utility has focused on identifying "rates" of sequence evolution within classes of data (gene classes, site-positional classes, structural and functional classes, etc.) and the application of such rates in identifying site-specific utility of particular markers. This focus is another outgrowth of the probabilistic practice of identifying classes of data, either to para-meterize models, to identify the appropriate level(s) for which a particular gene or gene region will be phylogenetically informative, or to identify situations when data sets should be analyzed separately rather than simultaneously.

Rate heterogeneity: swamping, saturation and simultaneous analysis of data sets

An early, popular, but apparently incorrect notion was that the sheer number of molecular characters would "swamp" morphological data in analyses (Miyamoto, 1985), obviating the combined analysis of morphological and molecular data sets. Such swamping has never been demonstrated empirically; to the contrary, many combined morphological and molecular matrices converge upon a solution supported by the morphology alone, but either add resolution (Miller et al., 1997) or support (Sullivan, 1995) for that tree. Nevertheless, the perception that molecular characters are *de facto* superior to morphological characters by virtue of sheer numbers persists to this day (e.g. Hedges and Maxson, 1996; Givnish and Sytsma, 1997b), as does the view that morphological and molecular data must be analyzed separately.

In theory, of course, even a single character can resolve any number of taxa, given unlimited character states. Character systems are rarely this rich, however, and molecular data in particular are characterized by a maximum of four possible states, excluding gaps and polymorphisms (which should not be coded as primary homologues). The immediate implication of this is a limit to the observed character variation or range for every character, no matter how many taxa are added. This four-state limit is the source of the assertion of saturation, the idea that homologous changes at a given site eventually become masked over time if point mutations occur either much more frequently than speciation events, or frequently relative to the time by which the taxa under analysis are separated. A given sequence is expected to confound the resolution of a group of taxa either when (i) it evolves too slowly for there to be enough informative sites, or (ii) when base changes occur so rapidly that the homologous character-state changes are masked by or saturated with subsequent mutations, so-called multiple hits.

Many authors have viewed saturation as the root of "long-branch attraction" (Felsenstein, 1978b, 1981), the notion that if a given site or gene evolves very fast, the distribution of its character states will result in topological attraction – due to congruent homoplasy – between unrelated taxa. Albert et al. (1993) showed mathematically that the Felsenstein zone may apply only under exceedingly limited circumstances, such as when rates are irrationally high. However, Whiting et al. (1997) have pointed out a more support-based flaw to Felsenstein's argument, namely the rejection of statements of relationship on the grounds that they are supported by too many synapomorphies, as though the nodal support in question is simply too good to be true. Sullivan et al. (1997), for example, retrieved identical topologies for a molecular data set from both parsimony and maximum likelihood, yet concluded that the nodal support for the parsimony tree was "artificially high", a conclusion about actual data that could derive only from a presumed model.

Because genes with different substitution rates are expected to yield substantially different phylogenetic information, many workers would advocate performing separate analyses for different data sets, or ignoring some data altogether. It is no accident of history that proponents of maximum likelihood methods also advocate partitioning data when rates vary "significantly"; such variability cannot be accommodated by a single model for the purposes of maximum likelihood estimation. We find the stated premise of partitioned analyses of data irreconcilable with the strategy of partitioning itself. If, indeed, rates among genes do vary, then one might expect simultaneous analysis of data sets to obtain a more resolved, better-supported tree than either analyzing the data separately or restricting their analysis to a subset of the taxa for which the data are uncontradicted. Sullivan (1995) found that two discordant molecular data sets converged upon the tree supported by morphology – and with greater nodal support – when combined. A growing number of authors are finding similar results (e.g. Remsen and DeSalle, 1998; Miller et al., 1997). Remsen and DeSalle's data, further, converged on a single best-supported tree despite *a priori* heterogeneity tests that indicated such data should not be combined (!). Thus, while it might be argued that rate heterogeneity obviates combining data in a likelihood analysis, it might well be viewed as an argument *for* combining data under parsimony.

The alternative, required by maximum likelihood, to combining data sets under simultaneous analysis is the computation of the consensus of trees derived from data sets analyzed separately, each according to set-specific models. That alternative involves another – and very relevant – kind of swamping: consensus trees, for their part, simply combine topologies (as sums of nodes) without regard to how well supported their nodes are (Mickevich and Farris, 1981; Wheeler, 1991). Hence, a node supported by one synapomorphy may collapse an alternative node supported by nine synapomorphies when the strict consensus tree is computed. Consensus trees may require more steps than any of the initial trees, in which case any of the initial trees is preferable to the consensus (Miyamoto, 1985).

While most workers are probably cautious of results derived from reconciling trees resulting from different data sets, it should be obvious that the inclusion of more taxa leads to greater potential for node collapse in computing consensus trees when the data are incongruent and decisive (*sensu* Goloboff, 1991a, b; Davis et al., in press). As those authors explain, a completely undecisive matrix does not contradict any topology, but when data sets are large and decisive, the potential for smaller matrices to be overruled regardless of their decisiveness is an issue. We will return to the issue of decisiveness in our discussion of tree support statistics.

Requisite data

Molecular systematists must make what are sometimes difficult choices of what and how many genes or gene regions to sequence. Of course, there can be no formula dictating the requisite amount of sequence data or the elusive "perfect gene" that will resolve an n-taxon matrix in every case. Typically, workers tend to view the characteristics of data affecting gene choice in terms of saturation, which is demonstrated by evaluating character-state change within a subset of characters (such as third positions) against summed state changes in the entire data set. But one really needs a tree to demonstrate saturation as homoplasy. It appears to be worthwhile to compare the accumulation of informative site changes per absolute site change (following cladogram construction) based on simultaneous analysis of multiple data sets (Mickevich and Farris, 1981; Baker and DeSalle, 1997). Lack of variability, for its part, is easy enough to detect by simple inspection of the data matrix. Perfectly incongruent character data, while appearing variable in the matrix, will of course result in an unresolved "bush".

Nonetheless, one might predict that the resolution of increasing numbers of taxa requires concomitant increases in character data (but see Hillis, 1996), an inference supported by Wheeler's (1992a) simulations. As more taxa are added to the matrix, not only does the possible number of topologies increase, but so does the number of possible optimizations for each character. Kim (1996), using a simulated data matrix for six "taxa", asserted that the probability of long-branch attraction actually increases with taxon sampling; that is to say, parsimony may be inconsistent in converging on the "wrong" tree as taxa are added. This claim runs counter to the expected result that adding taxa would, if anything, be likely to break up long branches (cf. Hillis, 1996). However, it is unclear what (if any) methodological contributions are made by such contrived data sets, or whether such simulations when limited in scope to only four or six taxa have any more bearing on phylogeny reconstruction from large data sets than do the chicken-frog-mouse "phylogenies". There is simply no way to predict generally whether the addition of taxa will increase, decrease or leave unchanged the number of most parsimonious trees obtained from the analysis, although some such generalizations may be possible in the arena of tree support, to which we shall return.

To the extent that the number of terminals impacts the actual algorithmic search for short trees, some workers advocate "pruning" the data set or comparing the output of pruned and unpruned analyses. Pruning may be achieved directly through taxon deletion or through nodal collapse and substitution. Siddall's (1994) Random Cladistics software package supports a unique operational function that allows the identification of problem taxa and critical taxa by displaying the number of most parsimonious trees analyzed under different taxon deletion schemes. Problem taxa are identified when their removal markedly reduces the number of equally short trees; critical taxa are identified when their removal increases that number. Of course, the interpretation of such output demands care, but the function does provide a frame of reference for identifying pivotal taxa.

Other authors have implied or suggested more indirect shortcuts to tree search compartmentalization of tree regions and localized analyses in order to simplify tree search. Most recently, Rice et al. (1997) advocate the "inferred ancestral states" or IAS approach, which collapses character data of taxa in a clade into a single presumed ancestral character set. This corresponds broadly to the taxic compartmentalization method (Mishler, 1994) and the placeholder method (Donoghue, 1994). While computationally attractive, such suggestions require that "known" monophyletic clades be hypothesized independently and then collapsed, in effect discarding primary data (Nixon and Carpenter, 1996a).

Support, congruence and decisiveness

Character and matrix support

As stated earlier, it is generally agreed that the more taxa there are in the matrix, the more sequence data may be needed to resolve them. But resolution is not all there is to phylogenetic inference, and workers have come up with numerous measures and indices to evaluate resolved phylogenies using one or more aspects of nodal support, conflict in or among data sets and character congruence. Several of these endeavors are sensitive to the absolute numbers of characters as well as taxa and to the relationship between characters and taxa. And, as with tree search, some measures become computationally cumbersome as data sets get very large. Many of the more popular measures appear to be retained in usage despite demonstrated shortcomings, and some novel algorithms may provide an efficient way of dealing with large matrices specifically.

The simplest measures of tree and character support are the consistency index (CI) (Kluge and Farris, 1969) and retention index (RI) (Farris, 1989), either of which may be calculated as unit indices for individual characters or as ensemble indices for an entire data set on a given tree. The CI is simply an expression for the fraction of actual character change on a given topology represented by the minimum possible amount of such character change as evaluated from the matrix. The complement of C is therefore the fraction of change attributable to homoplasy; the RI is simply the fraction of potential synapomorphy in the data set retained as realized synapomorphy on a given topology (Farris, 1989).

As Goloboff (1991a, p. 218) describes, the CI and RI are "completely adequate to compare trees from identical matrices"; the CI is also useful in comparing homoplasy across different data sets, whereas the RI reflects the fit of a fraction of a given matrix' informative data to a given topology[1]. However, neither index reflects the decisiveness of a matrix, i.e. the preferability of a single topology over others, because all the most parsimonious trees – whether there be two or two thousand – share common ensemble CI and RI values.

Some authors have applied such measures to the evaluation of molecular vs. morphological data. Givnish and Sytsma (1997a,b) indicate that molecular data generate trees with higher consistency indices, whereas Baker and DeSalle (1997) describe precisely the opposite trend. Regardless, a high consistency index is not necessarily a reason to prefer one form of data over another; nor do differences in any such indices militate against combining data under separate analyses. It is possible for a high consistency index and a low decisiveness index to obtain from the same tree.

The ensemble CI is augmented by the inclusion of autapomorphies in the calculation. Incongruence indices based on relative tree lengths are likewise affected by autapomorphies in the data set (Cunningham, 1997). Dependence on data set size of measures such as the CI and RI may bear indirectly on arbitration among competing topologies when such measures are the sources of iterative tree weights. Iterative or successive weighting (Farris, 1969) is a recursive procedure that often has the desirable effect of reducing the number of equally parsimonious topologies, facilitating choice among them (Carpenter, 1988). Superficially, one might expect such *a posteriori* weights to decrease with the addition of taxa. However, there is no empirical evidence to that effect (the issue has never been investigated), and upon close reflection, the utility of the rescaled consistency index may be such that this potential problem is corrected. In a way, successive weighting is a way of making the data more decisive, a quality that is not affected by data set size.

The consistency index is of course sensitive to data set size (Farris, 1972), but that is not in and of itself an indictment of the measure (Goloboff, 1991a). Goloboff (1991a,b) proposed the data decisiveness (DD) index as a unitary means of measuring robustness of a given data set, rather than on a per-tree basis. Data are more decisive when the most parsimonious tree(s) differ more in length from the mean tree length for the data. Contrary to popular notion, the level of homoplasy in a data set does not imply that the data are uninformative. In Goloboff's (1991a, p. 218) words: an "abundance of homoplasy can not shift the preference to less parsimonious alternatives." Theoretically, an increase in data set size is no more or less like-

---

[1] The RI is meaningless when counted from suboptimal trees.

ly to convey a more decisive data set. Furthermore, DD is not affected by uninformative characters; a tree can have a high ensemble CI without being decisive (i.e. incompatible with other possible trees). Decisiveness, as mentioned earlier, may be important to the interpretation of large data sets because it reflects potential incongruence (not incompatibility) with other data, something to consider when two matrices differ greatly in size (Davis et al., 1998). Data decisiveness is worthy of more investigation, particularly in the sphere of comparing morphological with molecular data.

Nodal support

So far as nodal support goes, the bootstrap (Felsenstein, 1985) has probably been revisited in the systematics literature more than any measure (e.g. Sanderson and Donoghue, 1989; Carpenter, 1992, 1996; Hedges, 1992; Kluge and Wolf, 1993; Brown, 1994; Harshman, 1994, Sanderson, 1995). The procedure relies on random sampling (with replacement) from the character matrix such that a contrived data set is created and analyzed for each iteration. Node-specific bootstrap values represent the percentage of such analyses in which a given node appears. From a purely statistical standpoint, some have argued that the bootstrap is inappropriate because it assumes that characters are independently sampled from identically distributed universes of possible characters (Farris, in Werdelin, 1989; Carpenter, 1992, 1996; Kluge and Wolf, 1993). But bootstrap values are also sensitive to uninformative data (autapomorphies and invariant columns; Carpenter, 1996), which may be a particular problem with molecular data. Among the bootstrap's other shortcomings are that the bootstrap appears to be topology-biased (M. Siddall, personal communication and in Wenzel, 1997). Nonetheless, bootstrapping continues to be practiced and is even required for publication in at least one prominent molecular journal.

Another framework for evaluating nodal support is Bremer's (1988) support index, which simply refers to the absolute minimum number of steps by which a tree's overall length must increase for a given node on that tree to collapse while holding all other nodes fixed. While many phylogeneticists (including ourselves) routinely use the Bremer index, there is no consensus as to what precisely the index refers. Wenzel (1997) has argued that Bremer values are potentially meaningless, if not misleading, because they are necessarily calculated by recourse to a topology that is not supported by the data (i.e. the tree that is equivalent to the shortest tree at every node except the collapsed one). Following Baker and DeSalle (1997), however, we view the Bremer index as a potentially useful tool for evaluating the contributions of different data partitions (e.g. gene sequences) to trees constructed by a simultaneous analysis of all data. Their partitioned Bremer support evaluates the relative contributions of separate matrices to the most parsimonious topologies derived from the simultaneous analysis of all the data. This method is analogous to Nixon and Carpenter's (1996b) clade concordance index, which evaluates the supported conflict against a strict consensus. Both Bremer support and bootstrap values can be prohibitively time-consuming when evaluating very large data matrices (Rice et al., 1997), but that is a function of data ambiguity, not necessarily data set size.

Further software implementation issues involving coding and tree search

Of course, the most immediate practical computational pitfalls in dealing with large data sets involve the computer implementation of tree-finding and tree-evaluation measures. There exist several phylogenetic inference programs for both Macintosh and IBM-compatible systems which vary in computational speed as well as accuracy of solution finding. While some of these programs differ more in size than in kind, several have unique functions worth employing and some appear to be entirely redundant.

Programs such as Hennig86 (1988) may yield trees with unsupported, although a newer program (NONA, Goloboff, 1993) appear to address this problem to some degree (Coddington and Scharff, 1994, 1996; Nixon and Carpenter, 1996b). To the extent such issues do exist, they appear to be exacerbated by the number of taxa under examination as well as the quality of sequence data. Platnick (1987, 1989) reviewed computer implementation, although his papers predated the publication of NONA (and a number of other programs with potentially useful applications, e.g. Random Cladistics; Siddall, 1994). More recently, Nixon and Carpenter (1996b) have emphasized some of the advantages of NONA, including its ability to filter out trees with branches possibly of zero length. We do not discuss alignment theory or software, as that topic is treated elsewhere in this volume (W.C. Wheeler, this volume).

Recalling the factorial function relating the number of taxa and the number of possible resolved topologies, it should come as no surprise that computational speed decreases markedly in most computer packages with the inclusion of more than 20 taxa or so; the effect is, of course, more exaggerated when one employs likelihood algorithms. The most useful programs offer exhaustive search options, but these too may become cumbersome for more than about 20 taxa. Programs therefore support approximate searches that are faster, but not guaranteed to find all (or any) of the shortest possible trees for the data. Such procedures operate first by generating a starting tree and then branch swapping to identify more parsimonious solutions or sets thereof. Terminals may be added according to their presented order in the matrix, according to some comparative optimality criterion, or randomly. When using random addition, however, one should perform multiple replicates. Chase et al. (1997) suggested that combination of separate data sets under simultaneous analysis might lessen search time by reducing the percent distance between starting trees and most parsimonious trees.

Farris et al. (1996) and Farris (1997) have suggested parsimony jacknifing specifically as a solution to the problem of analyzing large data sets in a way that employs a frequency distribution to arbitrate among competing nodes. As such it is not a nodal support measure *per se*, but does rely on a critical value (50% or 66% in their usage) to justify a given node. By resampling characters without replacement, the procedure expresses support as a frequency with which a given node appears in random subsamples of the character data, and uses that frequency as a means of finding short trees themselves. Parsimony jacknifing thus circumvents the bootstrap's sensitivity to uninformative data. In fairness, the bootstrap was not intended for actual tree search, but merely as a nodal support measure, although it too could be used for the former purpose. The jacknife procedure also has computational advantages, such as being orders of magnitude faster than bootstrap neighbor joining and avoiding problems associated with zero-length branches and input order sensitivity (Farris et al., 1996). Further, whereas it has been suggested that moderately large data sets with many taxa will tend to yield trees for which bootstrap values suggest low statistical support (Sanderson and Donoghue, 1989), jacknife frequencies may actually increase with addition of taxa [see Farris et al.'s (1996) Asterales examples]. Jacknifing may afford the first means of evaluating tree support that is logically and statistically sound, both as a support measure and as a way of going about tree search.

Some of the most serious practical difficulties in primary analysis involve the coding of missing applicable values and subset polymorphisms, whereas the most problematic postanalysis issues are the evaluation of trees with unsupported or ambiguously supported nodes (Coddington and Scharff, 1994, 1996), and the use of such trees to calculate tree measures or to arbitrate choices among equally well corroborated solutions. Missing data, polymorphic data and inapplicable data directly influence the resolution of tree output, yet are logically separate (Nixon and Davis, 1991). As mentioned earlier, the use of polymorphic characters can mask relationships if the (higher-level) taxa under investigation are not monophyletic. Missing data can drastically alter tree topology (e.g. Doyle and Donoghue, 1992; Novacek and Wheeler, 1992 and references therein; Nixon et al., 1994) and confound the efficacy of tree search.

Inapplicable data involve character systems unique to a subset of taxa under investigation. Characters involving scale structure in moths are obviously not codable in a higher-level multiordinal analysis, since the presence of scales themselves comprises a synapomorphy of the Lepidoptera. In such cases a separate matrix could be prepared for each level of the analysis, but that approach discards data, as mentioned. Clados (Nixon, 1992) allows the user to distinguish missing values from polymorphic values and inapplicable character states (?,*,-), but given that Clados operates as a shell program from within which Hennig86 performs the actual search, it operationally equates them all. Clados also shells to NONA; and although that program does distinguish polymorphisms, it too treats missing data and inapplicable data as equivalent computationally. Because of gaps in molecular data, there is also a risk that a given site may be informative but get coded as inapplicable in an analysis. One solution (Wheeler and Nixon, 1994) is to score the characters as gaps and add negatively weighted binary character states to the matrix that correct for any superfluous steps in the tree that obtains; this procedure fails under certain circumstances, however (Swofford and Siddall, 1997). Alternatively, one could simply score gaps as inapplicable, but that could be risky given the equation of "inapplicable" and "missing" in current software packages. It has further been pointed out that informative multistate characters can include autapomorphies that may go undetected by screening and removal commands (e.g. *mop* in DADA; Nixon, 1998). Thus autapomorphic states of informative characters should not necessarily be removed prior to analysis by scoring the taxa as missing for the character. This operation could decrease the length of the tree without affecting its topology, but might simply introduce ambiguities in the form of missing values. With the exception of Wheeler and Nixon's (1994) Sankoff procedure, which must be encoded manually and which may still fail in many cases, there exists no algorithm or software package to the best of our understanding that redresses this problem.

Meanwhile, sites informative at one hierarchic level but invariable in shallower areas (crown groups) of a tree present a more strategic dilemma to the molecular systematist: given the desirability of combining data sets under simultaneous analysis, the risks of coding data as "missing", and the equation of missing and inapplicable data in current software, should genetic data unlikely to vary at lower levels be gathered merely for the sake of comparability, that is to avoid coding large strings of missing data? Gathering invariant data is every bit as expensive and time-consuming as gathering informative data. Although philosophically we find the proposal to compartmentalize data risky at best, we cannot advocate sequencing slowly evolving genes for species-level work, for example, simply for the sake of maximizing comparability by eliminating inapplicable data.

Conclusion

Large data sets – especially those which are large by virtue of many taxa – present a daunting number of possible tree topologies and character optimizations. It is important, of course, that workers recognize and address the potential limitations of their taxon sampling schemes, especially insofar as evaluating optimized character evolution, and of a given gene to address a given question. But rather that concern ourselves exclusively with the absolute number of possible topologies, workers might do well to understand the different properties of phylogenetic inference methods and commonly used support statistics. It is not sufficient to justify a method by claiming statistical rigor; statistics must, more fundamentally, be appropriate for the questions they are employed to examine (e.g. Farris et al., 1996).

Many of the issues in the analysis of especially large data sets can be distilled to simple choices of rooting, taxon sampling and appropriate gene choice. In order to avoid confounding their own conclusions, molecular workers should be especially careful to avoid confusing nomenclatural validity with actual monophyly. Using paraphyletic or polythetic groups as terminals in higher-level analyses will simply perpetuate a cascade of errors in inferring relationships.

Finally, there appears to be a growing body of work suggesting that saturation and the long-branch problem may be mere epiphenomena and not necessarily obstacles to phylogenetic inference under parsimony. Studies such as Remsen and DeSalle's (1998) and Baker and DeSalle (1997) highlighting combinability speak directly to the irrelevance of *a priori* statistical comparisons of data, and suggest that molecular data may not only be less misleading than popularly believed, but indeed an outstanding complement to morphological data.

Acknowledgements

Victor Albert, Rick Baker, Jim Carpenter and Rob DeSalle provided helpful comments on early (but quite recent) versions of this paper. Any remaining ambiguities, inconsistencies, misrepresentations, faux arguments, unsupported statements and other errors are the sole responsibility of the authors. PZG was supported by an EPA graduate fellowship during the writing of this paper.

References

Albert, V. A., Chase, M. W. and Mishler, B. D. (1993) Character-state weighting for cladistic analysis of protein-coding DNA sequences. *Ann. Miss. Bot. Gard.* 80: 752–766.

Avise, J. C., Arnold, J., Ball, R. M., Bermingham, E., Lamb, T., Neigel, J. E., Reeb, C. A. and Saunders, N. C. (1987) Intraspecific phylogeography: The mitochondrial bridge between population genetics and systematics. *Annu. Rev. Ecol. Syst.* 18: 489–522.

Baker, R. H. and DeSalle, R. (1997) Multiple sources of character information and the phylogeny of Hawaiian Drosophilids. *Syst. Biol.* 46(4): 654–673.

Bremer, K. (1988) The limits of amino-acid sequence data in angiosperm phylogenetic reconstruction. *Evolution* 42: 795–803.

Brower, A. V. Z., DeSalle, R. and Vogler, A. (1996) Gene trees, species trees and systematics: a cladistic perspective. *Annu. Rev. Ecol. Syst.* 27: 423–450.

Brown, J. K. M. (1994) Bootstrap hypothesis tests for evolutionary trees and other dendrograms. *Proc. Natl. Acad. Sci. USA* 91: 12293–12297.

Carpenter, J. M. (1988) Choosing among multiple equally parsimonious cladograms. *Cladistics* 4: 291–296.

Carpenter, J. M. (1992) Random cladistics. *Cladistics* 8(2): 147–153.

Carpenter, J. M. (1994) Successive weighting, reliability and evidence. *Cladistics* 10(2): 215–220.

Carpenter, J. M. (1996) Uninformative bootstrapping. *Cladistics* 12(2): 177–181.

Cavalli-Sforza, L. L. and Edwards, A. W. F. (1967) Phylogenetic analysis. Models and estimation procedures. *Evolution* 21: 550–570.

Chase, M. W., Soltis, D. E., Olmstead, R. G., Morgan, D., Les, D. H., Mishler, B. D., Duvall, M. R., Price, R. A., Hillis, H. G., Qiu, Y.-L. et al. (1993) Phylogenetics of seed plants: an analysis of nucleotide sequences from the plastid gene *rbcL*. *Ann. Mo. Bot. Gard.* 80: 528–580.

Chase, M. W., Cox, A. V., Soltis, D. E., Soltis, P. S., Mort, M. E., Savolainen, V., Reeves, G., Hoot, S. B. and Morton, C. M. (1997) Large DNA sequence matrices, phylogenetic signal and feasibility: an empirical approach. *Am. J. Bot.* 84(6): 181.

Coddington, J. A. and Scharff, N. (1994) Problems with zero-length branches. *Cladistics* 10(4): 415–423.

Coddington, J. A. and Scharff, N. (1996) Problems with "soft" polytomies. *Cladistics* 12(2): 139–145.

Cunningham, C. W. (1997) Is congruence between data partitions a reliable predictor of phylogenetic accuracy? Empirically testing an iterative procedure for choosing among phylogenetic methods. *Syst. Biol.* 46(3): 464–478.

Dahlgren, R., Clifford, H. and Yeo, P. (1985) *The families of monocotyledons*. Springer-Verlag, Berlin, 530 pp.

Davis, J. I. (1995) A phylogenetic structure for the monocotelydons, as inferred from chloroplast DNA restriction site variation, and a comparison of measures of clade support. *Syst. Bot.* 20: 503–527.

Davis, J. and Nixon, K. C. (1992) Populations, genetic variation and the delimitation of phylogenetic species. *Syst. Biol.* 41(4): 421–435.

Davis, J. I, Simmons, StevensonM. P. and Wendel, J. F. (1998) Data decisiveness and data quality in phylogenetic analysis: an example from the monocots using mitochondrial <u>atpA</u> sequences. *Syst. Biol.* 47(2): 282–310.

Day, W. H. E. (1983) Computationally difficult parsimony problems in phylogenetic systematics. *J. Theor. Biol.* 103: 429–438.

Dobson, A. J. (1974) Unrooted trees for numerical taxonomy. *J. Appl. Prob.* 11: 32–42.

Donoghue, M. J. (1994) Progress and prospects in reconstructing plant phylogeny. *Ann. Mo. Bot. Gard.* 81: 405–418.

Doyle, J. A. and Donoghue, M. J. (1992) Fossils and seed plant phylogeny reanalyzed. *Brittonia* 44(2): 89–106.

Farris, J. S. (1969) A successive approximations approach to character weighting. *Syst. Biol.* 18: 374–385.

Farris, J. S. (1972) Estimating phylogenetic trees from distance matrices. *Amer. Naturalist.* 106: 645–668.

Farris, J. S. (1983) The logical basis of phylogenetic analysis. *In*: Platnick N. I. and Funk, V. A. (eds) *Advances in Cladistics* 2, Columbia University Press, New York, pp. 7–36.

Farris, J. S. (1989) The retention index and rescaled consistency index. *Cladistics* 5: 417–419.

Farris, J. S. (1997) The future of phylogeny reconstruction. *Zool. Scripta* 26: 303–311.

Farris, J. S. and Kluge, A. G. (1985) Parsimony, synapomorphy and explanatory power: A reply to Duncan. Taxon 34: 130–135.

Farris, J. S., Albert, V. A., Kallersjo, M., Lipscomb, D. and Kluge, A. G. (1996) Parsimony jacknifing outperforms neighbor-joining. *Cladistics* 12(2): 99–124.

Felsenstein, J. (1978a) The number of evolutionary trees. *Syst. Zool.* 27: 27–33.

271

Felsenstein, J. (1978b) Cases in which parsimony or compatibility methods will be positively misleading. *Syst. Zool.* 27: 401–410.

Felsenstein, J. (1981) A likelihood approach to character weighting and what it tells us about parsimony and compatibility. *Biol. J. Linn. Soc.* 16: 183–196.

Felsenstein, J. (1985) Confidence limits on phylogenies: an approach using the bootstrap. *Evolution* 39(4): 783–791.

Givnish, T. J. and Sytsma, K. J. (1997a) Consistency, characters, and the likelihood of correct phylogenetic inference. *Mol. Phylogenet. Evol.* 7(3): 320–330.

Givnish, T. J. and Sytsma, K. J. (1997b) Homoplasy in molecular vs. morphological data: The likelihood of correct phylogenetic inference. *In*: Givnish, T. and Sytsma, K. (eds) *Molecular Evolution and Adaptive Radiation*, Cambridge University Press, Cambridge, pp. 55–101.

Goloboff, P. A. (1991a) Homoplasy and the choice among cladograms. *Cladistics* 7: 215–232.

Goloboff, P. A. (1991b) Random data, homoplasy, and information. *Cladistics* 7: 395–406.

Goloboff, P. A. (1993) Nona, ver. 1.5.1., American Museum of Natural History, New York.

Gould, S. J. (1985) A clock of evolution. *Natur. Hist.* 94: 12–25.

Harshman, J. (1994) The effect of irrelevant characters on bootstrap values. *Syst. Biol.* 43(3): 419–424.

Hedges, S. B. (1992) The number of replications needed for accurate estimation of the bootstrap p value in phylogenetic studies. *Molec. Biol. Evol.* 9(2): 366–369.

Hedges, S. B. and Maxson, L. R. (1996) Re: Molecules and morphology in amniote phylogeny. *Mol. Phylogenet. Evol.* 6, 312–314.

Hillis, D. M. (1996) Inferring complex phylogenies. *Nature* 383: 130–131.

Huelsenbeck, J. P. (1995) Performance of phylogenetic methods in simulation. *Syst. Biol.* 44: 17–48.

Kim, J. (1996) General inconsistency conditions for maximum parsimony: effects of branch lengths and increasing numbers of taxa. *Syst. Biol.* 45: 363–374.

Kluge, A. G. and Farris, J. S. (1969) Quantitative phyletics and the evolution of Anurans. *Syst. Zool.* 18: 1–32.

Kluge, A. G. and Wolf, A. J. (1993) Cladistics: What's in a word? *Cladistics* 9(2): 183–199.

Luckow, M. (1995) Species concepts: assumptions, methods and applications. *Syst. Bot.* 20(4): 589–605.

Maddison, D. R. (1991) The discovery and importance of multiple islands of most-parsimonious trees. *Syst. Zool.* 40: 315–328.

Mickevich, M. F. (1983) Introduction. *In*: Platnick N. I. and Funk, V. A. (eds) *Advances in Cladistics* 2, Columbia University Press, New York, pp. 3–5.

Mickevich, M. and Farris, S. (1981) The implications of congruence in Menidia. *Syst. Zool.* 30: 351–370.

Miller, J. S., Brower, A. V. Z. and DeSalle, R. (1997) Phylogeny of the neotropical moth tribe Josiini (Notodontidae: Dioptinae): comparing and combining evidence from DNA sequences and morphology. *Biol. J. Linn. Soc.* 60: 297–316.

Mishler, B. D. (1994) Cladistic analysis of molecular and morphological data. *Amer. J. Phys. Anthropol.* 94: 143–156.

Miyamoto, M. M. (1985) Consensus cladograms and general classifications. *Cladistics* 1: 186–189.

Nixon, K. C. (1992) *Clados*, version 1.2, Trumansburg, New York.

Nixon, K. C. (1998) DADA, ver. 1.2. Bailey Hortorium, Cornell University, Ithaca, N.Y.

Nixon, K. C. and Carpenter, J. M. (1993) On outgroups. *Cladistics* 9(4): 413–426.

Nixon, K. C. and Carpenter, J. M. (1996a) On simultaneous analysis. *Cladistics* 12(3): 221–241.

Nixon, K. C. and Carpenter, J. M. (1996b) On consensus, collapsibility and clade concordance. *Cladistics* 12 (4): 305–321.

Nixon, K. C., Crepet, W. L., Stevenson, D. and Friis, E. M. (1994) A reevaluation of seed plant phylogeny. *Ann. Miss. Bot. Gard.* 81(3): 484–533.

Nixon, K. C. and Davis, J. I. (1991) Polymorphic taxa, missing values and cladistic analysis. *Cladistics* 7: 233–241.

Novacek, M. J. and Wheeler, Q. D. (eds) (1992) *Extinction and Phylogeny*, Columbia University Press, New York.

Phipps, J. B. (1976a) Dendrogram topology: capacity and retrieval. *Can. J. Bot.* 54: 679–685.

Phipps, J. B. (1976b) The numbers of classifications. *Can. J. Bot.* 54: 686–688.

Platnick, N. I. (1987) An empirical comparison of microcomputer parsimony programs. *Cladistics* 3(2): 121–144.

Platnick, N. I. (1989) An empirical comparison of microcomputer parsimony programs, II. *Cladistics* 5: 145–161.

Platnick, N. I., Griswold, C. E. and Coddington, J. A. (1991) On missing entries in cladistic analysis. *Cladistics* 7: 337–343.

Remsen, J. and DeSalle. (1998) Character congruence of multiple data partitions and the origin of the Hawaiian drosophilidae. *Molec. Pylogenet. Evol.* 9: 225–235.

Rice, R., Donoghue, M. J. and Olmstead, R. G. (1997) Analyzing large data sets: *rbcL* 500 revisited. *Syst. Biol.* 46(3): 554–563.

Sanderson, M. J. (1995) Objections to bootstrapping phylogenies: a critique. *Syst. Biol.* 44(3): 299–320.

Sanderson, M. and Donoghue, M. J. (1989) Patterns of variation in levels of homoplasy. *Evolution* 43: 1781–1795.

Siddall, M. E. (1994) *Random Cladistics*. version 2.1.1, Toronto, ON, Canada.

Struwe, L., Albert, V. A. and Bremmer, B. (1994) Cladistics and family level classification of the Gentianales. *Cladistics* 10: 175–206.

Sullivan, J. (1995) Combining data with different distributions of among-site rate variation. *Syst. Biol.* 45(3): 375–380.

Sullivan, J., Market, J. A. and Kirkpatrick, C. W. (1997) Phylogeography and molecular systematics of the *Peromyscus aztecus* species group (Rodentia: Muridae) inferred using parsimony and likelihood. *Syst. Biol.* 46(3): 426–440.

Swofford, D. L. (1991) When are phylogeny estimates from molecular and morphological data incongruent? *In*: Miyamoto, M. M. and Cracraft, J. (eds) *Phylogenetic Analysis of DNA Sequences*, Oxford University Press, New York, pp. 295–333.

Swofford, D. L. and Siddall, M. E. (1997) Uneconomical diagnosis of cladograms: comments on Wheeler and Nixon's method for Sankoff optimization. *Cladistics* 13(2): 153–159.

Wenzel, J. W. (1997) When is a phylogenetic test good enough? *In*: Grandcolas, P. (ed.) *The Origin of Biodiversity in Insects: Phylogenetic tests of Evolutionary Scenarios*, Mem. Mus. Natn. Hist. Nat., Paris, pp. 31–45.

Wenzel, J. W. and Carpenter, J. M. (1994) Comparing methods: adaptive traits and tests of adaptation. *In*: Eggleton, P. and Vane-Wright, R. I. (eds) *Phylogenetics and Evolution*, Academic Press, London, pp. 79–101.

Werdelin, L. (1989) We are not out of the woods yet – a report from a Nobel Symposium. *Cladistics* 5: 192–200.

Wheeler, W. C. (1991) Congruence among data sets: A Bayesian approach. *In*: Miyamoto, M. M. and Cracraft, J. (eds) *Phylogenetic Analysis of DNA Sequences*, Oxford University Press, New York, pp. 334–346.

Wheeler, W. C. (1992a) Extinction, sampling, and molecular phylogenetics. *In*: Novacek, M. J. and Wheeler, Q. D. (eds) *Extinction and Phylogeny*, Columbia University Press, New York, pp. 205–215.

Wheeler, W. C. (1992b) Quo vadis? *Cladistics* 8(1): 85–86.

Wheeler, W. C. and Nixon, K. C. (1994) A novel method for the economical diagnosis of cladograms under Sankoff optimization. *Cladistics* 10(2): 207–214.

Whiting, M. F., Carpenter, J. M., Wheeler, Q. D. and Wheeler, W. C. (1997) The Strepsiptera problem: phylogeny of the holometabolous insect orders inferred from 18S and 28S ribosomal DNA sequences and morphology. *Syst. Biol.* 46(1): 1–68.

The comparison of morphological and molecular data in phylogenetic systematics

Allan Larson

Department of Biology, Washington University, St. Louis, MO 63130–4899, USA

Summary

Analytical methods facilitating the use of molecular and morphological characters as complementary sources of phylogenetic information are explored. Separation of phylogenetically useful information from misleading patterns of character variation is most effective when the methods of "taxonomic congruence" and "character congruence" are used together. Statistical approaches for implementing both methods are described and illustrated with an example from the phylogeny of salamanders.

Introduction

The common evolutionary descent of species imposes on their morphological and molecular variation a nested hierarchical pattern of groups within groups. Phylogenetic analysis aims to recover the historical pattern of common descent among species by identifying congruent hierarchical patterns of variation in their morphological and molecular characters. Each taxonomic character constitutes a unit of phylogenetic information, and many characters are required to obtain a well-supported phylogenetic topology for a group of species. Because characters differ in how effectively they are formulated and how accurately their variation recovers components of phylogenetic history, conflicting patterns of character variation are expected both within and among different data sets. Congruence among many morphological and molecular characters implies a robust inference of phylogeny, whereas extensive conflict suggests that the phylogenetic history is not being faithfully recovered and that reassessment of the characters is necessary.

Reconstructing patterns of common descent among species requires maximizing the use of informative characters coupled with a systematic method for identifying misleading information. The purpose of this chapter is to evaluate how best to accomplish this goal when both morphological and molecular data are examined. Two general approaches to this problem have been advocated. One begins by measuring the "taxonomic congruence" between different data sets by constructing phylogenetic trees from each one separately, and then identifying points of agreement and conflict (Mickevich, 1978). This approach assumes that characters within data sets are more likely to be nonindependent estimators of phylogeny than are characters from different data sets (de Queiroz, 1993). In this case, systematic errors are more likely to affect multiple characters within rather than between data sets, and consensus methods that recognize only the groups supported by more than one data set are conservative (de Queiroz, 1993).

The alternative approach, called "character congruence", combines all morphological and molecular characters into a single data matrix from which a tree is constructed (Kluge, 1989). The effectiveness of different characters is then evaluated according to their congruence with the resulting tree. This approach assumes that the pattern prevailing upon combination of the data sets reflects the phylogeny of the species being compared, and that the separate analyses of the different data sets are individually less effective at recovering this pattern. This situation would occur when different data sets are individually effective at resolving different parts of the tree but not the entire tree, and when the phylogenetic signal-to-noise ratio is enhanced by the increased number of characters present in the combined analysis. The combined analysis is most effective when all characters are independent within and between data sets (de Queiroz, 1993).

The conflict between using taxonomic congruence vs. character congruence for comparing morphological and molecular data has generated much controversy (for summaries see Barrett et al., 1991; de Queiroz, 1993; Jones et al., 1993; Bull et al., 1993; Eernisse and Kluge, 1993; Chippindale and Wiens, 1994; de Queiroz et al., 1995; Miyamoto and Fitch, 1995; Huelsenbeck et al., 1996a). Each approach provides information on the structure of character variation using different assumptions, and each one can be effective in identifying how well different parts of a phylogenetic tree are supported by available evidence. I maintain that the most effective assessment of the phylogenetic information contained in morphological and molecular data is obtained when both approaches are used. Neither approach is difficult computationally given the available computer programs, making concurrent use of both approaches entirely feasible. I describe the implementation and assessment of both approaches using an example from my studies of the phylogeny of salamanders.

I make several assumptions in my comparison of these methods, some of which are controversial. It is beyond the scope of this chapter to evaluate each controversial issue in detail, so I will start by stating my assumptions to qualify my conclusions. I assume that both the molecular and morphological characters have discrete alternative states subject to cladistic analysis, that the states are polarized by outgroup comparison and that character conflicts are resolved using the criterion of maximum parsimony. Molecular characters constitute positional homologies from alignable protein or nucleic acid sequences (or sites on a restriction map of the latter). Morphological characters represent either instantaneous morphologies (those observed and measured at a single point of the organismal ontogeny) or sequences of ontogenetic transformation for morphological features, although the latter should be more effective for phylogenetic analysis (de Queiroz, 1985). Data sets are compared only on their ability to recover phylogenetic topology, not the lengths of branches. The phylogenetic analysis treats all characters initially as discrete and equivalent units of phylogenetic information with no differential *a priori* weighting. This procedure assumes that variability of the characters has been matched to a first approximation with the assumptions of parsimony analysis, specifically that substitutional saturation has not occurred for molecular data, that evolutionary rates are not highly dissimilar among lineages and that terminal branches are not excessively long relative to internal branches. If these assumptions are violated, misinformation caused by parallel changes is likely to exceed the useful phylogenetic signal, and parsimony analysis will be ineffective (Felsenstein, 1978; Hendy and Penny, 1989). Results of the analyses described below can identify violations of these assumptions and suggest alternative analytical methods where appropriate.

Character conflicts can have two sources barring systematic error imposed by the investigator: (i) homoplasy (nondivergent evolutionary change), which is common when a character has a limited number of alternative states and evolves rapidly relative to the phylogenetic events being evaluated, and (ii) incongruence between the phylogeny of the characters and that of the species in which they are measured (Pamilo and Nei, 1988). Both sources should produce primarily random error unless characters are nonindependent or the assumptions of parsimony presented above are violated (see de Queiroz et al., 1995). Congruence between the phylogeny of characters and the phylogeny of the species in which they are measured is guaranteed only when the characters being used are emergent at the species level. Neither the morphological nor the molecular characters standardly used to reconstruct phylogenies of species are emergent at the species level, and both kinds of characters can lead to erroneous conclusions if a "character phylogeny" is incongruent with the "species phylogeny" [see Roth (1991), for a detailed discussion of this issue].

Incongruence between a phylogeny of morphological characters and the phylogeny of species is particularly likely when instantaneous morphological characters are studied and ontogenetic criteria are used to polarize variation (de Queiroz, 1985). For molecular characters, intraspecific duplication and divergence of genes, horizontal gene transfer and lineage sorting (retention of molecular polymorphism through phylogenetic branching events followed by alternative fixations) may generate molecular phylogenies that are not congruent with the phylogeny of species. Statistical demonstration of incongruence between different sets of molecular characters (such as those from organellar and nuclear genomes) can identify cases where at least one set of characters recovers a gene phylogeny that does not match the phylogeny of the species being compared. A recent review (Brower et al., 1996) argues that the problem of incongruence between gene trees and species trees is less severe in practice than its extensive discussion in the literature suggests.

A preliminary step in the analysis of morphological and molecular data is to ask whether these data sets individually and in combination contain statistically significant phylogenetic structure. The method of Hillis and Huelsenbeck (1992) can be used to ask whether the frequency distribution of minimum lengths for all trees, or a random sample of trees, exhibits a significant leftward skew, suggesting that the shortest trees will be substantially more parsimonious than the majority (see also Faith and Cranston, 1991). Hillis and Huelsenbeck (1992) have shown that when data are effectively randomized with respect to phylogenetic structure, the most parsimonious tree bears no consistent relationship to the true tree. Only when significant phylogenetic structure is present will the most parsimonious tree estimate some portion of the true tree. This demonstration is one of the most effective arguments against the notion that the most parsimonious tree obtained from a data set is necessarily the best working hypothesis of phylogeny. The comparison of morphological and molecular data by either method described below is meaningful only if significant phylogenetic structure exists in both data sets.

Taxonomic congruence

Taxonomic congruence asks whether the phylogenetic topologies constructed from different data sets are congruent. In a large study, it is rare that the most parsimonious topologies obtained from morphological and molecular data sets are exactly the same, but topological discrepancies alone are not sufficient to support the conclusion that the data sets are substantially in conflict for some portion of the phylogeny. If the most parsimonious topology for the molecular data constitutes a near-optimal tree for the morphological data and *vice versa*, it is incorrect to conclude that the data sets conflict. We cannot reject the hypothesis that both data sets are estimating with error the same phylogenetic topology.

We need a statistical criterion for asking whether the most parsimonious topologies obtained from each data set constitute near-optimal topologies for the other one. For this purpose, I recommend the nonparametric test of Templeton (1983). A modified version of Templeton's test is included in the computer package PHYLIP (Felsenstein, 1993). This method evaluates paired phylogenetic topologies to ask whether character changes are significantly more parsimonious on one topology than on the other one for a given data set. Paired trees of any size can be evaluated. The test as applied here is formally a one-tailed test, because the most parsimonious tree for the data set being tested is identified before applying the test and the alternative topology is known to be less parsimonious. Felsenstein (1985) computed exact significance levels for a special case of Templeton's test, however, and found that the one-tailed values of this test are close to the exact values, but not always conservative; he recommended use of the more conservative two-tailed probabilities, which I use here.

The main impact of the statistical testing that I advocate has been to introduce caution in phylogenetic analysis and therefore to prevent overinterpretation of results. Templeton's (1983) test is particularly useful for asking whether the tree specified by a morphological data set constitutes a near-optimal tree for a molecular data set and *vice versa*. Data sets will not be judged significantly in conflict if their differences can be attributed to random error.

I illustrate the analytical approaches presented in this paper using a data set consisting of 10 species of salamanders (one from each taxonomic family) and two outgroups (one anuran and one caecilian) taken from my phylogenetic studies of the salamander families (Larson and Wilson. 1989; Larson, 1991a; Larson and Dimmick, 1993). The molecular characters consist of 99 informative positions from aligned ribosomal RNA sequences, and the morphological data consist of 30 informative characters from the head, trunk and cloacal glands (see Larson and Dimmick, 1993, for details). Species are identified in Figure 1. This condensed sampling of taxa and characters is intended only for illustration of methodology; for conclusions based on analysis of the entire data set, see Larson (1991a) and Larson and Dimmick (1993). Application of Hillis and Huelsenbeck's (1992) test documents significant phylogenetic structure in these morphological and molecular data sets analyzed both separately and jointly (Larson and Dimmick, 1993).

The first step in applying Templeton's (1983) test is to obtain the most parsimonious partitioning of characters in the data set being analyzed on each of the topologies being compared. The number of times that each character changes on each of the topologies is recorded. Only the characters that undergo different minimum numbers of changes on the topologies being compared are used in the statistical test (Tab. 1). This information can be obtained from the PAUP program (Swofford, 1993) by specifying an input topology, requesting the most parsimonious partitioning of changes on that topology and printing a "changelist" that enumerates the changes inferred for all characters analyzed. This procedure is done separately for the two topologies, and the changelist is inspected to find the characters that undergo different minimum numbers of changes on the different trees. The MacClade program (Maddison and Maddison, 1992) provides the additional useful step of identifying all characters that change different numbers of times on the two topologies (using the "compare trees" option).

The test is illustrated in Table 1, which lists by column (i) the characters, (ii) the minimum number of changes per character on tree 1, (iii) the minimum number of changes per character on tree 2, (iv) the difference between columns 2 and 3, and (v) a ranking of the numbers in column 4 by absolute value but retaining the sign (+ or −) of the corresponding entry in column 4. The smallest numbers receive the lowest ranks. Tied ranks are very common and receive the midpoint value that would result if the tied entries were ranked consecutively. The absolute values of the posi-

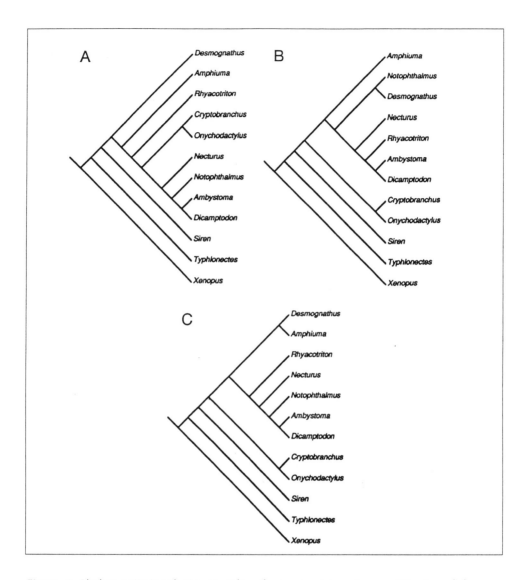

Figure 1. Phylogenetic topologies providing the most parsimonious partitioning of character changes for (A) molecular data alone, (B) morphological data alone, and (C) molecular and morphological data combined [all data from Figure 2 of Larson and Dimmick (1993)]. The amphibian taxa are as follows: Ambystoma californiense *(family Ambystomatidae),* Amphiuma means *(family Amphiumidae),* Cryptobranchus alleganiensis *(family Cryptobranchidae),* Dicamptodon aterrimus *(family Dicamptodontidae),* Desmognathus ochrophaeus *(family Plethodontidae),* Necturus beyeri *(family Proteidae),* Notophthalmus viridescens *(family Salamandridae),* Onychodactylus japonicus *(family Hynobiidae),* Rhyacotriton kezeri *(family Rhyacotritonidae),* Siren lacertina *(family Sirenidae),* Typhlonectes compressicauda *[order Gymnophiona (outgroup)],* Xenopus laevis *[order Anura (outgroup)].*

tive and negative ranks in column 5 are summed separately, and the smaller sum constitutes the test statistic, T_s. For the test to be significant, the test statistic must be less than the critical value for the Wilcoxon rank sum[1] (table B.11 of Zar, 1984) where n equals the number of characters listed in column 1 of the table described above.

In the sample data set, a single most-parsimonious tree was obtained from the molecular data alone (Fig. 1A), and two equally most parsimonious trees were found for the morphological characters (one of these appears in Fig. 1B). Table 1 illustrates an application of the Templeton (1983) test to ask whether one of the two trees maximally parsimonious for the morphological data (Fig. 1B; tree 1 of Tab. 1) is significantly more parsimonious for the morphological characters than the topology produced by the molecular analysis (Fig. 1A; tree 2 of Tab. 1). Nineteen of the morphological characters have different minimum numbers of changes on the topologies being compared. Fifteen of the characters differ by a single step between the two trees (in either direction) and receive a midpoint rank of 8 (1–15). Three characters differ by two steps between the trees and receive a midpoint rank of 17 (16–18). A single character differs by three steps between the trees and receives a rank of 19. All but three of the characters (Y, CCE, and CCL) favor the morphology-based topology (tree 1 of Tab. 2) over the one derived from molecular information (tree 2 of Tab. 2). The ranks with a positive sign (those favoring tree 2) yield the lower sum, which equals 35. For $T_s = 35$, the test is significant at the 0.02 level using a two-tailed test.

A comparable test applied to the molecular data (not shown) identifies 45 molecular characters that change different minimum numbers of times on the same two topologies (Figs 1A and 1B). The molecular-based topology (Fig. 1A) is favored over the morphology-based tree (Fig. 1B) at the 0.001 level ($T_s = 151$; $n = 45$). Thus, the topology derived from morphology is not close to optimal for the molecular data, nor is the topology derived from the molecular data near optimal for the morphological data; these data sets therefore may contain significant conflicts for at least part of the phylogenetic topology for the salamander species being compared.

When data sets are found to contain conflict, it is often of interest to determine which specific differences between the two topologies are individually significant (assuming that the topologies being compared differ by two or more nodes). An appropriate test can be done by rearranging individual branches of the less parsimo-

---

[1] Significance of T_s can be tested also by using a normal approximation (Zar, 1984), which is necessary if n > 100. Calculate $Z = (|T_s - \mu_T| - 0.5)/\sigma_T$ where $\mu_T = n(n + 1)/4$, and $\sigma_T = \{[n(n + 1)(2n + 1) - C]/24\}^{1/2}$ where C is a correction for tied ranks whose omission makes the test more conservative. $C = \sum(t_i^3 - t_i)/2$ where t_i is the number of ties in a group of tied ranks and the summation is performed over all groups of ties. Z is significant for a two-tailed test if it exceeds the following critical values (significance levels in parentheses): 1.96 (0.05), 2.3263 (0.02), 2.5758 (0.01), 2.8070 (0.005), 3.0902 (0.002), 3.2905 (0.001). For the worked example (Tab. 1), $Z = 2.544$ ($P < 0.02$).

Table 1. Templeton test of alternative trees

Character	Number of changes		Difference	Rank
	Tree 1	Tree 2		
A	2	3	−1	−8
BE	4	5	−1	−8
D	4	5	−1	−8
LA	3	4	−1	−8
LD	1	2	−1	−8
LE	1	2	−1	−8
Q	3	5	−2	−17
S	4	5	−1	−8
X	6	7	−1	−8
Y	3	2	1	8
CCE	4	3	1	8
CCH	1	2	−1	−8
CCL	4	1	3	19
CCM	1	2	−1	−8
CCR	1	3	−2	−17
CCS	1	2	−1	−8
CCT	1	2	−1	−8
CCU	2	4	−2	−17
CCW	1	2	−1	−8

$T_s = 35$, $n = 19$, $P < 0.02$. Nineteen phylogenetically informative morphological characters discriminate tree 1 (Fig. 1B), which is favored by the morphological data set, from tree 2 (Fig. 1A), which is favored by a molecular data set. For each character, the numbers of changes on the two trees are given, followed by their difference (tree 1−tree 2). The differences are then ranked in order of absolute value, but retaining the sign (+ or −) of the difference (tied ranks receive the midpoint value that would result if the entries were ranked consecutively). Positive and negative ranks are summed separately and the sum with the lower absolute value (in this case positive ranks) constitutes the test statistic (T_s). Tree 2 is significantly rejected by the morphological data [two-tailed probability from table B.11 of Zar (1984)]. Data and character designations are from Larson and Dimmick (1993).

nious topology to match the corresponding structure of the favored topology, and then asking whether the resulting tree is still significantly less parsimonious than the favored one using Templeton's (1983) test. Alternatively, the tree can be broken into subsets containing four taxa in which the positions of single internal branches differ among the three alternative possible topologies [however, see Hillis (1996) for

Table 2. *Homoplasy in morphological vs. molecular characters*

	Consistent	1 Homoplasy	2+ Homoplasies	Total
Morphology	8 (11.4)	14 (12.8)	8 (5.8)	30
Molecules	41 (37.6)	41 (42.2)	17 (19.2)	99
Total	49	55	25	129

Chi-square = 2.5, df = 9, n.s.

Contingency table evaluating homoplasy in morphological vs. molecular characters for the topology favored by parsimony analysis of the combined data (Fig. 1C). Expected values are in parentheses. Only characters phylogenetically informative for the 10 salamanders shown in Figure 1C are analyzed; changes occurring in outgroup taxa are excluded. Morphological and molecular characters do not differ significantly in levels of homoplasy for the taxa examined. Data are from Larson and Dimmick (1993).

cautions regarding the effectiveness of breaking phylogenetic problems into a series of four-taxon questions]. The test of Felsenstein (1985) can be used with the four-taxon statements to ask whether the presence of particular internal branches is significantly favored by parsimony [for worked examples using the salamander data, see Larson (1991a)]. Felsenstein's (1985) test was described for nucleic acid sequence data, but it is conceivably generalizable to other kinds of data, because all phylogenetically informative variation takes the form of binary characters in a four-taxon statement (although the more conservative significance levels must be used if approximate constancy of evolutionary rate cannot be assumed).

Several quantitative measures of the incongruence between data sets have been developed [see Swofford (1991) for review], and statistical testing can be used to evaluate their significance (Farris et al., 1994, 1995). Indices of incongruence contain summary information regarding the conflict between data sets, but they clearly do not substitute for thorough knowledge of how the topologies specified by data sets differ and how strongly various components of the alternative topologies are supported by the different data sets. Swofford (1991) suggests caution in the interpretation of indices of incongruence. These indices measure the character conflicts that occur between data sets in excess of the conflicts that occur within them. Low levels of incongruence will be measured if character conflict is high within data sets, even if characters also show substantial conflict between data sets. These low values are useful only in identifying situations where it is incorrect to attribute the majority of character conflict to the differences between data sets. For the study of the salamander families from which my worked example is taken, indices of incongruence suggest that only 4–26% of the total incongruence among characters occurs between the molecular and morphological data sets (Larson and Dimmick, 1993).

The taxonomic congruence between morphological and molecular data sets may be represented by "consensus trees" or "largest common pruned trees" (reviewed by Swofford, 1991). A "strict" consensus tree retains all groups of taxa that appear in trees specified by both the morphological and molecular data, whereas an "Adams" consensus tree attempts to retain all patterns of hierarchical nesting of taxa specified by both the morphological and molecular trees. Swofford (1991) reviews indices used to represent the amount of structure retained in a consensus tree as a measurement of the taxonomic congruence of the two data sets. The largest common pruned tree finds the largest subset of taxa for which the morphological and molecular trees specify identical relationships, and removes the other taxa from the tree. Figure 2 illustrates consensus trees and a pruned tree for the comparison of the molecular and morphological trees (Fig. 1A and B, respectively). Alternative methods for producing consensus trees and pruned trees are reviewed in detail by Swofford (1991), who finds none of them completely satisfactory. The difficulty of comparing trees using consensus and pruned trees increases when the question of whether particular topologies are statistically supported by a data set is considered. The statistical examination of specific, paired topologies described above is recommended as an alternative to the analysis of consensus trees and pruned trees for examining the relative support available in different data sets for particular taxonomic groupings.

A possible criticism of the statistical approach suggested above is that it could work against morphological data sets because they generally will contain fewer characters than aligned nucleic acid sequences, and often may contain too few character transformations for statistical discrimination of alternative branching patterns. It is also reasonable that the origin of a complex morphological novelty might be judged to contain more phylogenetic information than a positional substitution in a nucleic acid sequence. The statistical approach was designed for molecular data where it is assumed that many individually unimportant characters are analyzed and only the statistical tendencies of the collection are definitive. Morphological phylogenetics often uses small numbers of characters that are nonetheless considered individually important. One possible solution is to use differential weighting in the statistical test (as described by Templeton, 1983), although choice of specific weights often will be difficult to justify.

In my study of the salamander families (Larson, 1991a), I identified one character, evolution of internal fertilization, as a singularly important one whose conflict with the molecular analysis was difficult to explain. I did not conclude that the molecular data gave the definitive answer for branching events requiring homoplastic evolution of internal fertilization, although this result was statistically significant for the molecular data. I suggested further testing of the disputed relationships with additional molecular data and further examination of the fertilization character. The latter option was made available by the data of Sever (1991), which are incorporated in the example worked above (Tab. 1) as characters CCE, CCH, CCL, CCM,

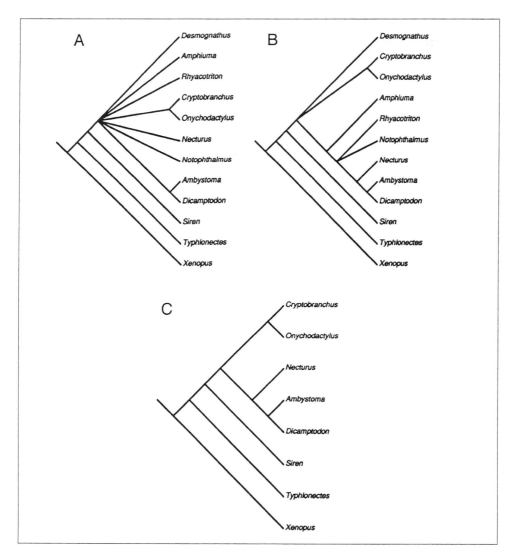

Figure 2. Phylogenetic topologies representing (A) the strict consensus tree, (B) the Adams consensus tree and (C) the largest common pruned tree for the molecular and morphological trees shown in Figure 1A and B, respectively. Taxa are designated as described in Figure 1.

CCR, CCS, CCT, CCU and CCW. The question of the phylogeny of the salamander families was then reconsidered, this time taking the approach of character congruence (Larson and Dimmick, 1993).

Character congruence

This approach begins by seeking the topology that is maximally parsimonious for variation observed in all relevant characters considered simultaneously, regardless of whether they are morphological or molecular. Each character is treated as an independent unit of phylogenetic information, and all characters are weighted equally in the initial phylogenetic analysis. Incongruence is evaluated for individual characters (rather than for predetermined sets of characters) against the topology generated from all relevant data. The character congruence approach is most useful if it directs the researcher toward knowledge of which characters are most effectively recovering the phylogeny of species and which ones are contributing primarily misinformation. The favored hypothesis of phylogeny then results from examining all characters that are informative for phylogenetic inference regardless of whether they are morphological or molecular. Characters may be misleading because (i) alternative states are misdiagnosed, (ii) rates of character evolution are mismatched to the divergence events being investigated or (iii) a character phylogeny is not congruent with the species phylogeny being estimated. Ideally, this approach would identify characters that are unreliable for any of these reasons.

Characters that are exceptionally homoplastic are highlighted for reevaluation. For molecular characters, reassessment of the alignment that produced the positional homologies on which the characters were based might be required. For morphological characters, hypothetically equivalent states that are required by the tree to have had separate phylogenetic origins should be reassessed. For example, investigation of the ontogenetic trajectory of a character may reveal developmental constraints that restrict transformations to a limited number of alternative states, thereby explaining homoplastic patterns; alternatively, character states considered equivalent when only the adult stage is observed may be found to have different ontogenetic origins.

For my example evaluating the relationships among the salamander families, the tree constructed from the combined data has a topology different from the trees generated using either the morphological or molecular data alone (Fig. 1; see also Larson and Dimmick, 1993). The resulting topology (Fig. 1C) closely resembles the structure of the molecular tree (Fig. 1A) except that the externally fertilizing taxa are moved toward the root, making the internally fertilizing salamanders monophyletic. This result lessens some concern that when morphological and molecular data sets are of different size, as they usually are (with the molecular characters being more numerous), the larger data set will entirely overwhelm the smaller one. Despite its smaller size, the morphological data in my example predominated in producing monophyly of the internally fertilizing salamanders contrasting with the results obtained using the molecular data alone. The molecular data contributed the majority of the information producing the remaining structure of the tree.

The analysis of individual characters for excessive homoplasy revealed three morphological characters that were judged unreliable for phylogenetic inference (Larson and Dimmick, 1993). Removal of these characters improved the consistency of the tree without altering its basic structure (Larson and Dimmick, 1993). These characters demonstrated paedomorphic loss of differentiation that presumably was generated in parallel by different ontogenetic perturbations that produced developmental truncation. Paedomorphosis was found by Kluge (1989) to explain inconsistency of characters in a phylogenetic study of snakes. Paedomorphosis does not automatically disqualify characters from phylogenetic analysis; other characters influenced by paedomorphosis were judged useful in my study (Larson and Dimmick, 1993) and might have been even more useful if they had been described as ontogenetic transformations rather than as instantaneous morphologies (de Queiroz, 1985). In contrast to the morphological characters, removal of the most homoplastic molecular characters lessened the resolution of the tree, and Larson and Dimmick (1993) concluded that these characters retained considerable phylogenetic information despite their relatively high homoplasy. Patterns of variation in these characters produced groupings that were sufficiently congruent with other characters that strong structure emerged from their combined analysis. A similar finding was made by Titus and Larson (1995) in the use of morphological and molecular data to examine intergeneric relationships within the salamander family Salamandridae.

The tree constructed from combined data sets provides a framework for examining the relative levels of consistency of characters categorized in different ways. I used a contingency table and the chi-square distribution to examine relative levels of homoplasy for different categories of characters (de Queiroz, 1989). Alternatively, statistical tests for differences in consistency indices among different kinds of characters have been used (McKitrick, 1992; de Queiroz and Wimberger, 1993). Numerous criteria can be used to circumscribe subsets of characters or character states for these tests. In molecular data, one may be interested in testing whether different structural or functional regions of a molecular sequence demonstrate different levels of homoplasy. Molecular sites undergoing transversions vs. those undergoing transitions might be compared. For morphological data, characters influenced by paedomorphic trends or losses of differentiation might be compared with those whose derived states involve primarily developmentally progressive changes.

The contingency chi-square analysis is illustrated in Tables 2 and 3 using the same data discussed above (Fig. 1 and Tab. 1). Levels of homoplasy observed in phylogenetically informative morphological vs. molecular characters are examined using the topology (Fig. 1C) generated by parsimony analysis of both data sets combined. All characters are categorized as being consistent with Figure 1C (no homoplasy), having a single homoplastic change (convergence, parallelism or reversal) or more than one homoplastic change among the 10 salamander taxa examined. The

Table 3. Patterns of homoplasy in molecular characters

	Consistent	1 Homoplasy	2+ Homoplasies	Total
Transitions	21 (20.4)	24 (20.9)	4 (7.8)	49
Transversions	10 (7.9)	5 (8.1)	4 (3.0)	19
Length changes	11 (13.7)	14 (14.0)	8 (5.2)	33
Total	42	43	16	101

Chi-square = 6.4, df = 4, n.s. Contingency table evaluating patterns of homoplasy observed for 99 molecular characters containing 101 phylogenetically informative derived states using the phylogenetic topology shown in Figure 1C. The phylogenetically informative derived states are categorized according to the kinds of changes that produced them (transitions, transversions, length mutations) and the level of homoplasy observed (consistent, one homoplastic change, two or more homoplastic changes). Expected values are in parentheses. Only the derived states that are phylogenetically informative for the 10 salamanders shown in Figure 1C are analyzed; changes occurring in outgroup taxa are excluded. The different categories of molecular change do not differ significantly in observed levels of homoplasy for the taxa examined. Data are from Larson and Dimmick (1993).

statistical test (Tab. 2) indicates that the morphological and molecular data do not differ significantly in amount of homoplasy and that most characters show no more than a single homoplastic change on the 10-taxon tree.

The molecular characters are examined further according to the kinds of changes observed at the different sites. Phylogenetically informative derived states (those shared by two or more taxa) are grouped into three categories: (i) those arising by transitional substitutions ("transitions"), (ii) those arising by transversional substitutions ("transversions") and (iii) those arising by length mutational changes ("length changes"). Transitions usually occur approximately twice as frequently as transversions in nuclear genomes (as observed in Tab. 3), with the consequence that they are expected to show greater homoplasy and substitutional saturation when comparisons cover long periods of evolutionary time. This observation has caused some workers to weight them *a priori* less heavily than transversional substitutions. The statistical analysis in Table 3 suggests that levels of homoplasy are not significantly different between transitional and transversional substitutions in this data set, nor are length changes significantly more prone to homoplasy than substitutional changes. Differential weighting of these different kinds of changes therefore is not warranted.

It is interesting to ask whether the tree specified by the combined morphological and molecular data constitutes a near-optimal topology for each of the data sets analyzed separately. The statistical test of Templeton (1983) described above is used to answer this question. The tree specified by the combined analysis (Fig. 1C) is found to be near optimal for both the molecular and morphological data sets analyzed separately. The Templeton (1983) test identifies the topology produced by the molecular data alone (Fig. 1A) as near optimal for the combined data, but the tree specified by the morphological data alone (Fig. 1B) is significantly less parsimonious than the favored tree (Fig. 1C) for the combined data.

Discussion

Evaluation of the phylogenetic information and the concordance of morphological and molecular data is most effective when both taxonomic congruence and character congruence are assessed concurrently using statistical approaches. In the worked example, the taxonomic congruence approach alone indicated that the morphological and molecular data contained significant conflict, implying that at least one data set could be recovering a character phylogeny that is at least partially incongruent with the species phylogeny. The combination of the two data sets for an analysis of character congruence identified a third topology that is near optimal for both the morphological and molecular data sets analyzed separately. This latter analysis challenges the conclusion that the molecular and morphological data sets contain significant conflicts. Although each data set can reject with statistical significance the phylogenetic topology specified by the other one, it is reasonable to propose that both data sets are estimating with error the topology that results from their combined analysis, and that the combined analysis is most effective because the phylogenetic signal in both data sets becomes reinforced when the data are combined. This outcome counters the argument that different data sets should not be analyzed in combination just because a preliminary test indicates significant incongruence between them [see Cunningham (1997) for further discussion of this issue].

The topology produced by the combined data makes more sense biologically than the topologies produced by either the morphological or molecular data considered separately; the internally fertilizing salamanders are monophyletic as specified by the morphology-based tree, but the relative relationships of taxa within this clade follow those of the molecular-based tree [an arrangement upheld by subsequent studies of mitochondrial DNA variation by Titus and Larson (1995)].

Based upon my use of both taxonomic congruence and character congruence, I interpret the respective phylogenetic information content of the molecular and morphological data as follows. The molecular data provided fairly strong resolution of salamander phylogeny, but the parsimony analysis failed for a few very deep internal branches that probably represent short intervals of evolutionary time relative to the terminal branches (especially for the branch connecting *Cryptobranchus* and *Onychodactylus* to the internally fertilizing salamanders). The morphological data complement the molecular data by providing strong evidence for monophyly of the internally fertilizing salamanders (all salamanders in Fig 1 except *Cryptobranchus*,

Onychodactylus and *Siren*), but contribute relatively little information to the structure of the remainder of the tree. This situation favors the performance of the combined analysis of molecular and morphological data, the result of which (Fig. 1C) is far more informative than the consensus diagrams (Fig. 2). Use of both the taxonomic congruence and character congruence approaches as described above nonetheless was critical for understanding character evolution and thereby identifying these data sets as appropriate for the combined analysis. My interpretation of this character variation will be tested further as additional data are collected.

Analysis of combined morphological and molecular data sets can be misleading if either one specifies primarily a character phylogeny that is inconsistent with the species phylogeny (Bull et al., 1993). This problem is most likely to occur where molecular characters come from a sequence that is transmitted as a unit where horizontal transfer, lineage sorting or mistaken inclusion of paralogous comparisons affect many characters simultaneously. In this case, the morphological and molecular data sets are expected to generate significantly different topologies for the same taxa, which would be detected using the taxonomic congruence analysis described above. It could be detectable also by the analysis of character congruence, which should reveal higher levels of homoplasy in characters that are tracing a phylogeny other than the one supported by the majority of the combined morphological and molecular characters. Observation of extensive incongruence between subsets of characters would alert the investigator to this problem; however, a tree resulting from combined analysis of the different data sets would not be a reliable estimate of phylogeny until the problem is corrected. For most studies of higher-level phylogeny, including my worked example, the likelihood of extensive discordance between a gene phylogeny and the species phylogeny is probably very small (de Queiroz, 1993; Brower et al., 1996); such discordances are more likely to be a problem when very recent divergences are investigated.

The character congruence approach provides an assessment of whether differential weighting of characters is desirable in phylogenetic analysis. In the initial analysis, each character is assumed to be an independent observation containing phylogenetic information, and no differential weighting is imposed. The contingency chi-square analysis illustrated above shows that the morphological and molecular data used in my example do not differ in level of homoplastic change, nor do different subcategories of the molecular characters. Provided that the study effectively matches variation of each character to the evolutionary timescale in which it is phylogenetically informative, excessive homoplasy or saturation of change should not occur. Transitions and transversions should be equivalent in their informational content, as should be substitutions in silent vs. replacement sites in a protein-coding sequence, regardless of their relative frequency in the data set. It is only when observing broader times of divergence over which a more rapidly evolving set of characters or character states (such as transitions) has reached saturation while a more slowly evolving set (such as transversions) has not undergone saturation that differential weighting is potentially useful. There are, furthermore, different and overlapping categories of weighting (transitions vs. transversions, silent vs. replacement substitutions), and it is often unclear *a priori* how these should be resolved. For example, for nucleic acids that show differential rates of evolution among sites, weighting transversions over transitions might preferentially select sites that are very rapidly evolving and most likely to show substitutional saturation. The contingency chi-square test combined with the character congruence analysis can provide a useful way to ask whether different hypothetical weighting schemes are justified.

Some investigators prefer to assess saturation of character change and appropriateness of differential weighting of characters prior to conducting a phylogenetic analysis. In an earlier paper (Larson, 1991b), I recommended using the ratios of transitions to transversions among aligned nucleic acid sequences for identifying substitutional saturation. For recently diverged sequences, the ratio of transitions to transversions is approximately 2 for nuclear sequences and usually higher for animal mitochondrial DNA sequences. Substitutional saturation lowers this ratio until it approximates an equilibrium value determined by the base composition of the sequence (Holmquist, 1983). I applied this test to the ribosomal RNA data on which my worked example is based, and found evidence for substitutional saturation only in three small domains of the sequence (Larson, 1991b). An alternative approach is to plot maximum-likelihood estimates of numbers of substitutions occurring between paired sequences against sequence divergence; a linearly increasing relationship is expected if saturation has not occurred, and a plateau is formed at a level of divergence for which saturation has become extensive [see Titus and Larson (1996) for an example]. For sequences having extensive length variation, the method of Gatesy et al. (1993) could be used to ask which positions in an alignment convey phylogenetic information and which ones are ambiguous because of superimposed length-mutational changes. I know of no comparable *a priori* methods for asking whether morphological characters are saturated with change.

If evolutionary saturation is detected or strongly suspected *a priori* and differential weighting of characters or character states is to be explored as a possible analytical solution, I recommend trying a series of different weighting schemes for the categories of characters or character states hypothesized to contain different relative amounts of phylogenetic information vs. noise. If differential weighting of characters does enhance the ratio of signal to noise, the phylogenetic topology should be most stable near the point where weighting is optimal for resolving homoplasy. This point would be indicated by a decrease in the number of equally most parsimonious and near-optimal trees generated from the data. This procedure is analogous to the one described above under character congruence where removal of three highly homoplastic morphological characters stabilized the tree but removal of some relatively homoplastic molecular characters destabilized it. A topology chosen in this manner could then serve as the framework for analysis of character congruence as described above.

A possible complication with differential weighting of characters involves trees for which rapidly evolving characters or character states contribute most of the phylogenetic information for relatively recent branching events but contribute largely misinformation for more basal branches. A single differential-weighting scheme would not be optimal for all parts of the tree; it might increase resolution for deeper branches of the tree while decreasing resolution of more recent phylogenetic events. A possible solution to this problem is a two-step procedure in which different weighting schemes are used sequentially to examine first the most basal and then the most highly nested nodes in the phylogenetic tree (Moritz et al., 1992).

A major limitation to the application of all approaches described above involves situations where analysis of character variation using parsimony is inappropriate. Some observational conditions that are not entirely under the control of the investigator produce statistical inconsistency of parsimony for molecular sequence data (Felsenstein, 1978). As noted above, parsimony is expected to support an incorrect topology if rates of molecular evolution are large and/or highly uneven, especially where internal branches on the tree are very short compared with terminal branches (Felsenstein, 1978, 1988; Hillis et al., 1994).

Statistical approaches based upon maximum likelihood (reviewed by Swofford et al., 1996) may be useful for phylogenetic analyses where parsimony is expected to fail. A model of evolutionary change appropriate for the molecular sequences whose evolution violates assumptions of maximum parsimony could be incorporated. Phylogenetic analysis by maximum likelihood is applicable also to morphological data, although the model of evolution appropriate for the morphological characters would differ from the one used to analyze the molecular data. A modified taxonomic congruence approach would consist of using statistical analyses based upon maximum likelihood to ask whether a topology optimal for the morphological data is significantly rejected by the molecular data and *vice versa*. A maximum-likelihood test of alternative phylogenetic topologies is described by Kishino and Hasegawa (1989) and can be implemented using the PHYLIP computer package of Felsenstein (1993). Huelsenbeck and Bull (1996) and Huelsenbeck et al. (1996b) also present likelihood-ratio tests useful for examining conflicts occurring among phylogenetic data sets. Application of the character congruence approach using maximum likelihood would be possible only if the different evolutionary models appropriate for the morphological and molecular data could be incorporated into a single analysis of the combined data sets.

Another major limitation on application of the approaches discussed above arises because some phylogenetic data are collected as distances between paired taxa rather than as characters of individual taxa. These data include ΔT_m measurements from DNA-annealing studies and immunological distances. Allozymic data are frequently expressed as genetic distances between paired taxa, although some methods for discrete coding of allozymic data (see Mabee and Humphries, 1993; Wiens, 1995) permit potential use of these data in both the taxonomic congruence and character congruence approaches described above. The topology derived from a distance analysis using, for example, the neighbor-joining procedure (Saitou and Nei, 1987) can be tested against a topology derived from discrete character data using Templeton's (1983) test to ask whether the discrete-character data significantly discriminate the alternative topologies. Comparable tests for use with distance data are available (Rzhetsky and Nei, 1992; see also Swofford et al., 1996). A modified application of the statistical taxonomic congruence approach is therefore possible when one or more of the data sets being compared constitute distance matrices. Distance data are fundamentally incompatible with the character congruence analysis, because one does not examine individual characters separately using distance data, nor is there a straightforward way to combine molecular and morphological data into a single distance matrix that accommodates the different sources of information.

Taxonomic congruence and character congruence are ways of asking which parts of a phylogenetic tree are well supported by the available morphological and molecular characters, and which parts require additional investigation. Although I have concentrated here on comparisons of morphological and molecular data, the methods described above can be generalized to any kind of comparative data (behavioral, physiological) where discrete taxonomic characters are used [e.g. McKitrick's (1992) analysis of behavioral and morphological characters]. The relative effectiveness of morphological vs. molecular or other data will vary among studies and among taxonomic groups (e.g. Wimberger and de Queiroz, 1996; Givnish and Sytsma, 1997). Recent studies applying the methods advocated here to the analysis of different kinds of systematic characters in a variety of taxa include Lafay et al. (1995), Titus and Larson (1995, 1996), Dimmick and Larson (1996), Mason-Gamer and Kellogg (1996), Miyamoto (1996), Poe (1996), Sites et al. (1996) and Shaffer et al. (1997). The combination of methods described above should permit each investigation to maximize use of the diverse forms of taxonomic information available.

Acknowledgements

I thank Kevin de Queiroz, Todd R. Jackman, Mary C. McKitrick, Tom A. Titus, David B. Wake and his laboratory group for helpful comments incorporated into the original version of this paper (Larson, 1994), and Todd Jackman for further comments on a draft of this updated version. I thank Alan R. Templeton for suggesting the contingency-table analysis of character congruence. This research was supported by NSF grant BSR 9106898 to the author.

References

Barrett, M., Donoghue, M. J. and and Sober. E. (1991) Against consensus. *Syst. Zool.* 40: 486–493.

Brower, A. V. Z., DeSalle, R. and Vogler A. (1996) Gene trees, species trees and systematics: a cladistic perspective. *Annu. Rev. Ecol. Syst.* 27: 423–450.

Bull, J. J., Huelsenbeck. J. P., Cunningham, C. W., Swofford, D. L. and Waddell. P. J. (1993) Partitioning and combining data in phylogenetic analysis. *Syst. Biol.* 42: 384–397.

Chippindale, P. T. and Wiens, J. J. (1994) Weighting, partitioning and combining characters in phylogenetic analysis. *Syst. Biol.* 43: 278–287.

Cunningham, C. W. (1997) Can three incongruence tests predict when data should be combined? *Mol. Biol. Evol.* 14: 733–740.

de Queiroz, K. (1985) The ontogenetic method for determining character polarity and its relevance to phylogenetic systematics. *Syst. Zool.* 34: 280–299.

de Queiroz, K. (1989) *Morphological and biochemical evolution in the sand lizards*, Ph.D. dissertation, University of California, Berkeley.

de Queiroz, A. (1993) For consensus (sometimes). *Syst. Biol.* 42: 368–372.

de Queiroz, A. and Wimberger, P. H. (1993) The usefulness of behavior for phylogeny estimation: levels of homoplasy in behavioral and morphological characters. *Evolution* 47: 46–60.

de Queiroz, A., Donoghue, M. J. and Kim, J. (1995) Separate versus combined analysis of phylogenetic evidence. *Annu. Rev. Ecol. Syst.* 26: 657–681.

Dimmick, W. W. and Larson, A. (1996) A molecular and morphological perspective on the phylogenetic relationships of the otophysan fishes. *Mol. Phylogenet. Evol.* 6: 120–133.

Eernisse, D. J. and Kluge, A. G. (1993) Taxonomic congruence versus total evidence and amniote phylogeny inferred from fossils, molecules, and morphology. *Molec. Biol. Evol.* 10: 1170–1195.

Faith, D. P. and Cranston, P. S. (1991) Could a cladogram this short have arisen by chance alone?: on permutation tests for cladistic structure. *Cladistics* 7: 1–28.

Farris, J. S., Källersjö, M., Kluge, A. G. and Bult, C. (1994) Testing significance of incongruence. *Cladistics* 10: 315–319.

Farris, J. S., Källersjö, M., Kluge, A. G. and Bult, C. (1995) Constructing a significance test for incongruence. *Syst. Biol.* 44: 570–572.

Felsenstein, J. (1978) Cases in which parsimony or compatibility methods will be positively misleading. *Syst. Zool.* 27: 401–410.

Felsenstein, J. (1985) Confidence limits on phylogenies with a molecular clock. *Syst. Zool.* 34: 152–161.

Felsenstein, J. (1988) Phylogenies from molecular sequences: inference and reliability. *Annu. Rev. Genet.* 22: 521–565.

Felsenstein, J. (1995) PHYLIP: *Phylogenetic Inference Package*, version 3.57c, Department of Genetics, University of Washington, Seattle.

Gatesy, J., DeSalle, R. and Wheeler, W. (1993) Alignment-ambiguous nucleotide sites and the exclusion of systematic data. *Mol. Phylogenet. Evol.* 2: 152–157.

Givnish, T. J. and Sytsma, K. J. (1997) Consistency, characters and the likelihood of correct phylogenetic inference. *Mol. Phylogenet. Evol.* 7: 320–330.

Hendy, M. D. and Penny, D. (1989) A framework for the quantitative study of evolutionary trees. *Syst. Zool.* 38: 297–309.

Hillis, D. M. (1996) Inferring complex phylogenies. *Nature* 383: 130–131.

Hillis, D. M. and Huelsenbeck, J. P. (1992) Signal, noise and reliability in molecular phylogenetic analyses. *J. Hered.* 83: 189–195.

Hillis, D. M. and Huelsenbeck, J. P. and Cunningham, C. W. (1994) Application and accuracy of molecular phylogenies. *Science* 264: 671–677.

Holmquist, R. (1983) Transitions and transversions in evolutionary descent: an approach to understanding. *J. Mol. Evol.* 19: 134–144.

Huelsenbeck, J. P. and Bull, J. J. (1996) A likelihood ratio test to detect conflicting phylogenetic signal. *Syst. Biol.* 45: 92–98.

Huelsenbeck, J. P., Bull, J. J. and Cunningham, C. W. (1996a) Combining data in phylogenetic analysis. *Trends. Ecol. Evolut.* 11: 152–158.

Huelsenbeck, J. P., Hillis, D. M. and Nielsen, R. (1996b) A likelihood ratio test of monophyly. *Syst. Biol.* 45: 546–558.

Jones, T. R., Kluge, A. G. and Wolf, A. J. (1993) When theories and methodologies clash: a phylogenetic reanalysis of the North American ambystomatid salamanders (Caudata: Ambystomatidae). *Syst. Biol.* 42: 92–102.

Kishino, H. and Hasegawa, M. (1989) Evaluation of the maximum likelihood estimate of the evolutionary tree topologies from DNA sequence data, and the branching order in Hominoidea. *J. Mol. Evol.* 29: 170–179.

Kluge, A. G. (1989) A concern for evidence and a phylogenetic hypothesis of relationships among Epicrates (Boidae, Serpentes). *Syst. Zool.* 38: 7–25.

Lafay, B., Smith, A. B. and Christen, R. (1995) A combined morphological and molecular approach to the phylogeny of asteroids (Asteroidea: Echinodermata). *Syst. Biol.* 44: 190–208.

Larson, A. (1991a) A molecular perspective on the evolutionary relationships of the salamander families. *Evol. Biol.* 25: 211–277.

Larson, A. (1991b) Evolutionary analysis of length-variable sequences: divergent domains of ribosomal RNA. *In*: Miyamoto, M. M. and Cracraft, J. (eds) *Phylogenetic Analysis of DNA Sequences*, Oxford University Press, New York, pp. 221–248.

Larson, A. (1994) The comparison of morphological and molecular data in phylogenetic systematics. *In*: Schierwater, B., Streit, B., Wagner, G, P. and DeSalle, R. (eds) *Molecular Ecology and Evolution: Approaches and Applications*, Birkhäuser, Basel, pp. 371–390.

Larson, A. and Dimmick, W. W. (1993) Phylogenetic relationships of the salamander families: an analysis of congruence among morphological and molecular characters. *Herpetol. Monogr.* 7: 77–93.

Larson, A. and Wilson, A. C. (1989) Patterns of ribosomal RNA evolution in salamanders. *Molec. Biol. Evol.* 6: 131–154.

Mabee, P. M. and Humphries, J. (1993) Coding polymorphic data: examples from allozymes and ontogeny. *Syst. Biol.* 42: 166–181.

Maddison, W. P. and Maddison, D. R. (1992) *MacClade*, version 3.01, Sinauer, Sunderland, MA.

Mason-Gamer, R. J. and Kellogg, E. A. (1996) Testing for phylogenetic conflict among molecular data sets in the tribe Triticeae (Graminae). *Syst. Biol.* 45: 524–545.

McKitrick, M. C. (1992) Phylogenetic analysis of avian parental care. *Auk* 109: 828–846.

Mickevich, M. F. (1978) Taxonomic congruence. *Syst. Zool.* 27: 143–158.

Miyamoto, M. M. (1996) A congruence study of molecular and morphological data for eutherian mammals. *Mol. Phylogenet. Evol.* 6: 373–390.

Miyamoto, M. M. and Fitch, W. M. (1995) Testing species phylogenies and phylogenetic methods with congruence. *Syst. Biol.* 44: 64–76.

Moritz, C., Schneider, C. J. and Wake, D. B. (1992) Evolutionary relationships within the *Ensatina eschscholtzii* complex confirm the ring species interpretation. *Syst. Biol.* 41: 273–291.

Pamilo, P. and Nei, M. (1988) Relationships between gene trees and species trees. *Molec. Biol. Evol.* 5: 568–583.

Poe, S. (1996) Data set incongruence and the phylogeny of crocodilians. *Syst. Biol.* 45: 393–414.

Roth, V. L. (1991) Homology and hierarchies: problems solved and unresolved. *J. Evol. Biol.* 4: 167–194.

Rzhetsky, A. and Nei, M. (1992) A simple method for estimating and testing minimum-evolution trees. *Molec. Biol. Evol.* 9: 945–967.

Saitou, N. and Nei, M. (1987) The neighbor-joining method: a new method for reconstructing phylogenetic trees. *Molec. Biol. Evol.* 4: 406–425.

Sever, D. M. (1991) Comparative anatomy and phylogeny of the cloacae of salamanders (Amphibia: Caudata). I. Evolution at the family level. *Herpetologica* 47: 165–193.

Shaffer, H. B., Meylan, P. and McKnight, M. L. (1997) Tests of turtle phylogeny: molecular, morphological and paleontological approaches. *Syst. Biol.* 46: 235–268.

Sites, J. W. Jr., Davis, S. K., Guerra, T., Iverson, J. B. and Snell, H. L. (1996) Character congruence and phylogenetic signal in molecular and morphological data sets: a case study in the living iguanas (Squamata, Iguanidae). *Molec. Biol. Evol.* 13: 1087–1105.

Swofford. D. L. (1991) When are phylogeny estimates from molecular and morphological data incongruent? *In*: Miyamoto, M. M. and Cracraft, J. (eds) *Phylogenetic Analysis of DNA Sequences*, Oxford University Press, New York, pp. 295–333.

Swofford, D. L. (1993) *PAUP: Phylogenetic Analysis Using Parsimony*, version 3.1, Illinois Natural History Survey, Urbana.

Swofford, D. L., Olsen, G. J., Waddell, P. J. and Hillis, D. M. (1996) Phylogeny reconstruction. *In*: Hillis, D. M., Moritz, C. and Mable, B. K. (eds) *Molecular Systematics*, Sinauer Associates, Sunderland, MA, pp. 407–514.

Templeton, A. (1983) Phylogenetic inference from restriction endonuclease cleavage site maps with particular reference to the evolution of humans and the apes. *Evolution* 37: 221–244.

Titus, T. A. and Larson, A. (1995) A molecular phylogenetic perspective on the evolutionary radiation of the salamander family Salamandridae. *Syst. Biol.* 44: 125–151.

Titus, T. A. and Larson, A. (1996) Molecular phylogenetics of desmognathine salamanders (Caudata: Plethodontidae): a reevaluation of evolution in ecology, life history and morphology. *Syst. Biol.* 45: 451–472.

Wiens, J. J. (1995) Polymorphic characters in phylogenetic systematics. *Syst. Biol.* 44: 482–500.

Wimberger, P. H. and de Queiroz, A. (1996) Comparing behavioral and morphological characters as indicators of phylogeny. *In*: Martins, E. P. (ed.) *Phylogenies and the Comparative Method in Animal Behavior*, Oxford Univ. Press, Oxford, pp. 206–233.

Zar, J. H. (1984) *Biostatistical Analysis*, 2nd ed. Prentice-Hall, Englewood Cliffs, NJ.

Beyond area relationships: Extinction and recolonization in molecular marine biogeography

Clifford W. Cunningham[1] and Timothy M. Collins[2]

[1]Zoology Department, Duke University, Durham, NC 27708, USA
[2]Department of Biological Sciences, Florida International University, University Park, Miami, FL 33199, USA

Summary

In vicariance biogeography, the traditional focus on solely determining area relationships can obscure biologically interesting complexity. Even in the case of neighboring sister areas, the rise and fall of barriers to dispersal can yield a complex pattern of vicariance and interchange. Vicariance biogeographers view incongruent historical patterns as noise that must be filtered out. Here, we sharpen the focus of vicariance biogeography, and attempt to identify organismal characteristics that unite sets of taxa with congruent histories. Emphasizing examples from coastal marine invertebrates, we apply this perspective to two well-studied model systems: the southeastern United States, and the trans-Arctic interchange through the Bering Strait.

In both systems, populations from neighboring areas tend to show either great genetic similarity, usually inferred to result from continuing gene flow, or reciprocal monophyly accompanied by deep genetic divergence. Because dispersal ability is a poor predictor of which taxa fall in either category, we consider the possibility that genetic similarities often attributed to continuing gene flow may result instead from extinction in one area followed by recolonization from the other. Similarly, reciprocal monophyly between neighboring areas suggests that taxa in those areas have resisted recent local extinction. Our perspective shifts focus away from larval dispersal ability, which has long dominated molecular marine biogeography. Instead, we can focus on extinction itself, asking why taxa showing reciprocal monophyly have resisted local extinction.

A focus on extinction and recolonization is especially fruitful for understanding the trans-Arctic interchange. In group after group, researchers have found genetic evidence consistent with local extinction in the NW Atlantic followed by recolonization either from the NE Atlantic or from the North Pacific. In general, taxa which are restricted to rocky substrata appear to have been more prone to local extinction. The ability to recolonize the NW Atlantic from neighboring areas does not appear to depend on dispersal ability, although present-day geographical distribution does seem to be important. We conclude by reviewing some of the reasons why biogeographers have failed to find a consistent relationship between larval dispersal ability and patterns of geographical subdivision.

Introduction

It is widely recognized that historical information is essential to fully appreciate the evolutionary context of species interactions (Vermeij, 1978; Brooks, 1985; Brooks and McLennan, 1991; Futuyma and McCafferty, 1990). In many cases, placing this historical information in a geographical context may also be important. Is a strong interaction between two or more species in a particular area the result of a long shared history, or did one or more of the species arrive only recently? Did a particular character state evolve *in situ,* or does it also appear in related species that diverged long ago? Establishing the geographical and historical context for the members of a community is known as historical biogeography. The only direct evidence for the composition of past communities comes from the fossil record (Valentine and Jablonski, 1993), but because of the paucity of fossil evidence for many taxa, and because of developments in phylogenetic methods, phylogenetic information has become increasingly important.

The first systematic attempt to infer biotic history from phylogenetic and geographic information was vicariance biogeography (Croizat et al., 1974; Nelson and Platnick, 1981; Nelson and Rosen, 1981; Humphries and Parenti, 1986). Vicariance biogeographers recognized that the current distributions of extant organisms may, in some cases, reflect the fragmentation (vicariance) of widespread ancestral ranges. Moreover, they recognized that in the absence of extinction or dispersal, the order in which these areas became fragmented will be reflected in the phylogenies of taxa taken from those areas (Fig. 1). The development of methods for recovering this sequence of fragmentation, or area relationships, has been the subject of much discussion (Rosen, 1976; Nelson and Platnick, 1981; Brooks, 1985; Humphries and

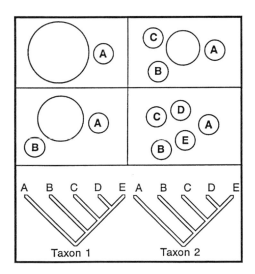

Figure 1. Geographic subdivision of a contiguous area by a sequence of vicariance events. If no postvicariance dispersal or extinction has taken place, the phylogenies of taxa taken from the subdivided areas will be perfectly congruent. In this case, the consensus area cladogram reflects the sequence of vicariant subdivision of the ancestral biota.

Parenti, 1986; Wiley, 1988; Page, 1988, 1991). All of these methods rely on finding congruence between phylogenies of unrelated species collected from the same areas (Fig. 1).

Even when area relationships can be readily determined, focusing only on area relationships obscures biologically interesting complexity. Consider the simplest possible case – neighboring sister areas. These areas share many pairs of closely related taxa and are currently divided by a barrier to dispersal. As is often the case, the barrier between these sister areas is transitory, driven by cyclical processes such as climate change (Briggs, 1974; Vermeij, 1978, 1989, 1991b). Even when barriers to migration are in place, they affect some taxa more strongly than others (Vermeij, 1978, 1989, 1991a, b). For these reasons, we would not expect all taxa in these neighboring areas to have experienced the same history. In the following section, we discuss two sets of taxa in these neighboring areas that have experienced distinct histories.

Vicariance, extinction and recolonization

Assume that one set of taxa in these neighboring sister areas has experienced little or no genetic connection (effective gene flow or recent recolonization) since the barrier first arose. As predicted by population genetic models, isolation has led to complete lineage sorting (Neigel and Avise, 1986; Avise, 1994), resulting in reciprocal monophyly and deep genetic divergence on either side of the barrier (Fig. 2A). Assume that a second set of taxa has experienced local extinction on one side of the barrier followed by recolonization across the barrier (Fig. 2B). Because of this recent recolonization event, there is little genetic divergence between these populations. Although considering either set of taxa would lead one to identify these as sister areas, the histories of the two sets of taxa differ in important ways. For example, every member of the second group of taxa experienced recent dispersal across the barrier. Does this mean that these taxa share characteristics making them better at dispersal than members of the first group?

A more subtle alternative scenario is that both sets of taxa share the same capability for dispersal and experienced approximately the same degree of intermittent dispersal across the barrier. In the first set of taxa, however, the genetic impact of dispersal was swamped by large standing populations on both sides of the barrier. In the second set of taxa, the genetic impact of the same level of dispersal was magnified because local extinction allowed the dispersers to found an entirely new population (e.g. Slatkin, 1977, 1985, 1987; but see Wade and McCauley, 1988). Therefore, an absence of genetic differentiation between neighboring populations may have as much to do with ecological opportunity created by local extinction as with dispersal ability.

A focus on extinction and recolonization shifts attention away from dispersal ability, and toward the conditions that underlie local extinction. Are taxa that show

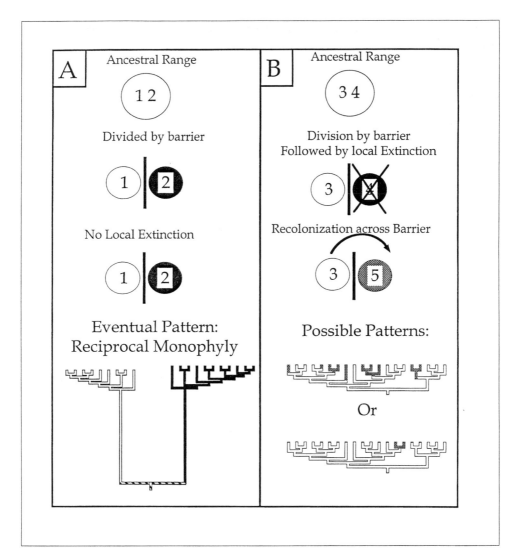

Figure 2. Two distinct histories experienced by taxa divided by the same vicariant event. (A) Cessation of gene flow resulting in reciprocal monophyly. (B) Local extinction on one side of the barrier followed by recolonization from the other side.

reciprocal monophyly between neighboring areas more resistant to extreme conditions? Is ability to disperse related to the probability of local extinction (Jablonski, 1986)? As Vermeij (1989) has argued, studying the factors that cause local extinction – or range restriction in his terminology – can help illuminate the factors that cause a species to become globally extinct.

Sharpening the focus of vicariance biogeography: searching for sets of congruent histories

Vicariance biogeographers attempt to identify a single set of area relationships from congruent elements in the phylogenies of taxa from those areas. If more than one set of congruent histories exists, these multiple histories are considered noise. In the preceding example, however, considering those multiple histories was more illuminating than simply identifying area relationships.

In this chapter, we argue that it is fruitful to focus and extend the goals of vicariance biogeography. Rather than focusing exclusively on identifying area relationships, we attempt to identify sets of taxa with congruent histories. This approach recognizes that regional biotas may have complex histories, and seeks to identify sets of taxa that have responded to the rise and fall of barriers to dispersal in similar ways. There are two advantages to this approach. First, it can help ecologists determine which interacting species may have had a long shared history (Brooks, 1985; Brooks and MacLennan, 1991). Second, and most important, it holds the promise of finding general lessons in biogeography. We may be able to learn if taxa with congruent histories share characteristics that can explain this congruence.

To explore characteristics that may underlie congruent histories, and to explore the implications of extinction and recolonization on geographic subdivision, we focus on genetic studies of coastal, benthic marine invertebrates. This is because they are reasonably well studied, have simple, largely one-dimensional ranges up and down coastlines, and often have excellent fossil records. The larvae of benthic marine invertebrates can be classified as planktotrophic (with pelagic, feeding larvae) or nonplanktotrophic [including pelagic nonfeeding, brooded or benthic crawling; Jablonski and Lutz (1983)]. Planktotrophic larvae are generally considered to have the greatest potential for dispersal (Scheltema, 1979, 1986). Although the population genetics and biogeography of marine invertebrates have been studied for nearly 30 years, much remains to be learned about the basic factors controlling the distribution of taxa in time and space. Even the organismal characteristic that has received the most attention, larval dispersal ability, is not consistently associated with patterns of geographical subdivision (Burton, 1983; Hedgecock, 1986; Ó Foighil, 1989; Palumbi, 1994, 1995).

We apply our perspective to two model biogeographic systems. The first system, the coast of the SE United States, represents the simplest possible case of two neighboring sister areas: the Gulf of Mexico and the western Atlantic. The second model system involves taxa that took part in the trans-Arctic interchange between the North Pacific and North Atlantic Oceans following the opening of the Bering Strait ca. 3.5 Ma. For both of these systems, we identify sets of taxa with congruent patterns of geographic subdivision. We find that neither an organism's geographic range nor its presumed larval dispersal ability, as predicted by larval type, can consistently explain patterns of geographic subdivision. We argue that in both model systems, local extinction and recolonization may be more important in explaining marine biogeographic patterns than has been previously thought.

Extinction and recolonization in the southeastern United States

Due largely to the efforts of John Avise and his colleagues, the coast of the SE United States has become one of the most intensively studied marine biogeographic systems (reviewed in Avise, 1992). The Gulf of Mexico and the Atlantic coast of the SE United States share many closely related species and subspecies, making them obvious sister areas (Fig. 3). This simple designation of sister areas, however, obscures a complex history between the Gulf of Mexico and the Atlantic. This complex history likely represents repeated episodes of vicariance and dispersal in response to climatically driven changes in sea temperature and sea level (reviewed in Felder and Staton, 1994). Two general patterns have emerged from studies of taxa collected from the SE United States – they either show reciprocal monophyly between the

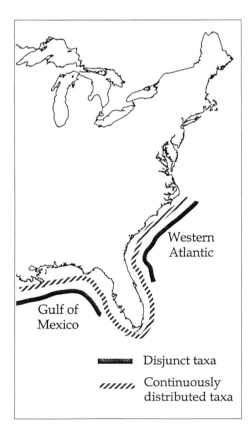

Figure 3. The first model system considered, the coast of the SE United States. Taxa found in both the Gulf and the Western Atlantic fall into two general categories: tropical species that are continuously distributed, and temperate species that are disjunct across southern Florida.

Gulf of Mexico and the Atlantic coast (Fig. 2A), or show clear evidence of recent genetic connection (Fig. 2B).

Are there organismal characteristics consistently associated with each of these biogeographic patterns? One easily defined characteristic is present-day geographic distribution. Some coastal taxa in the SE United States are continuously distributed, whereas other species found in both the Gulf of Mexico and the Carolinas are not found in the tropical waters of southern Florida (Fig. 3, Frey, 1965; Briggs, 1974; Cunningham et al., 1991; Felder and Staton, 1994). Have the same characteristics that prevent disjunct taxa from living in southern Florida increased the probability of vicariance during warmer periods? Probably not. Species with continuous distributions appear to be just as likely to show reciprocal monophyly between the Gulf of Mexico and the Atlantic as disjunct taxa (Tab. 1; Avise, 1992). Similarly, dispersal ability does not consistently predict patterns of geographical subdivision in the SE United States. Taxa with good dispersal ability are as likely to show reciprocal monophyly as they are to show evidence of recent genetic connection (Tab. 1).

Why is it difficult to explain these patterns of geographic subdivision? One obvious possibility is that a major vicariant event may have divided populations of taxa that now show reciprocal monophyly. Such an event is expected to affect all species regardless of their life history characteristics (Avise, 1992). However, this an-

Table 1. Genetic patterns in marine taxa found in the SE United States

Species	Distribution pattern	Dispersal ability	Reciprocal monophyly
Hydroid (*Hydractinia*)[†]	disjunct	poor	yes
Fish (toadfish)[§]	disjunct	good	yes
Fish (black sea bass)[§]	disjunct	good	yes
Hermit crab (*P. poll.*)[‡]	disjunct	good	yes
Crab (*Sesarma*)[ǁ]	disjunct	good	yes
Crab (*Uca*)[ǁ]	disjunct	good	yes*
Hermit crab (*P. long.*)[‡]	disjunct	good	no
Fish (sturgeon)[§]	disjunct	good	no
Fish (menhaden)[§]	disjunct	good	no
Horseshoe crab (*Limulus*)[§]	continuous	good	yes
Oyster (*Crassostrea*)[§]	continuous	good	yes
Mussel (*Geukensia*)[¶]	continuous	good	yes
Clam (*Mercenaria*)[%]	continuous	good	yes
Eel[§]	continuous	good	no
Fish (catfish)[§]	continuous	good	no

*monophyly not significant. [†]Cunningham et al., 1991. [‡]Cunningham et al., 1992. [§]Reviewed in Avise, 1992. [ǁ]Felder and Staton, 1994. [¶]Sarver et al., 1992. [%]Ó Foighil et al., 1996.

swer does not explain why some taxa show evidence of recent genetic connection. In some cases, continuing gene flow may account for the lack of genetic differentiation. In others it may have been caused by local extinction in one area followed by recolonization from another (Slatkin, 1977, 1985, 1987; but see Wade and McCauley, 1988). Whereas continuing gene flow may be enhanced by planktotrophic larvae, rare dispersal events in nonplanktotrophic species may be sufficient to colonize an area where the species is not currently found (Johannesson, 1988). If recolonization following local extinction occurs commonly in the SE United States, then vulnerability to local extinction may be as important as dispersal ability in determining which taxa show reciprocal monophyly.

For example, it is well known that isotherms become greatly compressed along the Atlantic coast of the SE United States during glaciation (e.g. Cronin, 1988). Such compression of isotherms may increase the probability of local extinction of temperate taxa in the Atlantic, followed by recolonization from the Gulf of Mexico. Using intraspecific allelic phylogenies, it may be possible to identify sets of taxa that were eradicated locally during the last glacial maximum and were later recolonized from the Gulf of Mexico. A newly colonized area is not only expected to show relatively low genetic diversity, but an intraspecific phylogeny is expected to show that alleles from the source population are paraphyletic with respect to alleles from the newly colonized area (Fig. 2B; Ortí et al., 1994; Templeton, 1993, 1994; Hellberg, 1994; Hewitt, 1996).

Just as taxa with little genetic differentiation may have experienced local extinction and recolonization, taxa with reciprocal monophyly on either side of a barrier must have resisted local extinction in the recent past. Genetic evidence for continuous residence is particularly important, because even a good fossil record may not provide unambiguous evidence for continuous residence. For example, if the population on the Atlantic side of the barrier repeatedly becomes extinct, but is always recolonized rapidly by the population on the other side of the barrier (relative to the rate of sedimentation or fossilization), the fossil record might appear continuous.

The preceding discussion suggests that genetic evidence might be an important source of information about whether local extinction has occurred in the recent past. If populations in neighboring sister areas have a genetic signature consistent with a recent range expansion (Ortí et al., 1994; Templeton, 1993, 1994; Hellberg, 1994), they may have undergone a local extinction followed by recolonization. This is especially likely if there was a long fossil history in the newly colonized area (e.g. Palumbi and Kessing, 1991; Ortí et al., 1994). On the other hand, taxa showing reciprocal monophyly are likely to have resisted extinction in the recent past.

The trans-Arctic interchange: a model system for marine biogeographic studies

The SE United States represents the simplest possible case – two neighboring sister areas. In this section, we consider a more complicated and interesting model system involving three neighboring areas: the North Pacific, the NW Atlantic and the NE Atlantic (Fig. 4). Before discussing biogeographic patterns, it is important to briefly sketch the geological background of a pivotal event in the marine northern hemisphere, the trans-Arctic interchange.

For most of the Cenozoic Era, the region that is now the Bering Strait between North America and Asia was a land bridge and was closed to marine migration. The isolation of the cold-water faunas of the North Pacific and North Atlantic was end-

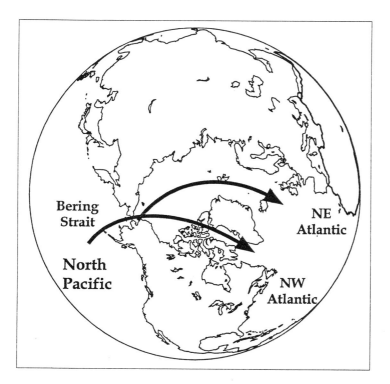

Figure 4. The second model system considered: three geographic areas that were involved in the trans-Arctic interchange of marine taxa through the Bering Strait ca. 3.5 Ma. As indicated by the arrows, this interchange was asymmetric, mainly proceeding from the Pacific to the Atlantic. We consider only taxa that currently have representatives in all three areas indicated: the North Pacific, NW Atlantic and NE Atlantic.

ed abruptly by the opening of the Bering Strait approximately 3.5 Ma. The exchange of taxa was asymmetric, with most invaders moving from the Pacific to the Atlantic (Fig. 4; Durham and MacNeil, 1967; Vermeij, 1989, 1991a,b). Because the opening of the Bering Strait during the late Pliocene took place when the earth's temperatures were considerably warmer than at any time during the Pleistocene (Herman and Hopkins, 1980; Shackleton et al., 1984; Carter et al., 1986; Andrews, 1988), many temperate and Arctic species took part in the interchange. The initial opening of the Bering Strait was soon followed by the onset of Northern Hemisphere glaciations between 2.5 Ma. and 3.1 Ma. (Shackleton et al., 1984). Pleistocene fluctuations of temperature and sea level have undoubtedly affected the likelihood of biotic interchange between the North Pacific and the North Atlantic.

Because the North Pacific and North Atlantic share many closely related taxa, they might be considered sister areas (Fig. 4). As with the SE United States, the simple designation of sister areas may obscure more interesting patterns. First, the major cause of this close area relationship is the dispersal associated with the trans-Arctic interchange rather than the vicariance of a widely distributed ancestral biota. Second, the repeated opportunities for vicariance and dispersal caused by glacial episodes have generated a complex history that is obscured by the designation of simple area relationships.

We consider the biogeography of the trans-Arctic interchange from two perspectives. We begin by considering the large-scale problem of determining which taxa took part in the exchange and in which direction they went. Although the identity of trans-Arctic invaders has been traditionally inferred from the fossil record, it is also possible to infer dispersal from phylogenetic information (Brundin, 1981; Bremer, 1992). We consider two gastropod genera with good fossil records and ask whether fossils and phylogenies give the same answers about their dispersal. Then we ask whether organismal characteristics can predict which species took part in the exchange. Like Vermeij (1991a), we conclude that the most important factor in determining which species took part in the interchange was ecological opportunity opened up by extinction in the North Atlantic.

Next, we consider at a finer scale the histories of species that invaded the North Atlantic. We identify four patterns that have emerged from genetic studies of the Northern Hemisphere marine fauna. These patterns are distinguished by the degree of genetic divergence between three areas – the North Pacific, the NW Atlantic and the NE Atlantic (Fig. 4). Finally, as before, we ask whether there are organismal characteristics that might explain the observed patterns.

Inferring dispersal from phylogenies and fossils: two case studies

Whether dispersal events can be inferred from the fossil record and from phylogenetic information is controversial (Croizat et al., 1974; Platnick and Nelson, 1978;

Platnick, 1981; Brundin, 1981; Patterson, 1981; Humphries and Parenti, 1986; Bremer, 1992). In the case of the trans-Arctic interchange, however, the simultaneous appearance of Pacific taxa in the North Atlantic (and *vice versa*) at the same time as the Bering Strait opened is strong evidence that a major biotic exchange took place. The marked asymmetry of the trans-Arctic interchange appears to be typical of invasions that follow the disappearance of major barriers (Webb, 1985; Vermeij, 1991a,b). In the following case studies, we ask whether the occurrence and directionality of dispersal events, as inferred from phylogenetic and fossil information, coincide.

The genus *Littorina* comprises 19 living species of marine rocky-intertidal herbivorous gastropods that are widely distributed through the Northern Hemisphere. Like many other genera that are thought to have participated in the trans-Arctic interchange, *Littorina* has a long history in the North Pacific but was not found in the North Atlantic fossil record until after the opening of the Bering Strait 3.5 Ma (Vermeij, 1991a; Reid, 1996). The fossil record indicates that two lineages of *Littorina* independently invaded the North Atlantic (Reid, 1996). A morphological cladogram of *Littorina* species is also consistent with two independent invasions of the North Atlantic and no dispersal in the opposite direction (Fig. 5A; Vermeij, 1991a; Reid, 1996). As with the morphological evidence, phylogenies derived from allozymes and mitochondrial DNA (mtDNA) both support the fossil-based hypothesis of two independent invasions of the North Atlantic (Zaslavskaya et al., 1992; Reid et al., 1996).

The genus *Nucella* is another group of marine rocky-shore gastropods. The fossil record suggests that there has been a single invasion of the Atlantic from the Pacific (Durham and MacNeil, 1967; Vermeij, 1991a). A mtDNA phylogeny also supports a single Pacific-to-Atlantic invasion (Fig. 5B; Collins et al., 1996). For *Nucella* and *Littorina,* fossil and phylogenetic information strongly support hypotheses of recent invasion of the North Atlantic from the Pacific. This agreement supports the assertion that phylogenetic information can illuminate the biogeographic histories of invaders that have no fossil record.

Why did only a subset of taxa take part in the trans-Arctic interchange?

Because so many taxa have recently invaded the North Atlantic, it is reasonable to ask if these invaders share organismal characteristics in common. In a statistical analysis of trans-Arctic invaders, Vermeij (1978, 1989, 1991a) found that the identity of invading species could not be predicted by larval dispersal ability. This is illustrated by *Littorina*, where one invading lineage has planktotrophic larvae, while the other, like *Nucella,* has nonplanktotrophic larvae.

On the other hand, Vermeij notes that invading species dominate rocky shore communities in the North Atlantic. This is evident in the shallow rocky shore of the

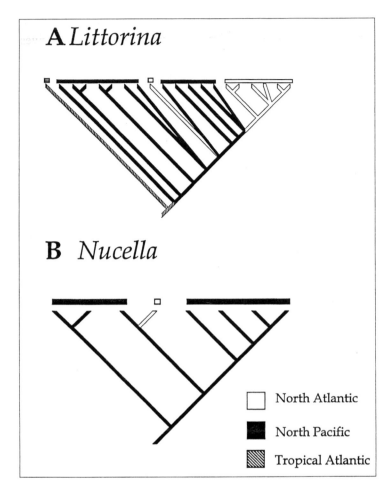

A *Littorina*

B *Nucella*

☐ North Atlantic

■ North Pacific

▨ Tropical Atlantic

Figure 5. (A) Morphological cladogram of the gastropod genus
Littorina, *after Reid (1990). This cladogram supports the fossil-based hypothesis of independent invasions of the Atlantic by two independent* Littorina *lineages. One of these invading lineages has planktotrophic larvae, while the other has non-planktotrophic larvae. (B) The mtDNA cladogram for the gastropod genus* Nucella, *based in the 740-bp mitochondrial cytochrome b gene (after Collins et al., 1996).*

NW Atlantic, where almost every major species is a recent invader from the North Pacific. The list of invaders includes 11 of 12 molluscan species, barnacles, hermit crabs, seastars, sea urchins and kelps (Vermeij, 1991a). Invading species compose much lower proportions of soft-bottom communities.

Does this mean that rocky-shore species in the North Pacific were more likely to have invaded the Atlantic than soft-bottom species? Apparently not. Close relatives of species that invaded the Atlantic make up similar proportions of present-day rocky-shore and soft-bottom communities of the North Pacific (Vermeij, 1991a). As discussed in the following section, the reason so many Pacific invaders are found on North Atlantic rocky shores appears to have more to do with ecological opportunity caused by extinction than dispersal ability (Vermeij, 1991a).

Glaciation and extinction in the NW Atlantic

The onset of Northern Hemispheric glaciation between 2.5 Ma and 3.1 Ma is believed to be responsible for large-scale extinction, especially in the North Atlantic (Raffi et al., 1985; Stanley, 1986). In the NW Atlantic, where the compression of isotherms was especially severe, cold-temperate species currently found north of Cape Cod appear to have been forced to the south of Cape Cod during the last glacial maximum (Vermeij, 1978, 1991a; Cronin, 1988). As a result, rocky-shore animals would have been especially likely to become extinct because there is virtually no hard substratum south of Cape Cod (Vermeij, 1978, 1991; Wethey, 1985; Ingólfsson, 1992). In an important paper, Ingólfsson (1992) has gone so far as to suggest that virtually the entire NW Atlantic rocky-shore fauna was exterminated during the most recent glacial maximum, and has subsequently been recolonized from elsewhere. If glaciations caused an especially high rate of extinction along the rocky shore, there would have been more ecological opportunities for invading species than for soft-bottom species.

A closer look at the histories of invading species

Because the Bering Strait opened before the onset of Northern Hemisphere glaciation, many temperate species were able to invade the Atlantic. Repeated opportunities for vicariant speciation during glacial maxima, combined with the continuing possibility of reinvasion from the Pacific, may have made the histories of invading taxa complex. In the following section, we consider only marine taxa that have invaded from the Pacific and that are currently found on both coasts of the North Atlantic. For these taxa, four broad classes of biogeographic histories can be defined according to the relative amount of trans-Arctic genetic divergence and trans-Atlantic genetic divergence (Fig. 6).

Large Pacific-Atlantic divergence: classes I and II

Taxa in classes I and II show clear reciprocal monophyly and deep genetic divergence between the Atlantic and Pacific (Fig. 6). Class I taxa are further defined by reciprocal monophyly between the NW and NE Atlantic, although the genetic divergence within the Atlantic is less than between the Atlantic and the Pacific (Fig. 6). This pattern is consistent not only with little or no recent genetic connection across the Atlantic, but also with a period of continuous residence on both sides of the Atlantic. By contrast, class II taxa also show deep Pacific-Atlantic divergence but

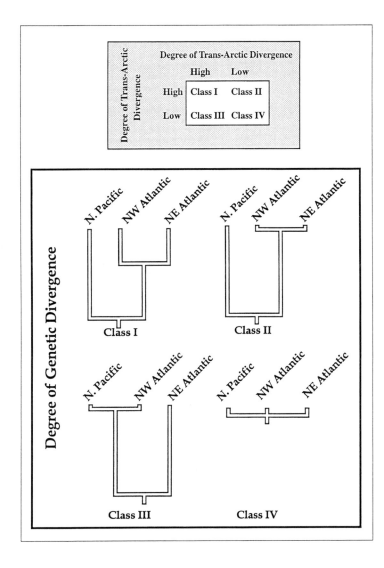

Figure 6. Four classes of biogeographic histories of Northern Hemisphere marine taxa. These classes are defined by the degree of genetic divergence across the Arctic and the across the Atlantic. See text for details.

very little differentiation between the NW and NE Atlantic (Fig. 6). The class II pattern is consistent with either continuing gene flow, or with a recent colonization event from one side of the Atlantic to the other (as in Fig. 2B).

As described above, in some cases it might be possible to distinguish between these alternatives (e.g. Templeton, 1993, 1994; Hellberg, 1994).

The class I pattern can be illustrated by a lineage of North Pacific hermit crabs that invaded the Atlantic during the trans-Arctic interchange. After their arrival in the Atlantic, these hermit crabs speciated on either side of the North Atlantic to become *Pagurus acadianus* (NW Atlantic) and *P. bernhardus* (NE Atlantic). The Atlantic species are monophyletic relative to the Pacific species (Cunningham et al., 1992; Cunningham, unpublished data). The Pacific-Atlantic divergence in a fragment of the mitochondrial 16S gene is about 3.2% across the Arctic and about 1% across the Atlantic (Cunningham et al., 1992; distances calculated according to Kimura, 1980). As a point of comparison, the same fragment was calculated to have diverged at a rate of about 1% per million years in several pairs of fiddler crabs now divided by the Isthmus of Panama (Sturmbauer et al., 1996). The degree of divergence observed in these hermit crabs is consistent not only with cessation of gene flow between the Pacific and Atlantic soon after the initial opening of the Bering Strait, but also with continuous residence on both sides of the Atlantic through abrupt climatic fluctuations over the past several hundred thousand years (Dansgaard et al., 1993).

The class II pattern can be illustrated by the gastropod genus *Nucella* discussed previously. As with *Pagurus*, an independent calibration of the rate of mitochondrial cytochrome *b* evolution in *Nucella* suggests a cessation of gene flow between the Pacific and Atlantic soon after the initial opening of the Bering Strait (Collins et al., 1996). Unlike *Pagurus*, however, there is virtually no divergence between populations of *N. lapillus* in the NW and NE Atlantic. European and North American populations share some identical mitochondrial haplotypes, and the average sequence divergence between the NW and NE Atlantic populations sampled over a 720-base pair region of cytochrome *b* is 0.05% (Collins, unpublished data). Another likely example of the class II pattern is of the barnacle *Semibalanus balanoides*. In a study of the AT-rich region of mitochondrial genome, *S. balanoides* exhibits a deep divergence between the North Pacific and the Atlantic, although the degree of divergence suggests a vicariance event in the past million or so years (Brown, 1995) Although there is considerable genetic diversity in the North Atlantic, there is no pattern of reciprocal monophyly between the NW and NE Atlantic (Brown, 1995; D. Rand, personal communication). This is consistent with a recent genetic connection across the Atlantic.

The class II pattern was also apparent in a study of the nuclear ribosomal intergenic spacer of the seaweed *Acrosiphonia arcta* (van Oppen et al., 1994). Like other class II species, there is deep divergence between Pacific and Atlantic populations and little divergence across the Atlantic. Unlike other class II species, however, high

Arctic populations of *A. arcta* in Greenland show great divergence from Pacific populations and from other Atlantic populations. Van Oppen et al. (1994) interpreted the high Arctic and temperate populations as having resulted from independent invasions from the Pacific.

What organismal characteristics might we expect to be associated with these patterns? Because both classes I and II show little or no genetic connection across the Arctic, they are likely to be largely absent from Arctic regions. Also, because class I taxa show deep genetic divergence across the North Atlantic, they are likely to have resisted local extinction through at least the most recent glacial episode.

By contrast, Atlantic members of class II taxa have diverged from Pacific taxa, but appear to have had a recent genetic connection across the Atlantic. Although this pattern can result from continuing gene flow, this is unlikely to be the only cause since at least one of the class II taxa, the gastropod *Nucella lapillus*, has non-planktotrophic larvae (Collins et al., 1996). A more likely possibility is that non-planktotrophic taxa went locally extinct on one coast or the other, only to be recolonized from the other coast, most likely by rafting of adults (Johannesson, 1988; Ingólfsson, 1992, 1995). As discussed previously, taxa found on the rocky shore of the NW Atlantic are especially likely candidates for local extinction. The absence of rocky shore south of Cape Cod may be fatal for *N. lapillus,* which is restricted to rocky shores, but not necessarily for *Pagurus,* which is more flexible in its habitat requirements (Ingólfsson, 1992).

Although habitat requirements and geographic range for these taxa seem consistent with expected class I and class II patterns, larval dispersal ability is clearly not a reliable predictor. It is *Pagurus,* with larvae that can spend 2–3 weeks in the plankton that shows deep trans-Atlantic divergence, whereas *Nucella*, with non-planktotrophic larvae, shows evidence of recent genetic connection.

Low Pacific-Atlantic divergence: classes III and IV

Taxa in classes III and IV show evidence of very recent genetic connection between the North Pacific and the North Atlantic (Fig. 6). The class III pattern is further characterized by a deep genetic divergence between the NW and NE Atlantic. This intriguing pattern is consistent with successive invasions of the Atlantic from the Pacific. The class III pattern has been observed in three taxa: the sea urchin *Strongylocentrotus droebachiensis* (Palumbi and Wilson, 1990; C. Biermann, personal communication), the red alga *Phycodrys rubens* (van Oppen et al., 1995) and the clam *Macoma balthica* (Meehan, 1985; Meehan et al., 1989). All three taxa show less divergence between the NW Atlantic and the North Pacific than either does with the NE Atlantic. The pattern is somewhat more complicated in *Macoma balthica* because a small population of NE Atlantic genotypes has recently been introduced into the San Francisco Bay. Introductions into the San Francisco Bay are relatively common, and are presumably caused by ballast water discharge (Meehan et al., 1989; Carlton and Geller, 1993).

Likewise, populations of the smelt *Osmerus* from the NW Atlantic show evidence of more recent genetic connection with the Pacific than with the NE Atlantic (Taylor and Dodson, 1994). Although *Osmerus* shows reciprocal monophyly among all three areas, the degree of genetic divergence between the NW Atlantic and the North Pacific is substantially lower than the degree of divergence across the Atlantic (6.6 vs. 15% in the cytochrome *b* gene).

Interestingly, all class III taxa only show evidence of recent genetic connection between the Pacific and the NW Atlantic, but not with the NE Atlantic. This is consistent with the hypothesis that class III taxa have been in continuous residence in the NE Atlantic for some time, but that the NW Atlantic has been recently colonized from the Pacific. In the case of the sea urchin *S. droebachiensis,* the presence of identical haplotypes in the Pacific and Western Atlantic indicates a recolonization within in the last 100,000 years (Palumbi and Kessing, 1991). Although NW Atlantic populations of the smelt *Osmerus* may have experienced relatively recent gene flow with populations from the Pacific, this connection appears to predate other class III taxa (Taylor and Dodson, 1994).

As with class II taxa, the class III pattern is consistent with local extinction in the NW Atlantic followed by recolonization from elsewhere. It may be that in some cases the only difference between class II and class III taxa is the location of the source population for recolonization, with a NE Atlantic source producing the class II pattern, and a Pacific source producing the class III pattern.

Class IV taxa show little genetic divergence either between the Pacific and the Atlantic, or across the Atlantic (Fig. 6). This pattern is seen in the sea urchin *S. pallidus* in which Pacific and Atlantic populations share identical mitochondrial haplotypes (Palumbi and Kessing, 1991; C. Biermann, personal communication). This pattern may result from recent, homogenizing gene flow, but is also consistent with a local extinction in the North Atlantic followed by recolonization from the Pacific. Palumbi and Kessing (1991) cite two factors supporting extinction followed by recolonization. First, the fossil record of *S. pallidus*, like *S. droebachiensis*, indicates that this taxon arrived soon after the initial opening of the Bering Strait ca. 3.5 Ma. Second, as with its congeneric *S. droebachiensis*, the genetic diversity of *S. pallidus* is extremely low, consistent with a recent and very extreme bottleneck (Palumbi and Kessing, 1991). Palumbi and Kessing argue that populations in both oceans were founded from a single refugium.

Another taxon showing the class IV pattern is the stickleback fish *Gasterosteus aculeatus* (Haglund et al., 1992; Ortí et al., 1995). The pattern of incomplete lineage sorting observed between the Pacific and the Atlantic strongly suggests a recent Pacific-Atlantic colonization event. Because Atlantic stickleback fossils go back 1.9 Ma, Ortí et al. (1995) have argued persuasively that the present pattern has resulted from a local extinction in the entire Atlantic followed by a recolonization from the Pacific. As predicted by a recent range expansion, not only is genetic diversity in the Atlantic lower than in the Pacific, but Pacific haplotypes are paraphyletic with respect to Atlantic haplotypes (Ortí et al., 1995).

Perhaps the best-studied taxon showing the class IV pattern is the mussel *Mytilus trossulus,* which shows little divergence across either the Atlantic or the Arctic (e.g. Varvio et al., 1988; McDonald et al., 1991; Rawson and Hilbish, 1995). Other *Mytilus* species in the North Atlantic appear to have arrived during an earlier invasion from the North Pacific. This earlier invasion gave rise to the endemic North Atlantic species *M. edulis* and *M. galloprovincialis.* Of these, only *M. edulis* is currently found on both coasts of the Atlantic, and it shows very little differentiation between the NW and NE Atlantic. This would place *M. edulis* in the class II category.

Summary of the histories of invading taxa

When we consider genetic patterns of invading Pacific taxa now found on both coasts of the Atlantic, complex patterns emerge. Some taxa appear to have had little or no genetic connection between the North Atlantic and the North Pacific since soon after the onset of Northern Hemisphere glaciation (class I and class II taxa, Fig. 6). Other taxa appear to have undergone multiple invasions of the North Atlantic (class III and some class IV taxa, Fig. 6). Interestingly, both class II and class III patterns are consistent with recent local extinction in the NW Atlantic followed by recolonization of the NW Atlantic from either the NE Atlantic (class II) or the North Pacific (class III). Class II and class III patterns are consistent with the hypothesis that the rocky-shore taxa of the NW Atlantic were especially hard hit during the last glacial maximum (Vermeij, 1978, 1991a; Wethey, 1985; Ingólfsson, 1992).

Several questions and concerns remain. Chief among these is the correct identification of taxa that have been introduced in historical times (e.g. Carlton and Geller, 1993). The identification of introduced taxa is somewhat easier in the Northern Hemisphere, due to relatively good historical records for macroinvertebrates. One promising approach is to document the genetic pattern observed in species known to have been introduced by human activity, and to treat with caution any taxa displaying similar patterns. The other outstanding question is whether taxa that show reciprocal monophyly across the same barrier were divided at the same time (true congruence) or at different times (pseudocongruence). This issue is beyond the scope of this chapter, but was considered in detail by Cunningham and Collins (1994). Suffice it to say that for some taxa in the trans-Arctic interchange, it should be possible to calibrate the rate of molecular evolution using fossil evidence (e.g. Cunningham and Collins, 1994; Collins et al., 1996; Reid et al., 1996).

Are there generalities in molecular marine biogeography? Or has the focus on larval type been a red herring?

Using examples from the SE United States and the trans-Arctic interchange, we argued that there is no consistent relationship between larval type, usually considered a proxy for dispersal ability, and geographic subdivision. This lack of consistency is not new and was noted in some of the earliest reviews of marine population genetics and biogeography (Burton and Feldman, 1982; Burton, 1983). Yet despite these troubling observations, the relationship between larval type and geographical subdivision has continued to draw more attention than any other pattern. As with the examples we considered here, some studies in other parts of the world have found the predicted relationship (e.g. McMillan et al., 1992; Kohn and Perron, 1994; Hellberg, 1996), while others have not (e.g. Hedgecock, 1986; Edmands and Potts, 1997).

In this chapter, we argued that in many cases, patterns of local extinction and recolonization may be as important as larval dispersal ability and may help to explain anomalous observations. We conclude this chapter by reviewing how larval type may influence the ability of dispersing organisms to colonize new habitats.

The paradox of Rockall and the efficacy of larval dispersal

Rockall is a tiny rocky island in the North Atlantic, 400 km west of the Outer Hebrides. The paradox of Rockall is that this tiny, remote island has 17 species of invertebrates, all of which have nonplanktotrophic larvae (Johannesson, 1988). If planktotrophic larvae are better at dispersal, why are there no planktonic species? Furthermore, how have these nonplanktotrophic species colonized this remote open ocean island? The pattern found on Rockall is not unique. Cobb seamount, 510 km west of the coast of Oregon in the NE Pacific, has a benthic community of 117 species that contains mainly species with nonplanktotrophic larvae (Parker and Tunnecliffe, 1994). South Georgia, a much larger but even more isolated island in the South Atlantic, is also dominated by nonplanktotrophic species (Davenport et al., 1997). These cases indicate that in certain circumstances, species with nonplanktotrophic larvae are able to disperse great distances and colonize habitats where planktotrophic larval species are unable to establish viable populations.

Most of these nonplanktotrophic species appear to raft, or arrive on floating debris. There is direct evidence that the brooding bivalve *Gaimardia trapesina* has rafted on kelp as far as 1300–2000 km (Helmuth et al., 1994). In one study of 41 floating clumps of seaweed around Iceland, 39 taxa were found (Ingólfsson, 1995), including hundreds of representatives of gastropods and bivalves. Other benthic invertebrates, such as corals, may disperse as adults on floating pumice (Jokiel, 1989, 1990). In the ascidian *Botrylloides*, adults rafting on eelgrass dispersed about 200 times farther than planktotrophic larvae of the same species, and had comparable success at recruitment (Worcestor, 1994).

These results indicate that dispersal by rafting may be effective for colonizing new and distant habitats. Planktotrophic larvae may more commonly disperse great distances, but rafting of adults in nonplanktotrophic species, while perhaps less common, may more often result in the establishment of viable populations. A single female that has brooding larvae, is impregnated, or is storing sperm may establish a new population [in some gastropod species females may store viable sperm for over 1 year (Giese and Pearse, 1977)]. In new habitats the poor dispersal ability of non-planktotrophic larvae may enhance the likelihood of establishment and survival of the population.

Therefore, it seems that the relationship between larval type and dispersal ability may be complex. In short, when local extinction takes place, recolonization from a neighboring area may be just as likely to take place by rafting as by dispersal of planktotrophic larvae. If this is true, then vulnerability to local extinction may be as important as dispersal ability in explaining patterns of geographic subdivision.

References

Andrews, J. T. (1988) Climatic evolution of the eastern Canadian Arctic and Baffin Bay during the past three million years. *In:* Shackleton, N. J,. West, R. G. and Owens, D. Q. (eds) *The Past Three Million Years: Evolution of Climatic Variability in the North Atlantic Region*, The Royal Society, London, pp. 235–250.

Avise, J. C. (1992) Molecular population structure and the biogeographic history of a regional fauna: a case history with lessons for conservation biology. *Oikos* 63: 62–76.

Avise, J. C. (1994) *Molecular Markers, Natural History and Evolution*, Chapman and Hall, New York.

Bremer, K. (1992) Ancestral areas: a cladistic reinterpretation of the center of origin concept. *Syst. Zool.* 41(4): 436–445.

Briggs, J. C. (1974) *Marine Zoogeography*, McGraw Hill, New York.

Brooks, D. R. (1985) Historical ecology: a new approach to studying the evolution of ecological associations. *Ann. Miss. Bot. Gard.* 72: 660–680.

Brooks, D. R. and McLennan, D. A. (1991) *Phylogeny, Ecology and Behavior*, The University of Chicago Press, Chicago and London.

Brown, A. F. (1995) The population genetics and biogeography of the barnacle, *Semibalanus balanoides*. Brown University.

Brundin, L. (1981) Croizat's panbiogeography versus phylogenetic biogeography. *In:* Nelson G, Rosen DE (eds) Vicariance biogeography: a critique, Columbia University Press, New York, pp. 94–158.

Burton, R. S. (1983) Protein polymorphisms and genetic differentiation of marine invertebrate populations. *Mar. Biol. Lett.* 4: 193–206.

Burton, R. S. and Feldman, M. W. (1982) Population genetics of coastal and estuarine invertebrates: does larval behavior influence population structure? *In:* Kennedy, V. S. (ed.)

Estuarine Comparisons, Academic Press, New York, pp. 537–551.

Carter, L. D., Brigham-Grette, J., Marincovich, L., Jr., Pease, V. L. and Hillhouse, J. W. (1986) Late Cenozoic Arctic Ocean sea ice and terrestrial paleoclimate. *Geology* 14: 675–678.

Carlton, J. T. and Geller, J. B. (1993) Ecological roulette: the global transport of nonindigenous marine organisms. *Science* 261: 78–82.

Collins, T. M., Frazer, K., Palmer, A. R., Vermeij, G. J. and Brown, W. M. (1996) Evolutionary history of northern hemisphere *Nucella* (Gastropoda, Muricidae): molecular, morphological, ecological, and paleontological evidence. *Evolution* 50(6): 2287–2304.

Croizat, L., Nelson, G. and Rosen, D. E. (1974) Centers of origin and related concepts. *Syst. Zool.* 23(2): 265–87.

Cronin, T. M. (1988) Evolution of marine climates of the U. S. Atlantic coast during the past four million years. *In:* Shackleton, N. J,. West, R. G. and Owens, D. Q. (eds) *The Past Three Million Years: Evolution of Climatic Variability in the North Atlantic Region*, The Royal Society, London, pp. 327–356.

Cunningham, C. W. and Buss, L. W. and Anderson, C. A. (1991) Molecular and geologic evidence of shared history between hermit crabs and the symbiotic genus *Hydractinia*. *Evolution* 458(6): 1301–1316.

Cunningham, C. W., Blackstone, N. W. and Buss, L. W. (1992) Evolution of king crabs from hermit crab ancestors. *Nature* 355: 539–542.

Cunningham, C. W. and Collins, T. (1994) Developing model systems for molecular biogeography: vicariance and interchange in marine invertebrates. *In:* Schierwater, B., Streit, B., Wagner, G, P. and DeSalle, R. (eds) *Molecular Ecology and Evolution: Approaches and Applications*, Birkhäuser, Basel, pp. 405–433.

Dansgaard, W., Johnsen, S. J., Clausen, H. B., Dahl-Jensen, D., Gundestrup, N. S., Hammer, C. U., Hvidberg, C. S., Steffensen, J. P., Jousel, Sveinbjörnsdottir, J. and Bond, G. (1993) Evidence for general instability of past climate from a 250-kyr ice-core record. *Nature* 364: 218–220.

Davenport, J., Barnett, P. R. O. and McAllen, R. J. (1997) Environmental tolerances of three species of the harpactacoid copepod genus *Tigriopus. Jour. Mar. Biol. Ass. U. K* 77: 3–16.

Durham, J. W. and MacNeil, F. S. (1967) Cenozoic migrations of marine invertebrates through the Bering Strait region. *In:* Hopkins, D. M. (ed.) *The Bering Land Bridge*, Stanford University Press, Stanford, CA, pp. 326–349.

Edmands, S. and D. C. Potts (1997) Population genetic structure in brooding sea anemones (*Epiactis* spp.) with contrasting reproductive modes. *Mar. Biol.* 127: 485–498.

Felder, D. L. and Staton, J. L. (1994) Genetic differentiation in trans-Floridian species complexes of *Sesarma* and *Uca* (Crustacea; Decapoda: Brachyura). *J. Crust. Biol.* 14(2): 191–209.

Frey, D. G. (1965) Other invertebrates – an essay in biogeography. *In:* Wright, J. and Frey, D. G. (eds) *The Quaternary of the United States*, Princeton University Press, Princeton, NJ, pp. 613–631.

Futuyma, D. J. and McCafferty, S. S. (1990) Phylogeny and the evolution of host plant associations in the leaf beetle genus *Ophraella* (Coleoptera, Chrysomelidae). *Evolution* 44(8): 1885–1913.

Giese, A. C. and J. S. Pearse (1977) *Reproduction of Marine Invertebrates*, vol. 4, *Molluscs: Gastropods and Cephalopods*, Academic Press, New York.

Haglund, T. R., Buth, D. G. and Lawson, R. (1992) Allozyme variation and phylogenetic relationships of Asian, North American, and European populations of the threespine stickleback, *Gasterosteus aculeatus*. *Copeia* 1992: 432–443.

Hedgecock, D. (1986) Is gene flow from pelagic larval dispersal important in the adaptation and evolution of marine invertebrates? *Bull. Mar. Sci.* 39(2): 550–564.

Hellberg, M. E. (1994) Relationships between inferred levels of gene flow and geographic distance in a philopatric coral, *Balanophyllia elegans*. *Evolution* 48: 1829–1854.

Hellberg, M. E. (1996) Dependence of gene flow on geographic distance in two solitary corals with different larval dispersal capabilities. *Evolution* 50: 1167–1175.

Helmuth, B., Veit, R. R. and Holberton, R. (1994) Long-distance dispersal of a subantarctic brooding bivalve (*Gaimardia trapesina*) by kelp rafting. *Mar. Biol.* 120: 421–426.

Herman, Y. and Hopkins, D. M. (1980) Arctic Ocean climate in late Cenozoic time. *Science* 209: 557–562.

Hewitt, G. M. (1996) Some genetic consequences of ice ages, and their role in divergence and speciation. *Biol. J. Linn. Soc.* 58: 247–276.

Humphries, C. J. and Parenti, L. R. (1986) *Cladistic Biogeography*, Clarendon Press, Oxford.

Ingólfsson, A. (1992) The origin of the rocky shore fauna of Iceland and the Canadian Maritimes. *J. Biog.* 19: 705–712.

Ingólfsson, A. (1995) Floating clumps of seaweed around Iceland: natural microcosms and a means of dispersal for shore fauna. *Mar. Biol.* 122: 13–21.

Jablonski, D. (1986) Larval ecology and macroevolution in marine invertebrates. *Bull. Mar. Sci.* 39: 565–587.

Jablonski, K. and Lutz, R. A. (1983) Larval ecology of marine benthic invertebrates: Paleobiological implications. *Biol. Rev.* 58: 21–89.

Johannesson, K. (1988) The paradox of Rockall: why is a brooding gastropod (*Littorina saxatilis*) more widespread than one having a planktonic larval dispersal stage (*L. littorea*)? *Mar. Biol.* 99: 507–513.

Jokeil, P. L. (1989) Rafting of reef corals and other organisms at Kwajalein Atoll. *Mar. Biol.* 101: 483–493.

Jokiel, P. L. (1990) Long-distance dispersal by rafting: reemergence of an old hypothesis. *Endeavour, New Series* 14: 66–73.

Kimura, M. (1980) A simple method for estimating evolutionary rates of base substitutions through comparative studies of nucleotide sequences. *J. Mol. Evol.* 16: 111–120.

Kohn, A. J. and Perron, F. E. (1994) *Life History and Biogeography: Patterns in Conus*, Clarendon Press, Oxford.

McDonald, J. H., Seed, R. and Koehn, R. K. (1991) Allozymes and morphometric characters of three species of Mytilus in the Northern and Southern Hemispheres. *Mar. Biol.* 111: 323–333.

McMillan, W. O., Raff, R. A. and Palumbi, S. R. (1992) Population genetic consequences of developmental evolution in sea urchins (Genus *Heliocidaris*). *Evolution* 46: 1299–1312.

Meehan, B. W. (1985) Genetic comparison of *Macoma balthica* (Bivalvia, Telinidae) from the eastern and western North Atlantic Ocean. *Mar. Ecol. Prog. Ser.* 22: 69–76.

Meehan, B. W., Carlton, J. T. and Wenne, R. (1989) Genetic affinities of the bivalve *Macoma balthica* from the Pacific coast of North America: evidence for recent introduction and historical distribution. *Mar. Biol.* 102: 235–241.

Neigel, J. E. and Avise, J. C. (1986) *Phylogenetic Relationships of Mitochondrial DNA Under Various Demographic Models of Speciation*, Academic Press, Orlando, FL.

Nelson, G. J. and Platnick, N. I. (1981) *Systematics and Biogeography: Cladistics and*

Vicariance, Columbia University Press, New York.

Nelson, G. J. and Rosen, D. E. (1981) *Vicariance Biogeography: A Critique*, Columbia University Press, New York City.

Ó Foighil, D. (1989) Planktotrophic larval development is associated with a restricted geographic range in *Lasaea*, a genus of brooding, hermaphroditic bivalve. *Mar. Biol.* 103: 349–358.

Ó Foighil, D., Hilbish, T. J. and Showman, R. S. (1996) Mitochondrial gene variation in *Mercenaria* clam sibling species reveals a relict secondary contact zone in the western Gulf of Mexico. *Mar. Biol.* 126: 675–683.

Orti, G., Bell, M. A., Reimchen, T. E. and Meyer, A. (1994) Global survey of mitochondrial DNA sequences in the threespine stickleback: evidence for recent migrations. *Evolution* 48(3): 608–622.

Page, R. D. M. (1988) Quantitative cladistic biogeography: constructing and comparing area cladograms. *Syst. Zool.* 37: 254–270.

Page, R. D. M. (1991) Clocks, clades, and cospeciation: comparing rates of evolution and timing of cospeciation events in host-parasite assemblages. *Syst. Zool.* 40(2): 188–198.

Palumbi, S. R. (1994) Genetic divergence, reproductive isolation and marine speciation. *Annu. Rev. Ecol. Syst.* 25: 547–572.

Palumbi, S. R. (1995) Macrospatial genetic structure and speciation in marine taxa with high dispersal abilities. *In*: Ferraris, J. D. and Palumbi, S. R. (eds) *Molecular Zoology: Advances, Strategies and Protocols*, Wiley-Liss, New York, pp. 101–117.

Palumbi, S. R. and Kessing, B. D. (1991) Population biology of the trans-Arctic interchange: mtDNA sequence similarity between Pacific and Atlantic sea urchins. *Evolution* 45(8): 1790–1805.

Palumbi, S. R. and Wilson, A. C. (1990) Mitochondrial DNA diversity in the sea urchins *Strongylocentrotus purpuratus* and *S. droebachiensis*. *Evolution* 44(2): 403–415.

Parker, T. and Tunnicliffe, V. (1994) Dispersal strategies of the biota of an oceanic seamount: implications for ecology and biogeography. *Biol. Bull.* 187: 336–345.

Patterson, C. (1981) Methods of paleobiogeography. *In*: Nelson, G. and Rosen, D. E. (eds) *Vicariance Biogeography: A Critique*, Columbia University Press, New York, pp. 446–489.

Platnick, N. I. (1981) The progression rule or progress beyond the rules of biogeography. *In*: Nelson, G. and Rosen, D. E. (eds) *Vicariance Biogeography: A Critique*, Columbia University Press, New York, pp. 144–150.

Platnick, N. I. and Nelson, G. (1978) A method of analysis for historical biogeography. *Syst. Zool.* 27: 1–16.

Raffi, S., Stanley, S. M. and Marasti, R. (1985) Biogeographic patterns and Plio-Pleistocene extinction of Bivalvia in the Mediterranean and southern North Sea. *Paleobiology* 11: 368–388.

Rawson, P. D. and Hilbish, T. J. (1995) Evolutionary relationships among the male and female mitochondrial DNA lineages in the *Mytilus edulis* species complex. *Molec. Biol. Evol.* 12(5): 893–901.

Reid, D. G. (1990) A cladistic phylogeny of the genus *Littorina* (Gastropoda): Implications for evolution of reproductive strategies and for classification. *Hydrobiology* 193: 1–19.

Reid, D. G. (1996) *Systematics and Evolution of Littorina*, The Ray Society, Dorchester, Dorset.

Reid, D. G., Rumbak, E. and Thomas, R. H. (1996) DNA, morphology and fosils: phylogeny and evolutionary rates of the gastropod genus *Littorina*. *Phil. Trans. R. Soc. Lond. B* 351: 877–895.

Rosen, D. E. (1976) A vicariance model of Caribbean biogeography. *Syst. Zool.* 27: 159–188.

Sarver, S. K., Landrum, M. C. and Foltz, D. W. (1992) Genetics and taxonomy of ribbed mussels (*Geukensia* spp.). *Mar. Biol.* 113: 385–390.

Scheltema, R. S. (1979) Dispersal of pelagic larvae and the zoogeography of Tertiary marine benthic gastropods. *In:* Gray, J. and Boucot, A. J. (eds) *Historical Biogeography, Plate Tectonics and the Changing Environment*, Oregon State University Press Corvallis, Oregon, pp. 391–397.

Scheltema, R. S. (1986) On dispersal and planktonic larvae of benthic invertebrates: an eclectic overview and summary of problems. *Bull. Mar. Sci.* 39(2): 290–322.

Shackleton, N. J., Backman, J., Zimmerman, H., Kent, D. V., Hall, M. A., Roberts, D. G., Schnitker, D., Baldauf, J. G., Desprairies, A., Homrighausen, R., Huddleston, P., Keene, J. B., Kaltenback, A. H., Krumsiek, K. A. O., Morton, A. C., Murray, J. W. and Westberg-Smith, J. (1984) Oxygen isotope calibration of the onset of ice-rafting and history of glaciation in the North Atlantic region. *Nature* 307: 620–623.

Slatkin, M. (1977) Gene flow and genetic drift in a species subject to frequent local extinctions. *Theor. Pop. Biol.* 12(3): 253–262.

Slatkin, M. (1985) Gene flow in natural populations. *Annu. Rev. Ecol. Syst.* 0(0): 393–430.

Slatkin, M. (1987) Gene flow and the geographic structure of natural populations. *Science* 236(4803): 787–792.

Stanley, S. M. (1986) Anatomy of a regional mass extinction: Plio-Pleistocene decimation of the western Atlantic bivalve fauna. *Palaios* 1: 17–36.

Sturmbauer, C., Levinton, J. S. and Christy, J. (1996) Molecular phylogeny analysis of fiddler crabs: test of the hypothesis of increasing behavioral complexity in evolution. *Proc. Natl. Acad. Sci. USA* 93(20): 10855–10857.

Taylor, E. B. and Dodson, J. J. (1994) A molecular analysis of relationships and biogeography within a species complex of Holarctic fish (genus *Osmerus*). *Molec. Ecol.* 3: 235–248.

Templeton, A. R. (1993) The "Eve" hypothesis: a genetic critique and reanalysis. *Amer. Anth.* 95: 51–72.

Templeton, A. R. (1994) The role of molecular genetics in speciation studies. *In:* Schierwater, B., Streit, B., Wagner, G, P. and DeSalle, R. (eds) *Molecular Ecology and Evolution: Approaches and Applications*, Birkhäuser, Basel, pp. 455–478.

Valentine, J. W. and Jablonski, D. (1993) Fossil communities: compositional variation at many time scales. *In:* Ricklefs, R. E. Schluter, D. (eds) *Species Diversity in Ecological Communities: Historical and Geographical Perspectives*, University of Chicago Press, Chicago, IL, pp. 341–349.

van Oppen M. J. H., Diekmann, O. E., Wiencke, C., Stam, W. T. and Olsen, J. L. (1994) Tracking dispersal routes: phylogeography of the Arctic-Antarctic disjunct seaweed *Acrosiphonia* (Chlorophyta). *J. Phycol.* 30: 67–80.

van Oppen, M. J. H., Draisma, S. G. A, Olsen, J. L. and Stam, W. T, (1995) Multiple trans-Arctic passages in the red alga *Phycodrys rubens*: evidence from nuclear rDNA ITS sequences. *Mar. Biol.* 123: 179–188.

Varvio, S.-L., Koehn, R. K. and Väinölä R (1988) Evolutionary genetics of the *Mytilus edulis* complex in the North Atlantic region. *Mar. Biol.* 98: 51–60.

Vermeij, G. J. (1978) *Biogeography and Adaptation*, Harvard University Press, Cambridge, MA.

Vermeij, G. J. (1989) Geographical restriction as a guide to the causes of extinction: the case of the cold northern oceans during the Neogene. *Paleobiology* 15: 335–356.

Vermeij, G. J. (1991a) Anatomy of an invasion: the trans-Arctic interchange. *Paleobiology* 17(3): 281–307.

Vermeij, G. J. (1991b) When biotas meet: understanding biotic interchange. *Science* 253: 1099–1104.

Wade, M. J. and McCauley, D. E. (1988) Extinction and recolonization: their effects on the genetic differentiation of local populations. *Evolution* 42: 995–1005.

Webb, S. D. (1985) Late Cenozoic mammal dispersals between the Americas. *In:* Stehli, G. G. and Webb, S. D. (eds). *The Great American Biotic Interchange*, Plenum Press, New York, pp. 357–386.

Wethey, D. S. (1985) Catastrophe, extinction and species diversity: a rocky intertidal example. *Ecology* 66(2): 445–456.

Wiley, E. O. (1988) Vicariance Biogeography. *Annu. Rev. Ecol. Syst.* 19: 513–42.

Worcestor, S. E. (1994) Adult rafting versus larval swimming: dispersal and recruitment of a botryllid ascidian on eelgrass. *Mar. Biol.* 121: 309–317.

Zaslavskaya, N. I., Sergievsky, S. O. and Tatarenkov, A. N. (1992) Allozyme similarity of Atlantic and Pacific species of *Littorina* (Gastropoda: Littorinidae). *J. Mollusc. Stud.* 58: 377–384.

The history of development through the evolution of molecules: gene trees, hearts, eyes and dorsoventral inversion

David K. Jacobs, Shannon E. Lee, Mike N. Dawson,
Joseph L. Staton and Kevin A. Raskoff

Department of Biology, University of California, 405 Hilgard Ave., Los Angeles, CA 90095-1606, USA

Summary

The initial surprise generated by the discovery of sequence-similar regulatory genes patterning the anterior/posterior axes of flies and mice has passed. Statements regarding the evolutionary implications of comparative developmental genetic data are now commonplace in the literature. However, despite the continued generation of developmental data, and citation of these data in reference to evolutionary issues, it is less than clear that there has been adequate integration of the developmental and evolutionary disciplines[1]. This is most evident in the language and examples chosen by developmental biologists when they turn to evolutionary themes. Here one finds frequent recourse to transcendentalism and typology. Thus, developmental workers seem to still hold Darwinism, with its unseemly aspects of natural selection, and the complexly branched evolutionary tree, at arm's length. Here, we examine some uses of typology and transcendentalism that pass for evolutionary analyses and apply phylogenetic methods to developmental genetic data. In particular, we use character tracing in phylogeny and gene trees of conserved homeobox regulators to examine the homology of hearts and eyes and the "inversion of the dorsoventral axis", topics of current interest in our rapidly increasing understanding of the evolution of metazoan development.

---

[1] How experimental embryology came to be so completely distinct from evolutionary studies is in itself an interesting question, treated in part by Maienschein (1991). It is thought that the dominance of Haeckel's biogenetic law, that development recapitulates the history of the evolutionary lineage (Haeckel, 1874), was influential. In this Haeckelian interpretation history of the lineage was the sole causal agent shaping development. No other causality need be sought. This perspective left no place for those who wished to explore causal links in the developmental process itself. Never mind that evolutionary workers were in the process of discarding Haeckel's biogenetic law (Garstang, 1922; Jacobs, 1992). The disciplines went their separate ways. Evolutionary biologists and experimental developmental biologists only now appear to be reestablishing the inherent links between their disciplines.

R. DeSalle, B. Schierwater (eds) Molecular Approaches to Ecology and Evolution
©1998, Birkhäuser Verlag Basel

Introduction

Most "evolutionary" analyses of developmental regulatory genes consist of a succession of comparisons and judgments of similarity of the compared entities. First, iterative comparsions of gene sequences result in alignment of sequence-similar molecules[2]. Aligned sequences can then be compared at the individual amino acid or nucleic acid positions along the molecule on the basis of identity of the amino acid or nucleic acid residue. In amino acids functional similarity can also be compared. Pairwise comparison of aligned sequences then allows the generation of a percent or frequency of similar or identical amino acid or nucleotide subunits in the molecule. These numbers can be used directly, or all pairwise similarities can be subjected to matrix operations resulting in grouping of the data on the basis of overall similarity. Alternatively and preferably, comparison can be made on the basis of each aligned position and an optimality criterion, such as parsimony, can be used to choose between tree topologies.

After sequence analyses, gene functions are assessed. In model organisms, such as flies, mice and nematodes, a wide variety of techniques can be applied. In these "systems" the conservation of complete sets of regulatory interactions involving cell-cell signaling, signal transduction from the cell surface to the nucleus and regulation of gene expression have been demonstrated. In flies, "rescue" experiments, where an exogenous gene from other taxa replaces the function of a deleted or mutated gene, are performed with increasing frequency. In less well studied organisms of interest in broader evolutionary analysis, comparison is largely restricted to expression of DNA-binding regulatory proteins, many of which contain a homeodomain. In such comparisons the structures expressing the gene are often hypothesized to derive from a structure in the shared ancestor of the two taxa (i.e. they are homologous).

Conservation of both sequence and gene expression or function can be used to support arguments of shared ancestry of morphologic features (Jacobs, 1994). For example, the expression of sequence-similar Hox genes in the same order along the anterio-posterior axes of flies and mice strongly supports preexisting hypotheses of shared ancestry of axial organization in these Bilateria (Duboule and Dolle, 1989). However, during the course of evolution, conserved protein motifs can also take on new roles. Signaling and DNA-binding proteins often have multiple discrete roles in development, suggesting a history of cooption of existing elements for new functions. One such cooptive event is the duplication of a suite of Hox genes in verte-

---

[2] Selection and alignment of sequences itself is a comparative process that usually involves the computer application of algorithms and optimization criteria often involving large data bases such as GenBank. In addition, in many cases sequences are aligned by eye. Clearly alignment algorithms and search algorithms are critical and provide a degree of objectivity. Regrettably, many workers fail to place their sequences in GenBank.

brates to specify positions along evolutionarily new axial structures, the limbs (Holland et al., 1992). This cooption for new roles of the same, or similar duplicate, regulatory molecules is the stuff of evolutionary change. Given this potential for assumption of new roles in development, expression studies that test clearly articulated hypotheses of morphologic evolution are preferred to those that *a posteriori* interpret or compare gene expression pattern in divergent taxa.

Developmental biologists often compare expression of sequence-similar genes between two divergent modern taxa and claim to have conducted an evolutionary analysis. However, such binary comparisons between modern taxa generate general statements of transformation that lack a precise interpretation of history. Such transformations are not evolutionary in that they don't explain change in a historical context, that is change over time from an ancestor to a descendent. Only through comparison with additional more distantly related "outgroup" taxa can the ancestral condition and direction of evolutionary change be inferred[3].

In terms of sequence, the use of the word *homology* as a proxy for *similarity* is problematic (Reeck et al., 1987). One often now sees the word *highly* modifying the word *homologous*. In the classic sense, homology is binary. Morphologic features are either homologous or not, and the use of modifiers to indicate degree of homology is meaningless. In fact, if one considers any amino acid or nucleic acid position that is identical in two aligned molecules, this identity can be the result of convergent evolution (homoplasy) as well as homology or shared ancestry. Given the constraints on evolution, even molecules that clearly share ancestry will often have numerous changes and reversions. Thus a certain fraction of the similarity observed is not necessarily due to shared ancestry in the sense of direct lineal descent of replicated DNA and is not homologous in that sense. Moreover, whole classes of molecules, such as homeodomains, presumably share ancestry, but have multiplied as a consequence of gene duplication. In the case of multiple sequence-similar genes, an explicit hypothesis of relationship among potentially relevant genes often needs to be developed. Such a hypotheses of relationship is most easily visualized as an evolutionary tree. In the absence of such an analysis, potentially related molecules may not be compared, and the evolutionary relationships between the molecules, their expression patterns, and the associated developmental processes and morphology may be misinterpreted.

---

[3] Taxa that are related to, but not members of, the group of interest are known as outgroups. For instance, one or more marsupials might provide a proper outgroup in an analysis of placental mammal phylogeny. Parsimony methods such as that of Fitch (1971) explicitly generate a hypothesis of the ancestral condition at each node in the reconstructed tree and thus make an explicit hypothesis of the character states present in the ancestor. This is done by recourse to the outgroups which permit determination of the ancestral or derived condition of character states at the succession of nodes in the tree [see Hillis et al. (1996) for a recent review of methodology]

Road map

In this work we explore the utility of phylogenetic methods in elucidating the evolutionary component of developmental genetic data. First we characterize the zootype as described by Slack et al. (1993) and contrast this with an analysis of development using an evolutionary tree. This leads to a discussion of the utility of gene trees (phylogenetic analyses of duplicated genes) for ascertaining the sequence of events in the evolution of development. These tree-based exercises are followed by consideration of the "master regulatory" gene *eyeless*. We then employ gene trees to examine the question of homology of the bilaterian heart and argue that this result is not as convincing as the argument for homology of eyes. This leads into a consideration of dorsal/ventral inversion arguments. Here we employ outgroups to constrain interpretation of the data and contrast this approach to reliance on pre-Darwinian transcendentalism.

Typology: the zootype vs. evolutionary trees

In their article entitled "The Zootype" Slack et al. (1993) discuss the concept of a suite of developmental genes characteristic of all animals. In so doing they focus on a singularity, a conceptual organism that has the shared molecular features of modern animal development. In this approach, as well as in the use of the term *zootype*, Slack et al. (1993) conform more to an Aristotelian typological or idealized approach than to a post-Darwinian, tree-based evolutionary perspective. In such a tree, multiple branches, not a single universal ancestor, are considered. The genes canonized as zootypical are primarily homeobox-containing genes. These genes are sequence-similar and are thought to have formed during a series of gene duplications or branching events. Clearly the 10 or so genes noted by Slack[4] were not acquired simultaneously; several duplication events were required to generate them. Thus assembly of the genetic program(s) controlling animal development involved a number, perhaps a very large number, of independent evolutionary events that occurred in the lineage leading to the group of animals under consideration. Subsuming these events under a single archaetypical entity obscures the details of this history. In particular, it obscures the succession of different organisms involved in, as well as the succession of gene duplication events necessary to generate, the developmental programs we observe today.

Given the branching nature of evolution, the vast majority of different kinds of organisms that have ever lived are extinct. Consequently, most branches along the

---

[4] If Slack et al. (1993) were to perform the same exercise today they would have to consider many more genes than originally examined.

stem of the tree of animals that might contain stages in the assembly of animal development are no longer available to be manipulated in the lab. Presumably this evolutionary assembly of the animal developmental program took place in the late Precambrian. How much time was involved is a matter of considerable discussion[5]. Given current interest in the early history of animal development and the fact that information is limited, we should use all the data available to us. This means we ought to: (i) consider the fossil record, and (ii) address all the nodes near the base of the animal tree for which we can reconstruct an ancestral condition. This will give a much clearer evolutionary picture than the compression of numerous nodes, where different developmental functions were gained, into a single "type" as was done by Slack et al. (1993).

Nodes, genes, and fossils

Although there is considerable argument about the relationships among animal phyla, there is reasonable agreement as to the coarsest level of animal relationships, as well as the relationship of animals with their immediate relatives (Fig. 1). Small ribosomal subunit DNA sequences suggest close relationships among the more complexly multicellular groups, the fungi, green plants and animals (Wainwright et al., 1993). However, morphologic and molecular data suggest that choanoflagellates are the sister taxon (closest relative) of the animals as a whole (Haeckel 1874; Wainwright et al., 1993). These ciliated, single-celled or colonial, organisms filter feed with a collar of microvilli and a flagellum in a fashion similar to the "collar" cells of sponges, the most basal branch among living animals. The next two branches above the sponges divide the radiates, the Ctenophora (comb jellies) and the Cnidaria (jellyfish, anemones, hydrozoans and corals), from the Bilateria. The Bilateria have bilateral symmetry at some stage in their life cycle. Ongoing disagreement as to the relationships amongst phyla is primarily focused within this bilaterian group.

In the fossil record of the late Precambrian, prior to a rapid radiation of animal phyla, a number of enigmatic forms that range from radial to bilateral in construction appear in the nearshore sedimentary rocks from around the world. These simple forms have limited internal structure, and generally lack evidence of a gut[6]. The details of their organization, their preservation and their evolutionary origin have

---

[5] Wray et al. (1996) recently argued for an ancient divergence of metazoa based on rate calculations involving sequence data. However, there are fundamental problems in extrapolating such data (Ayala et al., 1998). In addition trace fossil data strongly suggest that certain morphological features such as through guts did not appear until much later (Valentine, 1997).

[6] Recent reexamination of one vendazoan, *Kimberella,* suggests that it may indeed be a bilaterian (Fedonkin and Waggoner, 1997).

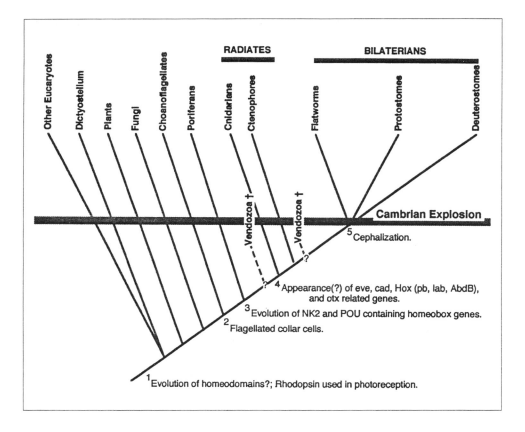

Figure 1. Basal relationships of the animals showing the choanoflagellate, fungal and green algal/plant outgroups. Note the uncertain placment of the Vendozoa, extinct forms with radiate and bilateral body plans that are found as fossils prior to the Cambrian radiation. The earliest appearance of features of developmental interest are noted at individual nodes on the tree. These include morphologic attributes as well as the first documented appearance of certain types of homeodomains.

been the subject of considerable discourse (Buss and Seilacher, 1994; Seilacher, 1988). However, these forms (vendozoans) are likely to represent extinct basal branches of the animal tree (Fig. 1), branching from one or more of the stems above the sponges but below the ctenophore, cnidarian and bilaterian crown groups of the Eumetazoan tree.

Considering the above information, the focus on a singularity suggested by the archaetypical analysis can be faulted from two additional perspectives. First, Slack et al. (1993) did not consider taxa more basal than Cnidaria. Yet sponges (Porifera) are generally considered animals, and have subsequently been found to contain homeobox genes of classes comparable to those found in the higher Metazoa. (Seimiya et al., 1994, 1997) (Figs 1 and 2). In addition, some animal homeodomains appear to be most closely related to fungal and plant homeodomains, suggesting still more basal antecedents of these genes in multicellular organisms[7]. Thus, even if one considers only the modern organisms and not the extinct fossil lineages basal on the tree of animal life, we should consider each of the living groups of organisms (plants, fungi, choanoflagellates, and radiates) that are related to the few, well-studied, bilaterian animals.

Sequence similarity and gene trees

The analysis of a newly recovered developmental molecule begins with an alignment. Often only sequences of a handful of genes known to be functionally similar are aligned. Until recently there may have been no more than a handful of molecules to compare. However, in some cases a huge richness of sequences are now available for comparison, e.g. homeodomain-containing genes. Given the developmental biologist's interest in function, there is a tendency to compare genes thought to have similar function in development. The breadth and complexity of available information suggests that the examination of such a small biased set of sequences may not be sufficient. Such selection of data precludes its use as an independent hypothesis of gene history on which the evolution of other aspects of gene function are mapped.

Once sequences are aligned, the analysis may extend no further than commentary on invariant amino acid positions that are unique to the group or subgroup of molecules in question. However, consideration of only those characters that uniquely diagnose a group excludes much of the phylogenetically useful information. In other cases trees are generated using sequence similarity. Such analyses involves pairwise comparison of all aligned sequences and the generation of matrices of similarities based on some coefficient of similarity or distance. Such coefficients can vary, but most simply are just the fraction of molecular units, aligned amino acid or nucleic acid, positions that are identical. These analyses usually employ classic dendrogram-generating programs such as UPGMA, where most similar sequences are grouped first and successively less similar sequences are chained on. However, analyses based on such simple similarity or distance matrices have their limitations. Distance or similarity methods accord changes at uninformative positions equal sta-

---

[7] Homeoboxes have now been described from the slime mold *Dictyostelium*. (Han and Firtel, 1998). Recent phylogenies based on elongation factor alpha suggest that the slime molds are a monophyletic clade related to the multicellular plant, animal, and fungal clades (Baldauf and Doolittle, 1997). These observations, taken at face value, suggest a correlation between the evolution of homeodomains and the evolution of multicellularity.

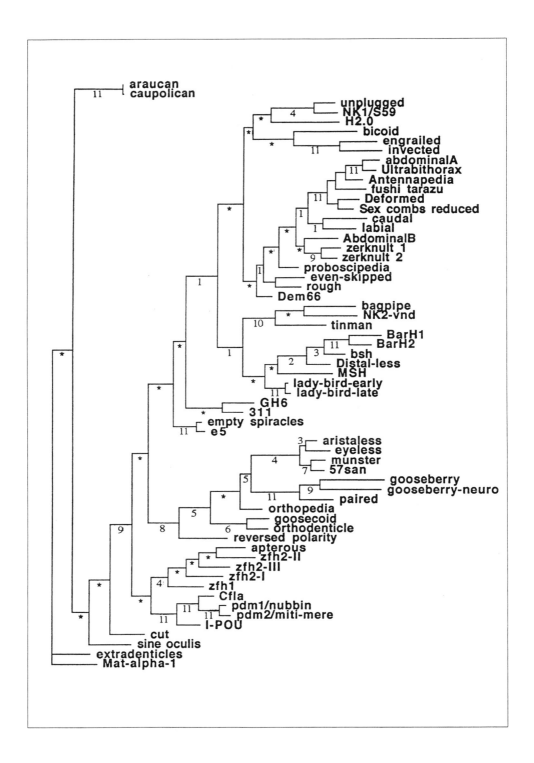

tus with those changes that actually document a shared evolutionary history. Only those instances where evolutionary change has generated new character states that are common to more than one descendant taxon, or sequence, have the power to resolve evolutionary history. For example, if there is rapid change of one sequence and it accumulates many novel character states, it will be less similar to the other sequences in the analyses and accorded a position near the base of the tree regardless of actual evolutionary relationship. These particular problems can be overcome by the use of parsimony methods to reconstruct evolutionary trees.

Reconstruction of an evolutionary tree using parsimony involves the examination of character changes rather than summation over all differences in pairwise comparisons. In parsimony analyses ancestral characters, and characters unique to particular taxa, do not generate different tree lengths given different tree topologies. Therefore they do not contribute to the selection of the shortest tree. In this regard these parsimony or cladistic tree reconstructions are preferable to similarity matrix-derived trees, which can be influenced by noninformative data.

We advocate the use of a particular parsimony method that includes both nucleic acid and amino acid data in the analysis (Agosti et al., 1996)[8]. Furthermore we argue for inclusion of available data where possible[9]. We present a parsimony analysis of *Drosophila* homeodomains as an illustration (Fig. 2). In the analyses present-

---

Figure 2. Gene tree of Drosophila *homeodomain-containing genes based on an analysis of first- and second-position nucleic acid data and amino acids from the homeodomain, paired-domain as well as the hexamer and octamer motifs are included in the analysis. Trees are the result of 500 heuristic searches using random addition sequences and tree bisection and reconnection in the program PAUP4.0d59. Branch length is proportional to the amount of character change. All these genes contain homeodomains, but many lack other motifs used in the analysis. Note that the gene* caudal, *which is not usually accorded Hox gene status due to its map position, is in the Hox gene clade on the basis of sequence. This kind of analysis has its uses as a first cut. It can be faulted for its lack of inclusion of genes from other taxa. Better resolution may be possible if more comprehensive analyses of individual classes of genes are performed as in Figure 4A, B. In addition the tree can be faulted for lack of inclusion of sufficient outgroup sequence. Mat-alpha-1, a yeast gene, is the only outgroup. Subsequent analyses suggest that some plant and fungal homeodomains fall inside the basal branches in the tree (cut, exd, sine oculis and POU-containing genes. Numbers adjacent to branches are decay indices. They document the number of steps longer a tree will have to be before the branch or node delimiting a certain set of taxa is no longer found in all trees recovered at that increased length. Low numbers indicate little support. As numbers increase above 2 they suggest independent character confirmation of the topology. Topologies that are still present in all trees half a dozen steps or so longer are very well supported. Decay indices were generated using successive sets of heuristic searches where trees of increasing length were obtained. In the analysis, more than one tree of minimal length was recovered. Nodes with no numbers are found in all minimal-length trees, but not in trees one step longer. Nodes with an asterisk were not present in all minimal-length trees.*

ed in this chapter, first and second nucleic acid positions of each codon as well as amino acids are included. First and second codon positions are critical in that they reflect the broader functional categories of amino acids[10]. There are additional reasons to use a parsimony (cladistic) approach rather than maximum-likelihood or other nucleotide-based approaches[11]. Parsimony performs well in circumstances where sets of character states are absent for suites of the taxa in the analysis. A point applicable to our homeodomain data: they all share one conserved motif, but only a subset of them share the additional conserved sequence motifs used in the analysis.

It is important to analyze relatively comprehensive sets of molecules if the relationships among them are to be adequately assessed. First, if closely related se-

---

[8] Many methods for generating phylogenetic trees are designed for application with ribosomal RNAs. Such sequences suffer from long-branch attraction. There are only four nucleic acid character states, so reversion to an original character state can occur relatively easily. In these circumstances, if there are differences in the rate of change in the molecules, the molecules with the most change will have the most reversions and tend to attract each other. Hence, the term *long-branch attraction*. It has been documented that simple distance methods perform the worst in this kind of circumstance. Parsimony methods also suffer. There are a range of additional methods that require various assumptions that appear to perform better under such circumstances. However, this problem and the analytical methods to correct for it are primarily a consequence of four-character-state nucleic acid data with the possibility of frequent reversion. In the analyses we perform, the inclusion of amino acid data helps buffer against long-branch attraction.

[9] Surprisingly, many initial reports of genes do not include a comprehensive search for sequence-similar genes.

[10] It is important to note that changes in amino acids result from change at either of the included nucleotide positions or at the uninduced third position. Change in either included nucleotide character from a codon will not map identically to *all* the changes in an individual amino acid character, as many amino acid changes may occur as a consequence of change in the nucleotide positions other than the one under consideration. Hence, there is a degree of independent character evolution, as well as a degree of character weighting, in such a scheme (Agosti et al., 1996) By excluding third positions, only those third-position changes that result in amino acid change, all of which are transversions rather than transitions, are acknowledged by the analytic approach, and they have a lower weight in the analysis relative to first- and second-position changes that change amino acids. This is consistent with the general functional similarity of amino acids that differ only in third position.

[11] The utility of parsimony has been demonstrated in the reconstruction of phylogenies of fossil and modern organisms where molecular data is only available for the modern organisms but morphological characters are available for both the fossil and modern taxa (Wheeler et al., 1993; Eernisse and Kluge, 1993). This is referred to as a "total evidence analysis" and is applicable in this context where all molecules share one conserved motif, whereas some of the genes have additional conserved regions not shared by all the taxa.

quences are excluded from the analysis, the detailed relationships between sequences will not be resolved. For example, previous analyses of Hox genes were designed to assess which of these genes are present in the Cnidaria, and predate the evolution of Bilateria; however, *caudal* was not included in these analyses, as *caudal* is not "linked" on a chromosome (e.g. Schubert et al., 1993). In our tree, the gene *caudal* is shown to group within the Hom/Hox cluster of genes (Fig. 2). Thus, our result suggests that some of the Cnidarian genes, previously identified as homologues of other Hox genes, may be related to *caudal* (unpublished data). The Hom/Hox genes are of general interest due to the parallel anterior/posterior pattern of expression and map position on the chromosome (Duboule and Dole, 1989). However, the gene trees document that *caudal*, although it does not map to the same chromosomal region, is a Hox gene by descent, an interpretation supported by the similar posterior pattern of expression in flies and mice (Gamer and Wright, 1993).

The above analysis involving *caudal* suggests exclusion of data based on *a priori* notions of relationship can have critical consequences for the analysis. It is also critical to include a range of sequences of presumed outgroups to the taxa. Inclusion of the most closely related sequences to the group in question provides the best polarization of the characters in the analysis[3]. If the outgroups are too few or too distantly related, they will not adequately constrain the set of character states possible at the base of the tree, optimal determination of which character states are shared and derived will not be possible, and the most informative tree topology may not be generated. In addition, inclusion of related sequences helps minimize the chance of accidental exclusion of ingroup taxa.

Given that only a small fraction of the relevant sequence information may have been discovered, avoidance of *a priori* biases based on expression data seems essential. This is especially true if sequence relationships are viewed as a test of such functional data. The best available hypothesis of relationship will likely result from continuous revision as more data become available. Some of the issues in tree reconstruction are evident in our analysis of *Drosophila* homeodomains (Fig. 2).

Master regulators

Several classes of information can be used to assess the similarity of gene product and function in disparate taxa. The genes themselves can be sequence-similar, especially as regards the conservation of functional protein-coding regions. Similarity of function can involve sequence similarity of functional components of molecules, such as DNA-binding motifs or transmembrane elements. Functional similarity can also be seen in the sets of interactions between suites of molecules involved in various aspects of the regulatory hierarchies controlling development. Furthermore, genes can play similar roles in development as determined by comparison of wild-type expression pattern and mutant phenotypes. Ultimately, in some cases, the mutant phenotype in one taxon can be rescued by the introduction of the gene product of the other taxon. All this, and more, has been demonstrated for the *eyeless* gene of *Drosophila* and its sequence-similar homologue, *Pax-6*, of mice (Callaerts et al., 1997; Halder et al., 1995). These genes contain similar homeodomain and paired-domain DNA-binding motifs, and both are required for eye development in the respective taxa. These similarities, in combination with additional results from flatworms and molluscs (Callaerts et al., 1997), have been used to argue for the homology of eyes, or at least light sensory structures, throughout the bilaterian animals.

Documentation of eye homology was quite a coup in that previous interpretations of eye evolution presumed multiple independent evolution of eyes in a number of animal lineages (Salvini-Plawen and Mayr, 1977). Yet the claims do not end there. It is also claimed that the same gene has the same station as the "master regulator", the ultimate arbiter of eyes, a claim supported by the generation of reasonably complete eye structures on antennae and limbs by the overexpression of the *eyeless* and *Pax-6* genes in *Drosophila* (Halder et al., 1995). It is argued that all other genes explicitly required for eye development such as *sine oculis*[12] operate downstream of *eyeless/Pax-6* and have a lesser status, as they are not master regulatory switches (Callaerts et al., 1997).

The *eyeless* result, and the interest in it, reveals as much about the thinking of developmentalists as it reveals about the evolution of eyes. The canonization of the *eyeless* gene as a "master regulator" is based on its "upstream" position. Other genes, such as *sine oculis*, are also required for eye development in both *Drosophila* and mouse (Cheyette et al., 1994; Oliver et al., 1995). The emphasis on *eyeless* may relate, in part, to the elegance of the experimentation. However, one also suspects that there is an implicit assumption that genes acting early in development must have evolved earlier (or be more conservative) because they occupy an upstream position in a set of regulatory interactions. Most developmental biologists perceive their role as unraveling the causal processes that result in development of the organism. In this context it may seem "natural" that upstream regulators, or more ultimate causes, be more conserved. However, this presumes that evolution acts by terminal additions to causal chains of interactions in development.

In many cases, evolution has not treated ultimate causes with special respect; new regulatory elements can appear upstream, downstream or in the middle of regulatory causal chains[13]. This point has been subsequently documented by work on the *Drosophila* gene *twin of eyeless*, which acts upstream of *eyeless* and is very similar in sequence to *eyeless*. Thus the "master" has undergone gene duplication and has a copy of itself (extramaster?) regulating it (Czerny et al., 1997). Clearly this

---

[12] *sine oculis* also has a vertebrate homologue, *Six-3* (Oliver et al. 1995).

suggests evolution of the early steps in the regulatory hierarchy controlling eye development. In addition, gene tree analyses that include *sine oculis* and *eyeless* (Fig. 2) suggest that *sine oculis* may have evolved earlier than *eyeless*. However, the functional evolution of these two molecules has yet to be constrained by data from organisms more basally branching than Bilateria. Thus, we argue that any expectation, or claim, that events earlier in causal chains are more evolutionarily important or more evolutionarily conserved than later events needs to be treated with caution until demonstrated empirically.

The expression of sequence-similar regulators of eye development in the eyes of *Drosophila* and vertebrates demonstrates shared ancestry of some aspects of visual sensory systems, entities that differ in their terminal differentiation and final optical structure. The distinct mechanisms of light gathering and image formation in compound eyes of *Drosophila* and the single-lens eyes of vertebrates seem quite different. Consequently, a number of workers have suggested that an eye spot, not a fully derived eye, is the shared ancestral condition. It has recently been determined that *eyeless* is a direct positive regulator of the visual pigment rhodopsin (Sheng et al., 1997), and suggested that this regulation of rhodopsin by *eyeless* was the ancestral shared function of the gene. In this interpretation *eyeless* would have been involved in photopigment production in a minimal ancestral metazoan "eye". Other aspects of eyes would then have evolved in parallel. However, genes besides *eyeless* such as *sine oculis* are also necessary for eye formation across a range of Bilateria. Thus, complex regulatory interactions in eye development, not just a simple relationship between a regulator and a photoreceptor, must have evolved prior to the diversification of the bilaterian taxa.

Two additional observations suggest that some component of the eye may have evolved prior to the diversification of the Bilateria. One is the eye, or eyelike structures, found in the medusoid forms of Cnidaria. In this regard the lens-containing eyes of the Cubomedusae are the particular focus of attention[14]. Second, it has re-

---

[13]This assumption about the evolution of causal chains closely parallels Haeckel's biogenetic law regarding the terminal addition of stages in development, which has long since been discarded in the light of the obvious evolution of early developmental stages (Garstang, 1922; DeBeer, 1940; Gould, 1977; see also Jacobs, 1992). Some of the recent work of Greg Wray and Rudy Raff (Wray and Raff, 1990; Raff, 1992) demonstrates the rapid and radical evolution of early development in the echinoderm genus *Heliocidaris* which strongly speaks to this point.

[14]With current knowledge, it appears that some aspects of eyes are derived from shared ancestry. The question, then, is how much of eye development is a shared ancestral feature and how much is derived individual lineages. This will vary depending on the lineages in question. So intriguing problems remain. Cubomedusae (Cnidaria), vertebrates and squids (Mollusca) have eyes with lenses. Are lenses, as well as photoreceptor function and/or innervation, homologous? *Nautilus*, the other living lineage of cephalopods, has pinhole eyes. Are pinhole eyes derived or ancestral in the cephalopods?

cently been ascertained that the motile sperm of fungi use rhodopsin in phototaxis (Saranak and Foster, 1997). Thus, some aspect of genes regulating rhodopsin in light sensory structures could conceivably precede the evolution of Metazoa (Fig. 1).

Hearts and eyes: Are hearts homologous?

Recent arguments for homology of eyes throughout the Bilateria are based on the similarity of homodomain and paired-domain sequence in the fly gene *eyeless* and genes referred to as *Pax-6* in other taxa (Callaerts et al., 1997). Parsimony analyses demonstrate the close relationship of these genes relative to other paired-like or homeobox-containing genes (Fig. 4A). This, in conjunction with other evidence, on the distribution of eyes in other taxa, reviewed above, and the gene tree analysis of pairedlike genes (including *eyeless/Pax-6*) discussed below, makes a convincing case for shared ancestry of eyes in bilaterian Metazoa. It has also been argued on the basis of similar function of sequence-similar regulators in heart development that hearts are homologous (e.g. Harvey, 1996). However, outgroup comparison and gene tree analysis do not support the homology of hearts in the same compelling way as similar analyses support the homology of eyes (Fig. 4B).

In terms of morphological considerations, arthropod circulatory systems are quite distinct from those of vertebrates. Vertebrates have a closed circulatory system; arteries, veins and capillaries – confine the circulation throughout its course through the body. The arthropods have an open circulatory system; the coelomic fluid is pumped to one end of the body and flows back through the coelomic cavity. The arthropod heart lacks the endothelial structures present in the vertebrate heart, and is in a dorsal position relative to the nerve chord and guts, as it is in other protostomes. The vertebrate heart is ventral. Thus, historically, there was little obvious morphologic basis to homologize these structures, other than a very general functional similarity. Shared ancestry of arthropod and vertebrate hearts was not anticipated on morphologic grounds (Fig. 3B); consequently, the potential homology of hearts is perhaps as exciting as the shared ancestry of eyes indicated by *eyeless* and *sine oculis*.

Hearts and eyes traced as characters on phylogeny

All the major clades of Bilateria contain representative taxa with eyes (Fig. 3). The Cnidaria, among the Radiates, also have eyelike structures (Salvini-Plawen and Mayr, 1977; Brusca and Brusca, 1990). Traditionally, flatworms are considered a basal group in the Bilateria, either as an outgroup to the protostomes and deuterostomes or near the base of the protostome clade (Eernisse et al., 1992). Free-living flatworms have well-developed eyespots that express the *eyeless* gene (Callaerts et

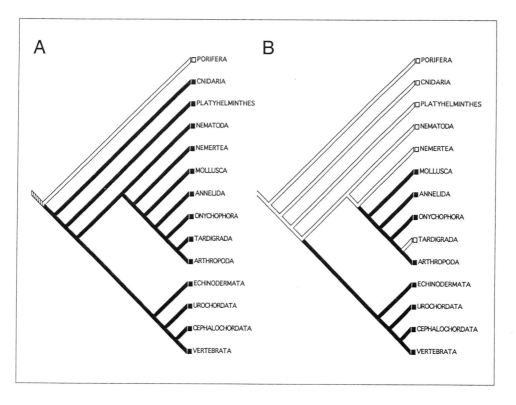

Figure 3. The evolution of eyes and hearts traced on a metazoan tree. (A) Presence or absence of eyelike innervated photoreceptors on a hypothesized set of metazoan relationships. Tracing of character evolution supports the presence of eyes in the ancestor of Bilateria. (B) Presence or absence of muscular heartlike circulatory system pumps mapped on the same hypothesized set of metazoan relationships. Note that the presence of a heart in the basal bilaterian is equivocal. Character states for phyla possessing heart structures were traced onto this tree using MacClade (ver 3.06, 1996). Basic tree topology and information on the distribution of heart and eyes are from Brusca and Brusca (1990). Note that although the most basal relationships of animals are generally agreed on (Fig. 1), many aspects of the relationships of the bilaterian phyla are controversial. Thus many of the details of the tree topology may vary between analyses. In addtion, even on fully resolved trees reconstruction of character evolution is often ambiguous in detail.

al., 1997). Other taxa often thought to represent basal branches of bilaterian groups such as nemerteans, nematodes and amphioxus all have eyes or eyespots. Thus eyes, or at least innervated light sensory structures of some type, occur in the outgroup to the Bilateria, the Cnidaria and in all the major Bilaterian phyla. Thus, given the above evidence, and recognizing all innervated light sensory structures as eyes, it becomes very difficult not to reconstruct the basal bilaterian with some type of eye

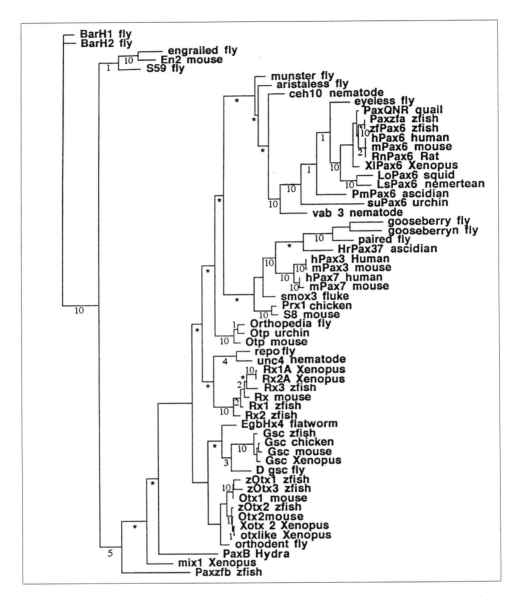

Figure 4. Parsimony analyses of the pairedlike clade of homeodomains (A) and the Nk2 homeobox-containing genes (B) conducted as described in Figure 2. In (A), the paired-like
gene analysis, based on the homeobox, paired-domain and octamer, note the very strong
decay indices for the eyeless/Pax-6 clade. In addition, note that several other clades of homeodomains MSH, RX, OTX and OTP are strongly supported by the decay indices. In (B),
the Nk2 analysis, based on homeodomain and Nk2-domain sequence, the bagpipe and putative vertebrate homologues form a clade that is poorly supported and that Nk2/vnd and
vertebrate Nk2.2 genes as well as some sequences from nematode and flatworms form a

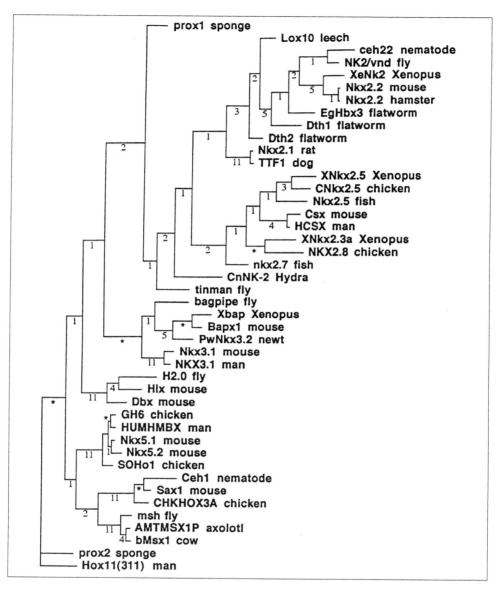

clade with some support. However, the vertebrate Nk2.2 and vnd genes are the only genes where obvious comparisons of expression pattern can be made (see text). In addition, although the vertebrate genes form a clade, the tinman gene is not the immediate sister taxon to this clade. Also, note the strong support for some other groups in the base of the tree, for example the MSH group. The H6 genes are strongly supported in a close relationship with the Nkx 5 genes, a relationship not noted by the papers reporting either group of genes. Sources of sequences are listed in Appendix 1.

(Fig. 3A)[15]. No such clear-cut case is evident when a muscular pumping heart is considered. Many of the groups thought to branch basally within the Bilateria lack muscles specific to a circulatory system; that is, they lack hearts. Cnidaria and flatworms lack circulatory systems. Amphioxus, the sister taxon to the vertebrates among the chordates, lacks a discrete muscular heart. Nemerteans are usually reconstructed near the base of the protostome clade, or in some cases reconstructed basal to the protostome deuterostome split (Eernisse et al., 1992). Nemerteans have a circulating coelomic fluid, but the fluid is moved by the contraction of the body muscles. There is no pumping heart *per se* (Brusca and Brusca, 1990). Nematodes, often considered basal in the Bilateria [but placed near the arthropods in the protostomes in recent analyses (Aguinaldo et al., 1997)], do not have a muscular heart. Given the lack of hearts, in the outgroup as well as in many of the taxa often reconstructed as basal within the Bilateria, it becomes more difficult to reconstruct the basal bilaterian node with a heart than an eye (Fig. 3B). Thus, this evidence does not strongly support homology of hearts in flies and mice, although it certainly does not preclude it.

eyeless/Pax-6 gene trees

Having examined the taxonomic distribution of hearts and eyes and traced these organs as single characters on a tree, we can consider the relationship of genes that function in the development of eyes and hearts. Genes sequence-similar to the *eyeless* gene of *Drosophila* have now been recovered from a range of bilaterian taxa. These form a strongly supported clade within an additional well-supported clade of homeodomain genes often referred to as pairedlike genes (Fig. 1). This *eyeless/Pax-6* clade contains genes exclusively involved in eye development[16]. This behavior of *eyeless* in combination with the presence of other conserved regulators such as *sine oculis* in *Drosophila* and sequence-similar genes in mice provide strong support for

---

[15] Salvini-Plawen and Mayr (1977) treated Metazoan eye evolution, but did not consider eyes or their shared features as single characters. They came to the view that eyes evolved independently in a large number of metazoan lineages. In their work they especially noted the ultrastructural differences in photoreceptor cells and argued, primarily on the basis of these cytological differences, for the absence of shared ancestry. However, negative arguments are especially treacherous. It seems that photoreceptor cell morphology must be more evolutionarily labile than Salvini and Mayr assumed. In addition, these authors did not use a consistent or comprehensive phylogenetic approach. They did argue for the absence of eyes in a molluscan ancestor on the grounds that several taxa within the Mollusca including scaphopod, monoplacophoran and aplacophoran taxa lack eyes. However, these taxa live buried in the sediment and/or live primarily in the deep sea, environments where a secondary loss of eyes would be expected and has occurred in numerous other lineages.

[16] Although the genes are involved in eye development, they also are likely to have other func-

homology of a developmental regulatory mechanism generating eyes, and for the eyes themselves in *Bilateria*[17].

Heart genes

With respect to hearts, the relevant *Drosophila* homeodomain is *tinman*, a gene involved in the differentiation of heart muscle and the development of the *Drosophila* heart (Bodmer, 1993; Azpiazu and Frasch, 1993). A putative homologue of *tinman*, *Csx/NK2.5* (e.g. Komuro and Izumo, 1993), was identified in the development of the murine heart. Subsequently, two additional NK2-type genes have been identified that function in heart development (Harvey, 1996). In *Drosophila*, three genes form a closely related *NK2* group of homeobox-containing genes (Fig. 2). Two of these genes, *tinman* and *bagpipe*, are involved in differentiating muscles in particular, and the development of mesoderm more generally. The third *NK2*-type gene in *Drosophila*, *ventral nervous system deformed (vnd)*, functions in the development of the ventral nervous system (Jiménez et al., 1995).

Of the dozen or more NK2-class genes characterized from vertebrates, a minimum of seven genes appear to have discrete functions in gnathostome vertebrates. These genes are often compared or grouped with one of the three *Drosophila* genes. However, it should be noted that *NK2*-type genes are known from sponges, cnidarians, flatworms, leeches and nematodes (Fig. 1), so clearly as a class they evolved coincident with, or prior to, the evolution of the Metazoa[18]. Thus the initial evolution of NK2 genes preceeds the evolution of many of the uniquely bilaterian structures in which genes in this group have developmental functions.

In our phylogenetic analysis of NK2 genes we come to the conclusion that there is minimal support for many of the relationships within the NK2 clade, including lack of support for a clade exclusively containing *tinman* and its putative vertebrate homologues (Fig. 4B). Thus, the sequences of these genes have yet to provide an in-

---

tions in the development of the nervous system. In the case of *Drosophila* there appear to be two genes, *eyeless* and *twin of eyeless* , that regulate each other in eye development. Vertebrates appear to have only one copy of the gene. This is unusual. Vertebrates frequently have multiple copies of single-copy *Drosophila* genes rather than the converse.

[17]Although eyes are present in some Cnidaria, a relationship to bilaterian eyes has not been demonstrated on the basis of these developmental molecules.

[18]Unlike photoreception, one of the primary functions of the *eyeless* gene, it is hard to imagine retention of a premetazoan function of an *NK2* gene, as they are involved in differentiating tissues thought to be unique to Bilateria. However, in sponges *prox 1*, an *NK2*-like gene, is expressed throughout all the cells during differentiation of the gemmule (Seimiya et al., 1994). Thus, it should be borne in mind that functions of these genes must be highly modified from those in these more distant outgroups which lack any of the structures under consideration in flies and mice.

dependent test of shared ancestry of heart gene function, and do not permit the same level of certainty of interpretation that is available for the *eyeless* gene. If we consider the *Drosophila* heart gene, *tinman*, different combinations of outgroups influence its placement relative to the suite of NK2 genes involved in development of the vertebrate heart. In most cases *tinman* is basal within the general group, outside both the group of vertebrate heart genes and a range of additional genes including the clade containing *vnd* (Fig. 4). Analyses employing distance rather than parsimony techniques generate a similar result (Harvey, 1996; Boettger et al., 1997). Analyses with different outgroups place *tinman* in a closer relationship to the vertebrate heart genes, but as an immediate sister taxon to the *CnNK2* gene, a gene expressed in the endoderm of the developing foot region of *Hydra*, a cnidarian (Grens et al., 1996). Thus, *tinman*, the *Drosophila* heart gene, is not strongly supported in its sister taxon relationship to the heart genes in vertebrates (Fig. 4)[19].

Given the lack of strong support of a relationship between *tinman* and the vertebrate heart genes from the sequence relationships (Fig. 4B), the presumed shared ancestry of *tinman* and the vertebrate heart genes rests to a larger degree on similarity of gene function. However, here again *tinman* differs significantly from the patterns observed in the NK2 gene functions in the development of the vertebrate heart. There are a minimum of three NK2 genes (NK2.3, 2.5 and 2.7) involved in patterning the vertebrate heart (Buchberger et al., 1996; Evans et al., 1995; Lee et al., 1996). *tinman* has additional functions not shared with any of these vertebrate heart genes. One such function of *tinman* early in development is an interaction with *bagpipe* to differentiate dorsal mesoderm in *Drosophila* development. *bagpipe* expression becomes restricted to a segmental registry of cells giving rise to the visceral mesoderm and ultimately the gut muscles (Azpiazu and Frasch, 1993). *tinman* becomes restricted to the most dorsal mesoderm and ultimately gives rise to the heart. No such pattern of regulation between the vertebrate heart genes and presumed vertebrate *bagpipe* homologues has been reported. In addition, *tinman* mutants are deficient in certain glial cells, indicating either a direct or an indirect function in nervous system development (Gorczyca et al., 1994).

The individual vertebrate NK2 heart genes (NK2.3, 2.5 and 2.7) do not appear to play the same critical role in early development that *tinman* does. With respect to the heart, the loss of *tinman* function results in complete loss of the heart. Such complete loss is not observed in the single knockout of the murine heart, NKx2.5 (Lyons

---

[19]The *NK2* genes as a class tend to have three conserved motifs: a 5' sequence similar to conserved octomer motifs found in a number of homeobox-containing developmental genes, a homeodomain and a 3' region that is found exclusively in this NK2 homeodomain-containing group (Price et al., 1992; Harvey, 1996). Although *bagpipe*, *vnd* and the vertebrate heart genes have the NK2-specific motif, *tinman* lacks this domain. Other workers have noted the absence of this motif in *tinman* and have concluded that *tinman* is degenerate in some way (Harvey, 1996).

et al., 1995), although, knockouts of all three genes have yet to be performed. The interactions of *bagpipe* and *tinman,* involved in early mesodermal patterning, are not evident in the putative vertebrate homologues of *tinman.* At the moment *tinman* is the only NK2 gene with a known heart-related function outside the vertebrates. Without more information from additional protostomes, such as polychaete annelids that have extremely well developed multiple hearts, it will be hard to assess which of the many *tinman* gene functions is ancestral and which may be more derived

If we turn our attention to expression patterns in the genes showing a sequence relationship to the *Drosophila* NK2 gene *vnd*, we find that these sequence-similar genes in other animals do not necessarily share aspects of the pattern of expression of *vnd* in *Drosophila. vnd* is required for ventral nervous system development, and genes in this general group are often referred to loosely as "nervous system-related". A gene, NK2.2, related to *vnd* is known from vertebrates (Guazzi et al., 1990), two aditional genes are known from flatworms (Garcia Fernàndez et al., 1993), and one each from nematodes (Okkema and Fire, 1994) and leech (Nardelli-Haefliger and Shankland, 1993). Of these genes, only the vertebrate gene, and potentially the leech gene, have "nerve-related" expression patterns[20].

An aspect of the *bagpipe* gene function may be conserved. The developing midgut musculature of *Drosophila* expresses *bagpipe*, and vertebrate gut musculature expresses a putative *bagpipe* homologue (Azpiazu and Frasch, 1993) during development. However, the earlier role of *bagpipe* in patterning the mesoderm found in *Drosophila* is not evident in vertebrate genes, such as the *Xbap* gene in "the frog" (Newman et al., 1997).

Taking the expression of the NK2 genes as a whole, many of the genes have important roles in patterning mesoderm. For example, two of the *Drosophila* genes,

---

[20]The vertebrate Nk2.2 genes are similar in nervous system expression to *vnd* (Price et al., 1992; Saha et al., 1993; Barth and Wilson, 1995; D'Alessio and Frasch, 1996; Hartigan and Rubenstein, 1996), and the leech gene *Lox 10* appears to have discrete expression patterns in the nervous system. However, the leech gene also has an endodermal function (Nardelli-Haefliger and Shankland, 1993), and the nematode gene *ceh22* functions in the pharynx and in the differentiation of gut-related muscle (Okkema and Fire, 1994), a role that seems more like *bagpipe* than *vnd*. The expression of sequence-similar genes in developing mesoderm and endoderm of the turbellarian flatworm *Dugesiat* further confounds the "neural" interpretation of this group of genes (Garcia-Fernàndez et al., 1993). Thus, the ancestral nervous system function of this group of genes is not confirmed by current knowledge of the expression pattern in one possible outgroup, the flatworms, despite the fact that flatworms have a well-developed nervous organization. The other vertebrate gene located in this part of the tree (Fig. 4B), Nk2.1/TTF1, is involved in differentiation of the thyroid and in its endocrine function, as well as in the expression of surfactant in the lung (Minoo et al., 1995, 1997). Cephalochordates have an endostyle, considered a possible homologue of the thyroid.

bagpipe and *tinman*, as well as the *ceh22* gene of nematodes, are directly involved in the regulatory cascade differentiating particular muscle types in development (Bodmer, 1993; Okkema and Fire, 1994)[21]. However, the NK2 genes appear to also play a wide range of roles in the differentiation of a complex suite of tissues not of mesodermal origin, or muscular in nature. Furthermore, NK2 genes have an ancestry that extends beyond the Eumetazoa back to at least the sponges (Seimeya et al., 1994) (Fig. 1). Thus, the gene family did not arise in the context of creatures with ectoderm, mesoderm, and endoderm and differentiated sets of muscles[17]. Given this evidence for change in gene function in the NK2 family, and the lack of strong support for many of the nodes in the NK2 clade, a range of evolutionary scenarios seem possible. These include:

1. Conservation of heart gene function from a gene ancestral to *tinman* and the vertebrate heart genes (NK2.3, 2.5, 2.7). This scenario requires divergent evolution of the non-heart-related aspects of *tinman* gene function, either the gain of a mesodermal patterning function in the *Drosophila* lineage or its loss in the vertebrate lineage, as well as duplication and divergence of roles in the three vertebrate heart genes. In addition, conservation of hearts postulates a heart-bearing ancestor of protostomes and deuterostomes which would require complex multiple loss of hearts in bilaterian evolution.

2. Independent derivation of hearts in separate protostome and deuterostome lineages. In this scenario, mesodermal tissue already expressing somewhat derived, but related NK2 genes would independently evolve details of circulatory system structure in protostomes and deuterostomes. Along these lines NK2 genes involved in muscle differentiation could have been independently coopted to specify heart muscle.

3. It is also possible that there has been a complex pattern of evolutionary gain and partial loss of heart function, but that the regulatory elements are all derived from a single ancestral gene with a heart gene function. This might explain the loss of the NK2 domain in the *tinman* gene.

It does seem relatively clear, from the tree analysis, that the NK2 genes that function in the development of the vertebrate heart are a product of gene duplication prior to the divergence of Osteicthyes and the lineage leading to tetrapods. To date, only one NK2 gene expressed in the heart, *tinman* from *Drosophila*, is known outside the vertebrates (Fig. 4B). A better sampling of NK2 genes from additional metazoa

---

[21]These genes interact with a set of myogenic genes that involve *bhlh* gene and *mef2*, a mads box regulator, that are characteristically involved in determining muscle cell lineages (Okkema and Fire, 1994; Olsen, 1997).

may help clarify the situation both in terms of the relationships among sequences and the evolution of expression patterns[22]. In comparison to the situation with *eyeless and* eyes, the documentation for homology of hearts in deuterostomes is more limited in terms of support from tracing of the morphologic feature itself, the strength of support for the relationships based on sequence analysis and the degree of consistency in expression pattern of the genes in the NK2 family. Thus although we cannot claim that hearts are definitively not homologous, the evidence assembled to date for the homology of bilaterian hearts is not as strong as that assembled for bilaterian eyes based on *eyeless/Pax-6*, among other genes.

Transcendentalism vs. the tree in arguments of dorsoventral inversion

Demonstration of homology of individual organs, such as "the eye", is of obvious critical interest in understanding the evolution of development and the evolution of animal form. Comparable excitement has accompanied initial observations of sequence-similar regulators expressed in inverse dorsoventral order in the development of flies and vertebrates. These observations have led to the claim that there has been an evolutionary inversion of the dorsoventral axis in the two different groups (De Robertis and Sasai, 1996; Bier, 1997). The regulatory molecules involved include *Drosophila decapentaplegic* (*dpp*) and vertebrate Bone Morphogenic Protein (BMP-4). BMP-4 is expressed ventrally in vertebrates and *dpp* is expressed dorsally in *Drosophila* (De Robertis and Sasai, 1996; Bier, 1997), precluding the conversion of ectoderm to neurectodermal fates in these portions of the embryo (Bier, 1997). *Short gastrulation (sog)* in the ventral region of *Drosophila*, and *chordin* in the dorsal region of vertebrates, inhibits the expression of BMP4/*dpp* in these regions, permitting the formation of neurectoderm (François and Bier, 1995; Bier, 1997). There is strong support for the interpretation that these molecules had a common ancestral function, that they are expressed in an inverted pattern along the dorsoventral axis of insects and vertebrates, and that they have similar functions in determining neural and neurectodermal fates. However, the evolutionary interpretation of these observations may not be as obvious as it appears to be at first glance.

It has been argued that molecular evidence for inverse dorsoventral patterning of flies and vertebrates "vindicates the hypothesis" first proposed by Geoffroy Saint-

---

[22] It is also possible that this region of this molecule could be better aligned or does not contain sufficient or appropriate information to resolve these issues or that the information in the NK2 and homeodomain information is in conflict. These interpretations can be further tested. Several clades at the base of the tree are well supported, including a strong relationship between H6 and NK5 homeodomains, a result not previously recognized. This resolution in adjacent clades suggests that it may be possible to resolve the relationships of the NK2 genes with additional data.

Hilaire (De Robertis and Sasai, 1996). Whether one refers to Geoffroy as having an antecedent and related idea is perhaps a matter of taste or style. However, to refer to "vindication of hypothesis" implies a much fuller endorsement of a precise mechanism of change. Geoffroy extended the typological or idealized aspect of comparative morphology of his day to include connections relating strikingly different forms, or archaetypes[23]. (Appel, 1987). Although Geoffroy was a contemporary of Lamarck and Buffon, who had explicit pre-Darwinian evolutionary ideas, he did not invoke these ideas of change over time in relating forms of vertebrates and protostome taxa. Appel (1987) argues that Geoffroy had a teratological mechanism in mind that would transform between different types in a single generation[24]. Is such a literal inversion a reasonable interpretation of the evidence?

Many organisms spend much of their lives in a dorsoventrally inverted position relative to *terra firma*. Horseshoe crabs swim upside down[25]; sloths hang upside down from trees. Flatfish lie on their sides and various structures, such as eyes, migrate during development to accommodate this new relationship between the sea bed and the side of the fish. Thus, change in behavior of the organism could lead to a relatively rapid evolution of a new organization. However, there are other differences between protostomes and deuterostomes that cannot be accommodated by the relatively trivial exercise of turning the organism over.

The gut of protostomes is dorsal to the nerve cord, while the gut of vertebrates is ventral to the nerve cord. Hearts are dorsal and ventral to the protostome and

---

[23] Geoffroy Saint-Hillaire endorsed, embellished or developed a huge range of transformational ideas (Appel, 1987). One of his more famous ideas involved the transformation of molluscs, rather than arthropods, to vertebrates. Here the cuttlebone or gladius of the cuttlefish or squid is "homologized" with the vertebral column. In this case the structure in question is a derivative of the molluscan shell. In cephalopods, and originally in the monoplacophoran ancestors of cephalopods, the shell is dorsal. This contrasts with the comparisons of ventral nervous organization of arthropods and the dorsal nerve cord in the vertebral column. In addition, Geoffroy endorsed the comparison or "homology" of the vertebral bones and the insect exoskeleton. Thus, Geoffroy endorsed at various times upsidedown, rightside-up and inside-out transformations. Ultimately, it was a comparison of vertebrates and cephalopods, not his earlier comparison of arthropods and vertebrates, that brought Geoffroy's conflict with Cuvier out in the open, leading to the famed "debate" in 1830 (see Appel, 1987).

[24] Teratologies were regularly accorded names in the Linnaean hierarchy earlier in the 19th century (Appel, 1987).

[25] This habit of *Limulus* the horseshoe crab appears to have inspired Patten's (1912) theory of arachnid origins of the vertebrates. In this evolutionary scenario agnathan fish derive from meristomate arachnid ancestors of *Limulus*. Detailed comparsion of cranial and neural architecture is used in the argument. Although this phylogenetic hypothesis is not tenable, our current knowledge of Hox genes suggests that some of the segmentally bounded neural structures compared in the analysis may be evolutionarily related.

deuterostome guts, respectively. On the face of it, then, a simple inversion seems possible. However, the topological relationships of gut to nerve cord are not so simple (Fig. 5). Mouths are ventral in both groups. In protostomes such as mollusks, annelids and arthropods, the gut proceeds from ventral to dorsal through a circumesophageal nerve ring and then continues toward the posterior, dorsal to the nerve cord. In the chordates the mouth and entire gut are ventral to the nerve cord. These

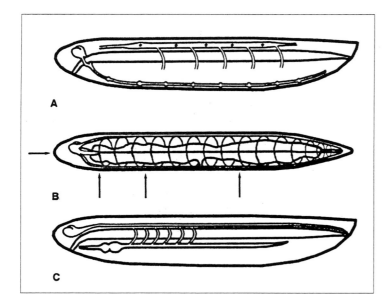

Figure 5. Arguments of dorsoventral inversion for protostomes (A) and deuterostomes (C) have generally not considered outgroup taxa, such as free-living flatworms (B). Flatworms and cnidarians lack a heart, and the relationship of the gut to the nervous system in flatworms is extremely variable. This suggests that the protostome gut, which goes from a ventral to a dorsal position through the circumesophogeal nerve ring, and the deuterosome gut, which remains below the nerve cord, are independently derived from flatworm-like ancestors that lacked a complete gut. The evolution of a centralized nervous system present in all three taxa would have preceded the evolution of the through gut.

differences in topology of gut and nerve cord cannot have resulted from simple dorsoventral inversion. Something else must have happened.

Several evolutionary hypotheses have been put forward that attempt to accommodate dorsoventral inversion with various preexisting evolutionary hypotheses

(Arendt and Nübler-Jung, 1994 a, b; Lacalli, 1995; Peterson, 1995; De Robertis and Sasai, 1996 among others). We do not intend to comment on these theories directly, but to step back and, through the use of outgroup comparison, develop a more general conception of the evolution of guts and nerves to address the question of evolutionary transformations at the base of the bilaterian tree. Again, flatworms[26] and the cnidarians provide potentially useful outgroups. These groups have nervous systems, but lack the through gut characteristic of deuterostomes and protostomes. Thus, comparison to outgroups suggests that the nervous system predates at least some details of gut evolution. This is further suggested by the different origins and orientations of gut development in protostomes and deuterostomes.

The flatworm nervous system is often conceived of as having an anterior nerve ring with four major nerves emanating posteriorly from it. These nerve axes then have well or poorly organized cross-commissures in different flatworm taxa. This whole generalized structure is often referred to as the orthogon (e.g. Reuter and Gustafsson, 1995). In some free-living flatworms, two of the ventral nerve cords have extensive development of cross-commissures. The other two nerves assume dorsal or lateral positions. In these cases, where the commissures of the ventral nerves are well developed, the nervous organization as a whole looks much like the nervous organization of many protostomes.

Relative to the nervous system, the gut in flatworms varies considerably in position. The gut opening can be anterior, or in a variety of positions on the venter. Guts can be absent, simple or complexly diverticulated. In some cases, multiple mouths and gut systems are present. Ringlike nervous structures are often associated with the gut. In the case of some flukes, the gut enters anteriorly through the nerve ring. In many free-living flatworms the pharynx and gut structures penetrate up through the nervous organization from various ventral positions (Fig. 5B). Here a secondary circumpharyngeal ring of nerves often develops from adjacent commissures of the ladderlike ventral component of the orthogon (Reuter and Gustafsson, 1995). Given the variety of different gut placements in flatworms, novel relationships between gut and nerve topology must have evolved multiple times. Moreover, given these topologies, it seems easy to envision a common ancestor of deuterostomes and protostomes that had a mouth and pharyngeal structures, but lacked a through gut. Independent evolution of the gut could then lead to different topologic relationships with the nervous system until a secondary opening and through gut reduced the evolutionary plasticity. Clearly such a hypothesis is testable if sufficient molecular

---

[26]It is probably not critical whether the modern flatworms are basal in the Bilateria, basal to the protostomes or treated as degenerate coelomate protostomes (Brusca and Brusca, 1990; Aguinaldo, et al. 1997). It may even be that the closest relatives of the modern deuterosome and protostome lineages are different groups of modern flatworms. Certainly, if one considers the tree, the condition in the stem basal to the crown group Bilateria will eventually lack a through gut, as vendozoans and radiates lack a through gut.

markers in gut development can be compared in deuterostomes, protostomes, flat-worms and possibly cnidarians.

Morphologically, the dorsoventral argument hinges on the relative placemant of nerve chord, gut and heart. However, molecular evidence focusses primarily on the formation of neurectoderm and the central nervous system. As mentioned above, *chordin/sog* and *dpp*/BMP-4 function in differentiation of neurectoderm. In essence they regulate interactions that determine the binary fate of ectoderm vs. neurecto-derm. The default condition of the ectoderm is neurectodermal (Bier, 1997). Downregulation of this default condition is then responsible for the presence of non-neural ectoderm. The regulation of particular conserved neural-specific genes tends to confirm the shared ancestry of this mechanism for defining neural fates. The genes in question (*msh/Msx* and *vnd/Nk2.2*) are downregulated by *dpp/BMP4*, and are expressed laterally and medially, respectively, in the neurectoderm of both flies and vertebrates (Bier, 1997; D'Alessio and Frasch, 1996). These observations tend to confirm the perspective that the conserved regulators *dpp/BMP4* and *sog/chd* are involved primarily in the ectoderm to define a concentrated nervous organization. Other aspects of dorsoventral organization can be thought of as coincident with this neural organization.

Cnidaria lack a centralized nervous system. Their relatively diffuse nerve nets are primarily associated with the gastroepithelia and lack the concentrated nervous or-ganization of the Bilateria. Here the innervated gastroepithelium may be compara-ble to neurectoderm, the default state without the repression of *dpp*/BMP-4. In such an epithelium the neurogenic genes[27], which determine the neural fate of cells des-tined to become neurons through interaction with and repression of the neural fate in adjacent cells, might be sufficient for the organization of a nerve net. If this is the case, *dpp*/BMP-4 repression might not be present or at least not involved in similar patterning of the ectoderm. In this scenario, the bilaterian central nervous system would be the product of concentrating the nervous organization in part of the ecto-derm, by eliminating it from other regions[28]. In a flatworm like ancestor of the Bilateria, more than one region of *dpp*/BMP-4 would be required to make the four large axial nerves of the orthogon. If this were the case, then the ventral nervous sys-tem in protostomes could derive from the ventral pair of nerves in the orthogon and the dorsal nervous system in vertebrates from the dorsal pair. In this scenario the signaling responsible for differentiating a nervous system would have operated dor-sally and ventrally in an ancestor. Evolutionary loss of a ventral nervous system in a chordate ancestor and the dorsal nervous system in a protostome ancestor would

---

[27]A proneural gene (CnASH) related to *achaete-scute* is present in Cnidaria (Grens et al., 1995).

[28]Alternatively, *dpp*/BMP-4 regulation might be responsible for differentiating innervated gastrodermis from the epidermis in Cnidaria.

leave homologous regulators operating at opposite ends of the dorsoventral axis. The above scenario explains the available data without invoking an instantaneous dorsoventral inversion as envisioned in the transcendental scheme of Geoffroy. In addition, by using outgroups to infer an ancestral condition, we have generated a hypothesis of evolution that makes specific testible predications. One such predication is that molecules homologous to *dpp* should be involved in making multiple gradients in flatworms that have an orthogonal nervous system.

Conclusion

In this work we have argued that:

1. Conceiving of single, archaetypical ancestors for groups of processes shared in the development of different taxa obscures the multiple gene duplications and multiple phylogenetic branches involved in the evolution of a complex developmental system. It is far better to consider all the branches near the base of the tree of animal life that we can resolve. Thus we can begin to consider the evolutionary assembly of developmental systems.

2. Tracing the evolution of important developments of character states, even the presence or absence of such organs as eyes, can aid in understanding the probability of evolutionary continuity and therefore of homology of these structures.

3. The generation of trees based on all the available sequences of whole classes of developmental regulatory molecules, here referred to as gene trees, can aid in assessing the degree of support for a hypothesis of homology of a structure controlled by putatively similar developmental regulators. We apply this approach to assess homology of hearts and of eyes. In the case of homology of bilaterian eyes, character tracing, gene trees, as well as the conservation of other developmental regulators in combination make a very strong case for the homology of eyes. The homology of hearts is not supported by character tracing, and gene trees are inconclusive as to the precise relationship of the NK2 trees. With additional data, such as more NK2 gene sequences, this issue may be resolved.

4. Many workers make comparisons between two taxa. Such arguments demonstrate transformation, but not the direction of evolutionary change. Perhaps appropriately workers who make make transformational arguments have invoked transcendental thinkers, such as Geoffroy Saint-Hilaire, who believed in direct morphologic transformations without consideration of ancestors. We, however, prefer to consider evolutionary arguments which require the reconstruction of ancestral character states through the use of outgroups. Such an approach leads

to hypothesis of character change between taxa and can lead to additional testable hypotheses. We apply such inferences to the question of dorsoventral inversion using flatworms and cnidarians as outgroups. This approach leads to evolutionary scenarios that do not require literal dorsoventral inversion and are subject to further test by examination of the regulatory molecules in the outgroup taxa.

In all these exercises, from our arguments against archaetypes and transcendental thinking, to our use of gene trees to shed light on homology, we are advocating the integration of phylogenetic methods into studies of development and developmental genes.

Acknowledgements

We thank Jesus Maldonado and Roxanne Tourabian for stimulating discussions. We also thank Diane Bridge, Doug Erwin, Nick Holland, Tom Roos, Judy Lengyel and an unidentified reviewer for their comments on earlier drafts of the manuscript. Use of the prerelease PAUP 4.0 provided to us by Dave Swofford was also greatly appreciated.

References

Agosti, D., Jacobs, D. and DeSalle, R. (1996) On combining protein sequences and nucleic acid sequences in phylogenetic analysis: the homeobox protein case. *Cladistics* 12(1): 65–82.

Aguinaldo, A. M. A., Turbeville, J. M., Linford, L. S., Rivera, M. C., Garey, J. R., Raff, R. A. and Lake, J. A. (1997) Evidence for a clade of nematodes, arthropods and other moulting animals. *Nature* 387: 489–492.

Appel, T. A. (1987) *The Cuvier-Geoffroy Debate: French Biology in the Decades before Darwin*, Oxford University Press, New York.

Arendt, D. and Nübler-Jung, K. (1994a) Inversion of the dorsoventral axis? *Nature* 371: 26.

Arendt, D. and Nübler-Jung,K. (1994b) Is ventral in insects dorsal in vertebrates? A history of embryological arguments favouring axis inversion in chordate ancestors. *Roux's Arch. Dev. Biol.* 203: 357–366.

Ayala, F. J., Rzhetsky, A. and Ayala, F. J. (1998) Origin of the metazoan phyla: Molecular clocks confirm palaeontological estimates. *Proc. Natl. Acad. Sci. USA* 95: 606–611.

Azpiazu, N. and Frasch, M. (1993) Tinman and bagpipe: two homeobox genes that determine cell fates in the dorsal mesoderm of *Drosophila*. *Gene. Develop.* 7: 1325–1340.

Baldauf, S. L. and Doolittle, W. F. (1997) Origin and evolution of the slime molds (Mycetozoa). *Proc. Natl. Acad. Sci. USA* 94: 12007–12012.

Barth, K. A. and Wilson, S. W. (1995) Expression of zebrafish nk2.2 is influenced by sonic hedgehog/vertebrate hedgehog-1 and demarcates a zone of neuronal differentiation in the embryonic forebrain. *Development* 121: 1755–1768.

Bier, E. (1997) Anti-neural-inhibition: a conserved mechanism for neural induction. *Cell* 89(5): 681–684.

Bodmer, R. (1993) The gene tinman is required for specification of the heart and visceral muscles in *Drosophila*. *Development*. 118: 719–729.

Boettger, T., Stein, S. and Kessel, M. (1997) The chicken NKX2.8 homeobox gene: a novel member of the NK-2 gene family. *Dev. Genes Evol.* 207(1): 65–70.

Brusca, R. C. and Brusca, G. J. (1990) *Invertebrates*, Sinauer, Sunderland, MA.

Buchberger, A., Pabst, O., Brand, T., Seidl, K. and Arnold, H.-H. (1996) Chick NKX-2.3 represents a novel family member of vertebrate homologues to the *Drosophila* homeobox gene tinman: differential expression of CNKX-2.3 and CNK-2.5 during heart and gut development. *Mech. Dev.* 56(1–2): 151–163.

Buss, L. W. and Seilacher, A. (1994) The Phylum Vendobionta: a sister group of the Eumetazoa? *Paleobiology.* 20(1): 1–5.

Callaerts, P., Halder, G. and Gehring, W. J. (1997) PAX-6 in development and evolution. *Ann. Rev. Neurosci.* 20: 483–532.

Chalepakis, G., Fritsch, R., Fickenscher, H., Deutsch, U., Goulding, M. and Gruss, P. (1991) The molecular basis of the undulated/Pax-1 mutation. *Cell* 66: 873–884.

Cheyette, B. N. R., Green, P. J., Martin, K., Garren, H., Hartenstein, V. and Zipursky, S. L. (1994) The *Drosophila sine oculis* locus encodes a homeodomain-containing protein required for the development of the entire visual system. *Neuron* 12: 977–996.

Czerny, T., Halder, G., Callaerts, P., Kloter, U., Gehring, W. J. and Busslinger, M. (1997) *Twin of Eyeless,* a second *Pax-6* gene of *Drosophila*, acts upstream of *Eyeless* in the control of eye development. *Abstracts of papers presented at the LXII Cold Spring Harbor Symposium on Quantitative Biology, Pattern Formation During Development. Cold Spring Harbor Laboratory, Cold Spring Harbor, New York.* Abstract #51.

D'Alessio, M. and Frasch, M. (1996) *msh* may play a conserved role in dorsoventral patterning of the neuroectoderm and mesoderm *Mech. Dev.* 58(1–2): 217–231.

DeBeer, G. R. (1940) *Embryos and Ancestors*, Oxford University Press, London.

De Robertis, E. M. and Sasai, Y. (1996) A common plan for dorsoventral patterning in Bilateria. *Nature* 380: 37–40.

Duboule, D. and Dolle, P. (1989) The structural and functional organization of the murine HOX gene family resembles that of *Drosophila* homeotic genes. *EMBO J.* 8: 1497–1505.

Eernisse, D. J. and Kluge, A. G. (1993) Taxonomic congruence versus total evidence, and amniote phylogeny inferred from fossils, molecules and morphology. *Molec. Biol. Evol.* 10(6): 1170–1195.

Eernisse, D. J., Albert, J. S. and Anderson, F. E. (1992) Annelida and Arthropoda are not sister taxa: a phylogenetic analysis of spiralian metazoan morphology. *Syst. Biol.* 41(3): 305–330.

Evans, S. M., Yan, W., Murillo, M. P., Ponce, J. and Papalopulu, N. (1995) tinman, a *Drosophila* homeobox gene required for heart and visceral mesoderm specification, may be represented by a family of genes in vertebrates: XNkx-2.3, a second vertebrate homologue of tinman. *Development* 121: 3889–3899.

Fedonkin, M. A. and Waggoner, B. M. (1997) The late precambrian fossil *Kimberella* is a mollusc-like bilaterian organism. *Nature* 388: 868–871.

Fitch, W. M. (1971) Toward defining the course of evolution: minimal change for a specific tree topology. *Syst. Zool.* 20: 406–416.

François, V. and Bier, E. (1995) *Xenopus* chordin and *Drosophila* short gastrulation genes encode homologous proteins functioning in dorsal ventral axis formation. *Cell* 80(1): 19–20.

Gamer, L. W. and Wright, C. V. E. (1993) Murine Cdx-4 bears striking similarities to the *Drosophila* caudal gene in its homeodomain sequence and early expression pattern. *Mech. Develop.* 43: 71–81.

Garcia-Fernàndez, J., Baguña, J. and Saló, E. (1993) Genomic organization and expression of the planarian homeobox genes Dth-1 and Dth-2. *Development.* 118: 241–253.

Garstang, W. (1922) The theory of recapitulation: a critical restatement of the biogenetic law. *J. Linn. Soc. Zool.* 35: 81–101.

Gorczyca, M. G., Phillis, R. W. and Budnik, V. (1994) The role of *tinman*, a mesodermal cell fate gene, in axon pathfinding during the development of the transverse nerve in *Drosophila. Development.* 120(8): 2143–2152.

Goriely, A., Stella, M., Coffinier C., Kessler, D., Mailhos, C., Dessain, S. and Desplan, C. (1996) A functional homologue of goosecoid in *Drosophila. Development.* 122(5): 1641–1650.

Gould, S. J. (1977) *Ontogeny and Phylogeny*, The Belknap Press of Harvard University Press, Cambridge, MA.

Grens, A., Mason, E., Marsh, J. L. and Bode, H. R. (1995) Evolutionary conservation of a cell fate specification gene – the hydra *achaete-scute* homolog has proneural activity in *Drosophila. Development* 121: 4027–4035.

Grens, A., Gee, L., Fisher, D. A. and Bode, H. R. (1996) CnNK-2, an NK-2 homeobox gene, has a role in patterning the basal end of the axis in *Hydra. Dev. Biol.* 180: 473–488.

Guazzi, S., Price, M., Felice, M. D., Damante, G., Mattei, M.-G. and DiLauro, R. (1990) Thyroid nuclear factor (TTF-1) contains a homeodomain and displays a novel DNA binding specificity. *EMBO J.* 9(11): 3631.

Haeckel, E. (1874) The gastraea-theory, the phylogenetic classification of the animal kingdom and the homology of the germ-lamellae. *Quart. J. Microsc. Soc.* 14: 142–165.

Halder, G., Callaerts, P. and Gehring, W. J. (1995) Induction of ectopic eyes by targeted expression of the *eyeless* gene in *Drosophila. Science* 267: 1788–1792.

Han, Z. and Firtel, R. A. (1998) The homeobox-containing gene *Wariai* regulates anterior-posterior patterning and cell-type homeostasis in *Dictyostelium. Development* 125: 313–325.

Hartigan, D. J. and Rubenstein, J. L. R. (1996) The cDNA sequence of murine NKX-2.2. *Gene* 168(2): 271–272.

Harvey, R. P. (1996) NK-2 homeobox genes and heart development. *Dev. Biol.* 178: 203–216.

Hillis, D. M., Moritz, C. and Mable, B. K. (eds) (1996) *Molecular Systematics*, Sinauer Associates, Sunderland, MA.

Holland, P. W. H., Holland, L. Z., Williams, N. A. and Holland, N. D. (1992) An *Amphioxus* homeobox gene: sequence conservation, spatial expression during development and insights into vertebrate evolution. *Development.* 116(3): 653–661.

Jacobs, D. K. (1992) The support of hydrostatic load in cephalopod shells: adaptive and ontogenetic explanations of shell form and evolution from Hooke 1695 to the present. *In:* Hecht, M. K., Wallace, B. and MacIntyre, R. J. (eds) *Evoluionary Biology*, vol. 26, Plenum Press, New York, pp. 287–349.

Jacobs, D. K. (1994) Developmental genes and the origin and evolution of Metazoa. *In:* Schierwater, B., Streit, B., Wagner, G, P. and DeSalle, R. (eds) *Molecular Ecology and Evolution: Approaches and Applications*, Birkhäuser, Basel, pp. 537–549.

Jiménez, F., Martin-Morris, L. E., Velasco, L., Chu, H., Sierra, J., Rosen, D. R. and White, K. (1995) vnd, a gene required for early neurogensis of *Drosophila*, encodes a home-odomain protein. *EMBO J.* 14(14): 3487–3495.

Komuro, I. and Izumo, S. (1993) Csx: a murine homeobox-containing gene specifically expressed in the developing heart. *Proc. Natl. Acad. Sci. USA* 90: 8145–8149.

Lacalli, T. C. (1995) Dorsoventral axis inversion. *Nature* 373: 110–111.

Lee, K.-H., Xu, Q. and Breitbart, R. E. (1996) A new *tinman*-related gene, nkx2.7, anticipates the expression of nkx2.5 and nkx2.3 in zebrafish heart and pharyngeal endoderm. *Dev. Biol.* 180: 722–731.

Lyons, I., Parsons, L. M., Hartley, L., Li, R., Andrews, J. E., Robb, L. and Harvey, R. P. (1995) Myogenic and morphogenetic defects in the heart tubes of murine embros lacking the homeo box gene Nkx2-5. *Genes Develop.* 9: 1654–1666.

Maienschein, J. (1991) *Transforming Traditions in American Biology 1880–1915*, Johns Hopkins University Press, Baltimore.

Minoo, P., Hamdan, H., Bu, D., Warburton, D., Stepanik, P. and deLemos, R. (1995) TTF-1 regulates lung epithelial morphogenesis. *Develop. Biol.* 172(2): 694–698.

Minoo, P., Li, C. G., Liu, H. B., Hamdan, H. and deLemos, R. (1997) TTF-1 is an epithelial morphoregulatory transcriptional factor. *Chest* 111(6): S135-S137.

Nardelli-Haefliger, D. and Shankland, M. (1993) Lox 10, a member of the NK-2 homeobox gene class, is expressed in a segmental pattern in the endoderm and in the cephalic nervous system of the leech *Helobdella*. *Development* 188: 877–892.

Newman, C. S., Grow, M. W., Cleaver, O., Chia, F. and Krieg, P. (1997) Xbap, a vertebrate gene related to *bagpipe*, is expressed in developing craniofacial structures and in anterior gut muscle. *Dev. Biol.* 181: 223–233.

Okkema, P. G. and Fire, A. (1994) The *Caenorhabditis elegans* NK-2 class homeoprotein CEH-22 is involved in combinatorial activation of gene expression in pharyngeal muscle. *Development* 120: 2175–2186.

Oliver, G., Mailhos, A., Wehr, R., Copeland, N. G., Jenkins, N. A. and Gruss, P. (1995) SIX3, a murine homologue of the *sine oculis* gene, demarcates the most anterior border of the developing neural plate and is expressed during eye development. *Development* 121(12): 4045–4055.

Olsen, E. N. (1997) Transcriptional control of cardiovascualr development. *In*: *Pattern Formation during Development*, abstracts of papers presented at the LXII Cold Spring Harbor Symposium on Quantitative Biology, Cold Spring Harbor Laboratory Press, New York, p. 10.

Patten, W. (1912) *The Evolution of the Vertebrates and Their Kin*, Blakiston, Philadelphia, USA.

Peterson, K. J. (1995) Dorsoventral axis inversion. *Nature* 373: 111–112.

Price, M., Lazzaro, D., Pohl, T., Mattei, M.-G., Rüther, U., Olivo, J.-C., Duboule, D. and DiLauro, R. (1992) Regional expression of the homeobox gene Nkx-2.2 in the developing mammalian forebrain. *Neuron* 8: 241–255.

Raff, R. A. (1992) Direct-developing sea urchins and the evolutionary reorganization of early development. *Biological Essays.* 14: 211–218.

Reeck, G. R., Haën, C. D., Teller, D. C., Doolittle, R. F., Fitch, W. M., Dickerson, R. E., Chambon, P., McLachlan, A. D., Margoliash, E., Jukes, T. H. and Zuckerkandl, E. (1987) "Homology" in proteins and nucleic acids: a terminology muddle and a way out of it. *Cell* 50: 667.

Reuter, M and Gustafsson, M. K. S. (1995) The flatworm nervous system: pattern and phylongey. *In*: Breidbach, O. and Kutsch, W. (eds) *The Nervous Sytstem of Invertebrates: An Evolutionary and Comparative Approach*, Birkhäuser Verlag, Basal, Switzerland, pp. 25–59.

Saha, M. S., Michel, R. B., Gulding, K. M. and Grainger, R. M. (1993) A *Xenopus* homeobox gene defines dorsal-ventral domains in the developing brain. *Development* 118: 193–202.

Salvini-Plawen, L. V. and Mayr, E. (1977) On the evolution of photoreceptors and eyes. *Evolut. Biol.* 10: 207–263.

Saranak, J. and Foster, K. W. (1997) Rhodopsin guides fungal phototaxis. *Nature* 387: 465–466.

Schubert, F. R., Nieselt-Struwe, K. and Gruss, P. (1993) The Antennapedia-type homeobox genes have evolved from three precursors separated early in metazoan evolution. *Proc. Natl. Acad. Sci. USA* 86: 7067–7071.

Seimiya, M., Ishiguro, H., Miura, K., Watanabe, Y. and Kurosawa, Y. (1994) Homeobox-containing genes in the most primitive metazoa, the sponges. *J. Biochem.* 221: 219–225.

Seimiya, M., Watanabe, Y. and Kurosawa, Y. (1997) Identification of POU-class homeobox genes in a freshwater sponge and the specific expression of these genes during differentiation. *Eur. J. Biochem.* 243(1–2): 27–31.

Seilacher, A. (1988) Vendozoa: organismic construction in the Phanerozoic biosphere. *Lethaia* 22: 229–239.

Sheng, G., Thouvent, E., Schmucker, D., Wilson, D. S. and Desplan, C. (1997) Direct regulation of rhodopsin 1 by Pax-6/eyeless in *Drosophila*: evidence for a conserved function in photoreceptors. *Genes Develop.* 11: 1122–1131.

Simeone A., Dapice M. R., Nigro V., Casanova J., F. Graziani, Acampora, D. and Avantaggiato, V. (1994) *Orthopedia*, a novel homeobox-containing gene expressed in the developing CNS of both mouse and *Drosophila*. *Neuron* 13: 83–101.

Slack, J. M., Holland, P. W. and Graham, C. F. (1993) The zootype and the phylotypic stage. *Nature* 361: 490–492.

Valentine, J. W. (1997) Cleavage patterns and the topology of the metazoan tree of life. *Proc. Natl. Acad. Sci. USA* 94(15): 8001–8005.

Wainright, P. O., Hinkle, G., Sogin, M. L. and Sickel, S. K. (1993) Monophyletic origins of Metazoa: an evolutionary link with fungi. *Science* 260(5106): 340–342.

Wheeler, W. C., Cartwright, P. and Hayashi, C. Y. (1993) Arthropod phylogeny: a combined approach. *Cladistics* 9: 1–39.

Wray, G. A. and Raff, R. A. (1990) Novel origins of lineage founder cells in the direct-developing sea urchin *Heliocidaris erythrogramma*. *Develop. Biol.* 141: 41–54.

Wray, G. A., Levinton, J. S. and Shapiro, L. H. (1996) Molecular evidence for deep Precambrian divergences among metazoan phyla. *Science* 274(5287): 568–573.

Appendix 1a.

GenBank accession numbers, or alternative sources, of sequences used in the "eye genes" analysis.

Sequence name	Accession no.	Sequence name	Accession no.
aristaless fly	L08401	hPax-3 human	U02368
BarH1 fly	M59963	mPax-3 mouse	X59358
BarH2 fly	M82884	zfPax-6 zfish	X63183
ceh10 nematode	X52812	hPax-6 human	M93650
EgbHx4 flatworm	X66820	LoPax-6 squid	U59830
engrailed fly	M10017	LsPax-6 nemertean	X95594
En2 mouse	L12705	mPax-6 mouse	X63963
eyeless fly	X79493	suPax-6 urchin	G984799
gooseberry fly	M14942/4	PmPax-6 ascidian	Y09975
gooseberryn fly	M14941/3	RnPax-6 Rat	U69644
D gsc fly	Goriely et al. (1996)	XlPax-6 Xenopus	U67887
Gsc zfish	L03395	hPax-7 human	Z35141
Gsc chicken	X70471	mPax-7 mouse	U20792
Gsc mouse	M85271	PaxB Hydra	G2102727
Gsc Xenopus	M81481	PaxQNR quail	X70475
mix1 Xenopus	M27063	paired fly	M14548
munster	W. McGinnis (personal communication)	Prx1 chicken	D13433
Otp mouse	Simeone et al. (1994).	repo fly	X78218
Otp urchin	Simeone et al. (1994).	Rx1A Xenopus	AF001048
Orthopedia fly	Simeone et al. (1994).	Rx1 zfish	AF001907
otxlike Xenopus	L26509	Rx1A Xenopus	AF001048
zOtx1 zfish	D26172	Rx2A Xenopus	AF001049
zOtx2 zfish	D26173	Rx2 zfish	AF001908
zOtx3 zfish	D26174	Rx3 zfish	AF001909
orthodent fly	X58983	Rx mouse	AF001906
Otx1 mouse	X68883	s59 fly	X55393
Otx2 mouse	X68884	S8 mouse	X52875
Xotx 2 Xenopus	U19813	smox3 fluke	M85303
Paxzfa zfish	X61389	unc4 nematode	Miller et al. (1992)
Paxzfb zfish	X63961	vab 3 nematode	Z49937
HrPax-37 ascidian	D84254		

Appendix 1b

GenBank accession numbers, or alternative sources, of sequences used in the "heart genes" analysis.

Sequence name	Accession no.	Sequence name	Accession no.
AMTMSX1P axolotl	D82577	NK2/vnd fly	S78691
bagpipe fly	L17133	nkx2.1 rat	D38035
Bapx1 mouse	U87957	Nkx2.2 mouse	U31566
bMsx1 cow	D30750	Nkx2.2 hamster	X81408
ceh1 nematode	X52810	Nkx-2.5 fish	U66572
Ceh-22 nematode	U10081	nkx-2.7 fish	U66573
CHKHOX3A chicken	M23065	NKX2.8 chicken	Y10655
CnNK-2 Hydra	Grens et al. (1996)	NKX3.1 man	U80669
Cnkx-2.5 chicken	X91838	Nkx3.1 mouse	U73460
Csx mouse	L20300	Nkx5.1 mouse	X75330
Dbx mouse	U61853	Nkx5.2 mouse	S80989
Dth-1 flatworm	X56499	prox1 sponge	L10984
Dth-2 flatworm	X69202	prox2 sponge	L10985
EgHbx3 flatworm	X66819	PwNkx3.2 newt	U88714
GH6 chicken	L13786	Sax1 mouse	X75384
H2.0 fly	Y00843	SOHo1 chicken	U35815
HCSX man	U34962	tinman fly	X55192
Hlx mouse	X58250	TTF-1 dog	X77910
Hox11(311) man	M75952	Xbap Xenopus	U75487
HUMHMBX man	M99587	XeNk2 Xenopus	L10327
Lox10 leech	Z22635	Xnkx-2.3a Xenopus	L38674
msh fly	U33319	XNKx-2.5 Xenopus	L25600

Subject index

Vetvicka V., University of Louisville, USA / **Sima P.,** Czech Academy of
Sciences, Prague, Czech Republic

Evolutionary Mechanisms of Defense Reaction

1998. 206 pages. Hardcover
ISBN 3-7643-5813-0

This book represents an evolutionary approach to defense mechanisms of all living
organisms. The results achieved in developmental and comparative immunology are
among the most interesting data in immunology. These results have great impact
on our understanding fundamental problems of the pathology of the human immune
system. In addition, the health questions related to commercially important animals
such as shrimps or fish have become a serious problem. At the same time, the field
of evolutionary immunology provides not only inspiration for further investigation in
biomedicine, but also a number of results applicable in clinical and commercial
practice.

This book evaluates the advantages and limitations of studying the development of
defense reactions. In addition to reviewing the major and crucial achievements of
the past, the book offers a comprehensive state-of-the-art treatise focused primarily
on the latest experiments described in the last few years. The well-defined mixture
of analyses of older and new data makes this book a valuable reference for all scientists
and scholars seeking detailed information on this important field even without preliminary knowledge.

The book is geared for scientists involved in immunology, cell biology, developmental biology, immunology and cancer research. Besides, this book will be particularly
beneficial for teachers of biology and/or developmental courses as well as both graduate
and undergraduate students.

Check our Highlights for new and notable titles selected monthly in each field.

http://www.birkhauser.ch

BioSciences with Birkhäuser

(Prices are subject to change without notice. 8/98)

For orders originating from all over the
world except USA and Canada:

For orders originating in the USA and
Canada:

Birkhäuser Verlag AG
P.O. Box 133
CH-4010 Basel / Switzerland
Fax: +41 / 61 / 205 07 92
e-mail: orders@birkhauser.ch

Birkhäuser Boston, Inc.
333 Meadowland Parkway
USA-Secaucus, NJ 07094-2491
Fax: +1 / 201 348 4033
e-mail: orders@birkhauser.com

Birkhäuser

EXS 86

Braunbeck T., Institute of Zoology, University of Heidelberg, Germany / **Streit B.,** Institute of Zoology, University of Frankfurt, Germany / **Hinton D.,** School of Veterinary Medicine, University of California at Davis, USA (Ed.)

Fish Ecotoxicology

1998. Approx. 300 pages. Hardcover
ISBN 3-7643-5819X
Due in September 1998

In the last twenty years, ecotoxicology has successfully established its place as an interdisciplinary science concerned with the effects of chemicals on populations and ecosystems, thus bridging the gap between biological and environmental sciences, ecology, chemistry and traditional toxicology. In modern ecotoxicology, fish have become the major vertebrate model, and a tremendous body of information has been accumulated. This volume attempts to summarize our present knowledge in several fields of primary ecotoxicological interest ranging from the use of (ultra)structural modifications of selected cell systems as sources of biomarkers for environmental impact over novel approaches to monitoring the impact of xenobiotics with fish in vitro systems such as primary and permanent fish cell cultures, the importance of early life-stage tests with fish, the bioaccumulation of xenobiotics in fish, the origin of liver neoplastic lesions in small fish species, immunocytochemical approaches to monitoring effects in cytochrome P450-related biotransformation, the impact of heavy metals in soft water systems, the environmental toxicology of organotin compounds, oxidative stress in fish by environmental pollutants to effects by estrogenic substances in aquatic systems. This collection of up-to-date reviews thus covers a broad range of topics important in current ecotoxicological research and should be of interest not only to specialists in ecotoxicology, but also to scientists and students in related fields of environmental sciences.

BioSciences with Birkhäuser

(Prices are subject to change without notice. 8/98)

For orders originating from all over the world except USA and Canada:

For orders originating in the USA and Canada:

Birkhäuser Verlag AG
P.O. Box 133
CH-4010 Basel / Switzerland
Fax: +41 / 61 / 205 07 92
e-mail: orders@birkhauser.ch

Birkhäuser Boston, Inc.
333 Meadowland Parkway
USA-Secaucus, NJ 07094-2491
Fax: +1 / 201 348 4033
e-mail: orders@birkhauser.com

Birkhäuser

EXS 82

Streit B. / Städler T., University of Frankfurt, Germany / **Lively C.M.,**
Indiana University, Bloomington, IN, USA (Ed.)

Evolutionary Ecology of Freshwater Animals
Concepts and Case Studies

1997. 384 pages. Hardcover
ISBN 3-7643-5694-4

Evolutionary ecology includes aspects of community structure, trophic interactions, life-history tactics, and reproductive modes, analyzed from an evolutionary perspective. Freshwater environments often impose spatial structure on populations, e.g. within large lakes or among habitat patches, facilitating genetic and phenotypic divergence. Traditionally, freshwater systems have featured prominently in ecological research and population biology.

This book brings together information on diverse freshwater taxa, with a mix of critical review, synthesis, and case studies. Using examples from bryozoans, rotifers, cladocerans, molluscs, teleosts and others, the authors cover current conceptual issues of evolutionary ecology in considerable depth.
The book can serve as a source of critically evaluated ideas, detailed case studies, and open problems in the field of evolutionary ecology. It is recommended for students and researchers in ecology, limnology, population biology, and evolutionary biology.

Check our Highlights for new and notable titles selected monthly in each field.

http://www.birkhauser.ch

BioSciences with Birkhäuser

(Prices are subject to change without notice. 8/98)

For orders originating from all over the world except USA and Canada:

Birkhäuser Verlag AG
P.O. Box 133
CH-4010 Basel / Switzerland
Fax: +41 / 61 / 205 07 92
e-mail: orders@birkhauser.ch

For orders originating in the USA and Canada:

Birkhäuser Boston, Inc.
333 Meadowland Parkway
USA-Secaucus, NJ 07094-2491
Fax: +1 / 201 348 4033
e-mail: orders@birkhauser.com

Birkhäuser

Molecular Approaches
to Ecology and Evolution

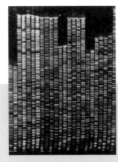

Rob DeSalle
Bernd Schierwater (Editors)

The last ten years have seen an explosion
of activity in the application of molecular
biological techniques to evolutionary and
ecological studies. This volume attempts to
summarize advances in the field and place into context the wide
variety of methods available to ecologists and evolutionary biologists using molecular techniques. Both the molecular techniques
and the variety of methods available for the analysis of such data
are presented in the text.

The book has three major sections – populations, species and
higher taxa. Each of these sections contains chapters by leading
scientists working at these levels, where clear and concise discussion of technology and implication of results are presented.

The volume is intended for advanced students of ecology and
evolution and would be a suitable textbook for advanced undergraduate and graduate student seminar courses.

ISBN 3-7643-5725-8

9 783764 357252